Utilization and Environmental Impact Assessment of Red Mud Resources

Yu Cheng · Xuemei Zhu

Utilization and Environmental Impact Assessment of Red Mud Resources

Springer

Yu Cheng
Shandong University of Science
and Technology
Qingdao, Shandong, China

Xuemei Zhu
Chinese Research Academy
of Environmental Sciences
Beijing, China

ISBN 978-981-96-7410-7 ISBN 978-981-96-7411-4 (eBook)
https://doi.org/10.1007/978-981-96-7411-4

The original submitted manuscript has been translated into English. The translation was done using artificial intelligence. A subsequent revision was performed by the author(s) to further refine the work and to ensure that the translation is appropriate concerning content and scientific correctness. It may, however, read stylistically different from a conventional translation.

This Springer imprint is published by the registered company Springer Nature Singapore Pte Ltd.
The registered company address is: 152 Beach Road, #21-01/04 Gateway East, Singapore 189721, Singapore

Contents

Chapter 1
Preface

1.1 Overview

Red mud is an alkaline solid waste inevitably produced in the process of alumina production (a step in refining metallic aluminum). According to statistics, every 1.0 ton of alumina produced generates 1.0 ~ 2.0 tons of dry red mud. By the end of 2021, the stockpile of red mud in China has exceeded 1.5 billion tons, and with the development of the economy, it is rapidly growing at a rate of 120 million tons per year. Shandong Province, as the largest aluminum industry province in the country, has an alumina production capacity of over 30 million tons/year, a red mud stockpile of over 500 million tons, covering more than 35,000 acres, and an annual increase of over 45 million tons of red mud emissions, accounting for about 40% of the country's alumina production capacity, is the province with the largest scale of red mud production in the country. The red mud stockyards of 47 alumina enterprises nationwide (4 temporarily suspended production in 2020) have mostly approached the stockpile limit, and in the next few years, they will face the situation of nowhere to pile up red mud, which will have a great impact on the development of China's aluminum industry and strategic security. Due to the complex and variable physical and chemical properties of red mud, it is difficult to be recycled and can only be stockpiled on site (as shown in Fig. 1.1 Red Mud Stockyard), however, the harm caused by the large amount of red mud emissions and accumulation to the environment and safety is becoming increasingly serious. Therefore, it is urgent to effectively recycle and utilize red mud resources, which has also become a universally recognized global problem. Exploring large-scale red mud absorption technology has become the most core and urgent problem of red mud resource utilization.

In the actual production of alumina, after the bauxite is mined, it is first crushed and ground, and the large ore becomes very small powder. Next, the bauxite powder is added to a high concentration of sodium hydroxide aqueous solution. Under high temperature and pressure, the alumina in the bauxite reacts with sodium hydroxide to form water-soluble sodium aluminate, while other components cannot react with

Y. Cheng and X. Zhu, *Utilization and Environmental Impact Assessment of Red Mud Resources*, https://doi.org/10.1007/978-981-96-7411-4_1

Fig. 1.1 Red mud stockyard

alkali and still exist in solid form. After filtration, alumina can be separated from other compounds in bauxite. As the temperature drops, the alkali solution containing sodium aluminate gradually cools. The solubility of sodium aluminate will decrease as the temperature decreases. By adjusting the ratio of bauxite and NaOH, we can make sodium aluminate nearly saturated in the alkali solution at high temperature. Once the temperature drops, the alkali solution cannot dissolve so much sodium aluminate. By adding a small amount of $Al(OH)_3$ as a "seed crystal" to the solution, the sodium aluminate in the solution will grow around the crystal of $Al(OH)_3$, so a large amount of $Al(OH)_3$ precipitates from the solution. These $Al(OH)_3$ will be calcined at high temperature to become alumina again, and then electrolyzed to obtain metallic aluminum. This process of separating alumina from bauxite was invented by Austrian Karl Josef Bayer more than a hundred years ago, so it is called the Bayer process. Today, most of the alumina factories in the world are using the Bayer process to produce alumina.

In summary, a crucial step in the Bayer process is to treat alumina with alkali solution to separate alumina from other components of bauxite. The compounds such as SiO_2, Fe_2 O_3, TiO_2 that cannot be dissolved by alkali are of no help to aluminum production and must be separated from the solution by filtration. During the filtration process, some NaOH solution is also separated out with the solid, so the final waste is usually a dark red or brick red fluid like mud. These wastes appear red because their main component is dark red Fe_2O_3 Therefore, people have given this waste a very vivid name—red mud. After being separated from bauxite, aluminum factories usually discharge it into specially excavated ponds, then let the water in it evaporate naturally, leaving behind solid powder, which is the so-called wet stacking. Some factories will first try to remove the water from the red mud, and then pile the dry powder in the open air, this method is called dry stacking.

At present, the most basic disposal methods for red mud are stacking, dumping into the sea, filling in depressions, etc., and the disposal cost accounts for about 5% of the value of alumina products. Most alumina enterprises in our country adopt stacking treatment. This disposal method not only requires a large amount of farmland and land, but also costs a lot in the construction and maintenance of the stacking site, and it also has a certain impact on the environment: red mud has a high alkali content and also contains metals such as titanium and iron. If leakage occurs during the stacking process and enters groundwater, surface water, and other bodies of water, it can form sediments, suspended matter, soluble matter, causing heavy metal pollution and pH value rise in the water body, and other adverse ecological impacts. In the karst area of Guizhou, this problem is particularly serious, making the selection of red mud stacking sites very difficult.

Although every ton of alumina produced will produce 1.0–2.0 tons of dry red mud, this weight is only the dry weight of its solid powder. If the alkali solution carried out during the separation process of red mud is included, the actual weight is much higher than this number. In recent years, China's annual alumina production capacity exceeds 70 million tons, and the production of red mud exceeds 100 million tons, and this number is just the annual discharge of red mud. If the accumulation over the years is included, the total amount of red mud produced is frighteningly high. This means that a large amount of precious land has to be used to store these wastes, and once these lands are used to store red mud, it is difficult to use them again for farming or construction. At the same time, red mud contains a large amount of highly corrosive NaOH, and sometimes heavy metals, etc. Once these harmful substances seep into the soil or rivers, they will cause serious pollution.

If the place where red mud is stored leaks due to natural or human reasons, the harm caused will be even more serious. On October 4, 2010, an artificial lake storing red mud at an aluminum factory in the western Hungarian city of Ajka suddenly burst, causing up to 1 million cubic meters of red mud to leak out, affecting several surrounding villages and eventually flowing into the Danube. This accident caused 10 deaths, more than 100 injuries, and affected 40 square kilometers of land, causing a serious environmental disaster. Therefore, how to resolve the contradictions between technology, economy, and environment, and find an economical and environmentally friendly way to dispose of red mud on a large scale is the key to the comprehensive utilization of red mud.

As a type of industrial waste with a large stock in our country, red mud has great application potential, but the current utilization rate of red mud nationwide is extremely low, and it has hardly been effectively utilized. The application of red mud in highway engineering mainly involves research on the combination of red mud with lime, fly ash, and cement as subbase and base materials. Although red mud has shown many excellent characteristics in engineering and construction, because of its strong alkalinity, its application in engineering is greatly restricted, and its research is still in the experimental research stage, with only a few small experimental sections being paved, but there are no reports on the follow-up evaluation of the experimental sections. If red mud can be modified to meet the strength and environmental protection requirements of road construction under the premise of affordable cost, it can provide a good way for the large-scale application treatment of red mud.

The large-scale application of red mud in infrastructure construction is in line with the basic national policy of protecting the environment and saving resources. The project research can not only consume a large amount of red mud, reduce the environmental pollution and safety hazards brought by red mud, but also save land resources, protect farmland and the environment, save energy and reduce emissions, and promote the value-added utilization of red mud. This is also an important manifestation of implementing the scientific development concept, developing a green, circular economy, building a resource-saving and environmentally friendly society, is the fundamental solution to the environmental pollution and safety hazards caused by red mud stacking, and is an important measure for the sustainable development of the economy and society.

1.2 Domestic and International Research Status

1.2.1 Current Status of Resource Utilization of Red Mud

Researchers and practitioners around the world have conducted extensive research on the comprehensive utilization of red mud, involving multiple fields and disciplines. In summary, the research on the comprehensive utilization of red mud mainly includes: reclamation of red mud dumps, production of building materials from red mud, recovery of iron, aluminum, sodium and rare metals from red mud, production of functional materials from red mud, and alkali removal from red mud, etc. However, there has been little progress in the large-scale industrial utilization of red mud. In summary, the more concentrated research directions for the utilization of red mud can be summarized into the following three aspects:

① Recycling: The research on recycling mainly includes the recovery of valuable metals, rare elements and the leaching and leaching of alkali from red mud. The chemical composition and mineral composition of red mud produced by different regions and enterprises vary greatly, but generally speaking, red mud is rich in iron, aluminum, calcium, silicon, and also contains titanium, scandium, niobium, tantalum, zirconium, gallium, vanadium, thorium and uranium and other rare metal elements. Research on extracting valuable metal elements from red mud has been ongoing, and experiments have proven that it is technically feasible to extract and recover valuable metals from red mud. However, how to economically and effectively concentrate and extract, increase the extraction rate, avoid the production of hazardous waste and secondary pollution, and implement industrialization are all issues that restrict the implementation of red mud recycling technology, which currently remains in the laboratory.

② Production of building materials: The production of building materials is the most likely way to consume red mud on a large scale and effectively solve the problem of large amounts of red mud accumulation. Its main feature is to use red mud as a whole without causing secondary pollution. The utilization methods include preparing building materials (such as bricks, tiles, cement, concrete, etc.), ceramic products (such as microcrystalline glass, refractory materials, pottery, etc.), roadbed and pavement materials, composite material fillers, and using red mud to build dams, etc. Dozens of ways and methods for producing building materials from red mud have been proposed worldwide, but due to the high iron and alkali content in red mud, or high cost and complex processes, it is difficult for large-scale utilization of red mud to meet the requirements of industrial production, or it is restricted by factors such as small addition of red mud in industrial production. The way to use red mud on a large scale in the true sense still needs further exploration and research.

③ Environmental remediation and management: In the field of environmental remediation and management, the pollution absorption function of red mud has been a research hotspot in recent years. The principle is that the porous structure of red mud and its large specific surface area have strong adsorption effects, which can be used to clean gases, treat wastewater, or use the catalytic effects of iron or heavy metal elements in red mud to carry out comprehensive treatment of waste gas and waste liquid. The range of red mud adsorption treatment involves using red mud to adsorb inorganic anions (such as PO_4^{3-}, F^-, NO^{3-}), toxic heavy metal ions (such as Pb^{2+}, Cd^{2+}, Cu^{2+}, Zn^{2+}, As^{3+}, Cr^{3+}) and organic matter (including dyes, phenols, bacteria, viruses) etc. Treated and activated red mud can also be used as a gas cleaner to absorb toxic and harmful components in the gas (such as H_2S, NO_x, CO_2), and its effect of fixing CO_2 can contribute to reducing greenhouse gas emissions, and has certain application prospects. In addition, because red mud also contains various components, such as iron, nickel and some heavy metal elements, red mud can also be used as a catalyst for chemical reactions; most of the composition of red mud is clayey silicate, so red mud is also used as a soil amendment to promote crop growth and development. However, these environmental remediation and management technology researches are limited by low cost-effectiveness, and the consumption of red mud is too small in the face of huge stockpiles, and they have not yet been effectively promoted for industrial use.

Shandong Aluminum Factory drew on the experience of the Soviet Union, combined with its own research, and built a national large-scale cement plant in the early 1960s aimed at comprehensive utilization of red mud. Then, with the aim of increasing the proportion of red mud in raw materials, successful low-silicon high-iron cement batching experiments were carried out, providing a basis for increasing the proportion of red mud in raw materials to 40 ~ 60%. However, due to the high alkali content in red mud, which hindered the increase of the proportion of red mud in raw materials, "low-calcium dioxide method" and "high-calcium-sulfite pulp waste liquid method" red mud desalination laboratory experiments and industrial trials were carried out in the 60s and 70s, which significantly reduced the alkali content in red mud and the alkali consumption in alumina production. But because it also caused a decrease in the fluidity of red mud slurry, it could not be persisted in production. Although work continued afterwards, due to the high cost of wet cement and the impact of alkali in red mud on cement performance, it has not been applied to production to this day. In order to increase the proportion of red mud in raw materials without desalination, experiments on adding mineralizers to cement raw materials were carried out, and the effect of normal sintering under high alkali content (less than 1.8%) was achieved by adding 0.5% of fluorite to raw materials, providing conditions for improving the utilization rate of red mud, and it was applied to production in the early 1980s.

The most researched abroad is Bayer red mud, because it contains a higher amount of iron oxide, it is first considered as a raw material for ironmaking. The Ural Aluminum Plant and Bafurodar Aluminum Plant in the former Soviet Union, the Gebr. Kirini Company in Germany, the McDowell-Weirman Company in the United States, as well as Japan and Hungary have all conducted experiments on red mud ironmaking. But due to technical and economic reasons, most of them stayed at the experimental stage. The most successful use of red mud is the production of building materials. The former Soviet Union used Bayer red mud from the Dnieper Aluminum Plant to produce cement, and the proportion of red mud in raw materials could reach 14%. Mitsui Alumina Company in Japan cooperated with cement plants to use red mud as iron raw material for cement raw materials, and cement clinker could use 5 ~ 20 kg/t of red mud, and the utilization level of Bayer red mud was relatively low. The Volkhov, Achin, and Kalyev alumina plants in the former Soviet Union that used diaspore as raw material to produce alumina have successfully used diaspore slag (red mud) to produce cement, and conducted limestone and red mud two-component batching experiments, and cement can use 629 ~ 795 kg/t of red mud, which has opened up an effective way for the comprehensive utilization of sintered red mud.

In recent years, Kaiser Aluminum Chemical Company in the United States has also conducted experiments on adding binders to Bayer red mud to make waste cover layers and solidification, and drying to make desiccants. The Rugo Alumina Plant in Spain mixed Bayer red mud with fly ash from thermal power plants at a certain ratio, and after high-pressure molding, drying, and sintering, it conducted experiments to make wall bricks.

To solve the problem of comprehensive utilization of stockpiled red mud, Shandong Aluminum Factory used the strong resistance of red mud itself to sulfate erosion, and on the basis of red mud batching to produce clinker, it also used dried red mud as a mixed material to conduct experiments on sulfate-resistant red mud cement, and carried out several years of industrial trial production in the 1970s. This experiment opened up two ways to use red mud in cement production, batching and mixing materials, and the utilization of red mud in cement can reach 600 ~ 800 kg/t.

In addition, Shandong Aluminum Factory has also conducted experiments on red mud fluidized sand (for casting), red mud steam-cured bricks, red mud cold-pressed bricks, red mud sintered bricks, red mud glazed bricks, etc., but all of them have not been able to enter industrial production due to product performance, process conditions, and economic benefits. Using the characteristics of high calcium oxide/silicon dioxide ratio, strong alkalinity, and low melting point of sintered red mud, the factory also conducted experiments on properly adding materials with high silicon dioxide and alumina content and heat-insulating graphite to produce casting protective slag for steelmaking. The product was trial-produced in the 1970s (with a production volume of 100 t/a), and has been transferred to local enterprises for production along with equipment and technology in the 1980s.

Using the characteristics of red mud containing a certain amount of alkali, calcium, and trace elements, Shandong Aluminum Factory dried and ground it, and conducted field experiments on red mud silicon-calcium fertilizer. The experimental volume in the 1970s reached several hundred tons, which had a certain regulatory effect on acidic soil and a certain yield-increasing effect on land lacking calcium, silicon, and corresponding trace elements. But because it does not contain the main components needed by crops, its effect on the soil is limited, and it has been terminated for many years. The Stadt Alumina Plant in Germany used a large amount of discarded red mud and sludge to fill low-lying land and turn barren land into fertile land. But this technology is rarely used at present, because long-term use can easily cause leakage, which in turn pollutes groundwater.

At present, many experts abroad are dedicated to the research of red mud adsorbent for wastewater purification, using red mud adsorbent to remove arsenic, cadmium, phosphorus, etc. from wastewater. According to experimental data, when the pH of wastewater is 3.2, the adsorption capacity of red mud for As^{3+} is 4.31 μmol/g, for As^{5+} the adsorption capacity is 5.07 μmol/g. In addition, the adsorption capacity for cadmium is 30 ~ 45%. However, because red mud itself contains a large amount of chemicals, the red mud used for adsorption in wastewater will inevitably have a certain impact on the turbidity and toxicity of the water. Therefore, under the current technical conditions, it is not feasible to directly use red mud as a purifier for wastewater in practical applications.

Although these studies have made good progress, there are still many problems. Many methods of using red mud have high production costs, and the performance of the products obtained is not entirely satisfactory, so it is difficult to get large-scale promotion and application. In addition, many methods can only use a very small part of the red mud, for example, although the use of red mud for sewage treatment is effective, sewage treatment generally only requires a small amount of red mud, so it is unrealistic to hope to digest the hundreds of millions of tons of red mud produced each year with this method. Therefore, so far, a large amount of red mud can only be dealt with by stacking.

With the development of the economy, the advancement of science and technology, and the needs of our lives, more and more aluminum will be produced, which also means that the accompanying red mud will have an increasingly serious impact on the environment. Finding a suitable method to reuse red mud as a resource is urgent. Therefore, minimizing the storage of red mud and realizing the large-scale, harmless, and resourceful comprehensive use of red mud is the fundamental solution to its environmental pollution and safety hazards, and it is also the only way for the sustainable development of China's aluminum industry.

1.2.2 Current Status of Red Mud Resource Utilization Engineering

The ideal comprehensive utilization technology for red mud should be a technology with large processing capacity, simple process, convenient operation, and no secondary pollution. The use of red mud in roads, ports and other projects is undoubtedly one of the best ways. Research in this area is still in its infancy, and there is no mature technology and process, so the key to this technology is to research and adopt suitable solidification technology and road construction technology to meet the national standards required by road design. This technology has broad market prospects and huge environmental benefits. It is a perfect combination of environmental protection and infrastructure construction, and will bring the comprehensive utilization of red mud to a new stage.

Based on the test analysis of different red mud physical and mechanical properties, engineering characteristics and environmental protection monitoring, it is found that it is feasible to use red mud for highway road materials, and the sintering method has more advantages than the Bayer method red mud. Because the sintering method has a sintering process, the red mud contains more active ingredients such as active calcium oxide and silicon dioxide. The Bayer method red mud contains high iron and aluminum, low silicon and calcium, and the mineral composition is mainly needle iron ore and hematite, which is a special waste slag with high iron and low silicon. Compared with the sintering method red mud, the Bayer method red mud has a greater difficulty in solidification. Soviet scholars have carried out exploratory research on the regeneration of red mud for road use. The sintered red mud was first applied to the road embankment fill, and phosphogypsum was used as the active ingredient for the solidification of red mud. The research results show that the compressive strength of the red mud binder with added phosphogypsum is 25.2 ~ 33.5 MPa after three months of hardening, and the setting or hardening period varies with the type of phosphogypsum used.

In order to improve the road performance of red mud, the road performance of the mixed material was studied through the design of the cement lime stabilized red mud soil base layer and the cement red mud stabilized graded crushed stone base layer. Sabat and Mohantab studied the possibility of using dolomite powder and fly ash to stabilize red mud as road base material. The red mud was first stabilized by F-grade fly ash at different proportions from 0 to 25%, and then 0–12% of different proportions of dolomite powder was added to the red mud mixture stabilized by the optimal percentage of fly ash, and the optimal percentages of dolomite powder and fly ash for stabilizing red mud were found to be 8% and 15% respectively.

The Shandong Provincial Transportation Science Research Institute built a cement, fly ash stabilized red mud road base test section in the Zhangbo pavement test road in 1989. The base layer used a material combination of lime: fly ash:red mud:soil = 4:25:25:46, and the 7d strength value reached 1.2 MPa; the base layer used a material combination of lime: fly ash:red mud:crushed stone (mass ratio) = 6:12:12:70, and the 7d strength value reached 2.0 MPa. Shandong Aluminum

Industry Zibo City Zichuan District built a road about 15 m long and about 4. The highway is made of red mud subgrade material. Qi Jianzhao's experimental research on the red mud subgrade material of Shandong Aluminum Industry shows that the red mud road subgrade material mixed in the ratio of red mud:fly ash:cement = 80:10:10, the 7d compressive strength can reach more than 2.0 MPa, and the 28d strength can reach more than 3.0 MPa. The red mud road base has reached the strength requirements of the first level of lime industrial waste slag stabilized soil and expressways. Construction practice and use process show that cement fly ash stabilized red mud base has good dynamic stability, shrinkage and temperature shrinkage. Since the red mud base stabilized by two ashes is mainly composed of red mud and fly ash, it can greatly reduce the road construction cost and turn waste into treasure, with good social and economic benefits.

By utilizing the special properties of red mud, it can be mixed into cement slurry to adjust the performance of cement and reduce costs. Research shows that red mud can adjust the water demand of cement mortar, improve the fluidity of mortar, and significantly improve the later strength. The research on red mud cement concrete pavement material shows that when the finely ground red mud replaces the appropriate amount of cement (replacement amount < 1/4), the formed cement red mud concrete has good physical and mechanical properties. It can be used for cement concrete pavement. In the research of using red mud as a cement concrete admixture, 15% of the cement in the cement concrete is replaced with finished red mud with a fineness of 2%, and some finely ground raw gypsum is added. The research results show that $CaSO_2 \cdot 2H_2O$ can stimulate the potential activity of red mud, and the 28d age compressive strength of cement concrete reaches 93% of the strength of the reference cement concrete; the compressive strength after the 90d age exceeds the reference cement concrete, and it has not fallen. The literature studied the effect of red mud excessively replacing cement in cement concrete on compressive strength, and used red mud to excessively replace cement in cement concrete to improve the compressive strength of red mud cement concrete. The research results show that using red mud to excessively replace cement in cement concrete is better than replacing the same amount, the early strength of Bayer red mud cement concrete is improved, but the 7d age strength of sintered red mud cement concrete is almost unchanged, and it does not increase much in the later period; when using red mud to replace cement in cement concrete, it is best to use additives or other additives for compound use. Bao Huiming and others added red mud as a stabilizing material to prepare red mud stabilized asphalt in matrix asphalt, and used the viscosity-temperature relationship analysis method to discuss the effects of different shear rates and dosages on the viscosity of red mud stabilized asphalt, and used scanning electron microscopy to observe the micro-state changes between red mud and asphalt. H. Zhang and others used red mud as a filler to prepare porous asphalt and conducted road performance tests, believing that this stabilized asphalt has good rutting resistance and aging resistance.

In summary, the current research focus on red mud in roads is to improve the strength of stabilizing materials and prevent secondary pollution during application. Due to the strong alkalinity of Bayer red mud, low stability strength and related environmental issues, there is a lack of effective theoretical support and systematic summary, and there is no very perfect technical process. Therefore, it has not been widely used in highway construction, and there are some problems to be solved, mainly:

(1) The engineering properties of red mud itself are poor and the bearing capacity is low. It is necessary to use appropriate curing materials to stabilize it to improve its engineering properties. The cost of the curing agent and the excitation effect on red mud are important factors restricting the use of red mud as a subgrade and pavement material.
(2) Red mud is concentrated in the stockyard for a long time and solidified into large particles. How to solve the problem of red mud crushing is related to the large-scale utilization of red mud.
(3) Red mud particles are porous and have a high specific surface area. They are strongly water-absorbent and the water content is not easy to control. Dry red mud is extremely hygroscopic and soluble in water, and can be decomposed by carbon dioxide and low-concentration polymer acid. It is necessary to find a water content range that is conducive to the use of red mud.
(4) Although red mud contains a lot of β-dicalcium silicate, it is in an inert state. How to stimulate the activity of red mud is a key issue.
(5) The pH of red mud is high, and it contains trace heavy metal ions and radioactive elements, which pose certain hazards to the environment. How to reduce the pH value of red mud itself and stabilize the heavy metal ions and radioactive elements in the leachate of red mud while ensuring strength and reducing costs is a problem that needs to be solved in the application of red mud.

Furthermore, because red mud is concentrated in the stockyard, building roads away from the stockyard will increase transportation costs, which is also a major factor restricting the use of red mud as a road base material. Therefore, it is necessary to carry out suitable stabilization techniques and engineering methods research based on different red mud characteristics, seek the optimal mix design, ensure the strength of the stabilized red mud mixture, minimize the impact of the red mud mixture on the environment, and reduce engineering costs, to achieve the large-scale and value-added engineering application of red mud.

Chapter 2
Engineering Characteristics of Red Mud for Road Use

This chapter focuses on the basic physical, chemical properties, mechanical characteristics, and road performance of red mud, providing foundational data and basis for the composition design optimization, performance, and mechanism analysis of inorganic and composite stabilizers stabilizing red mud.

2.1 Characteristics of Red Mud Components

The main mineral phases of Bayer process red mud are gibbsite $Al(OH)_3$, boehmite γ-AlOOH, hematite Fe_2O_3, goethite FeO(OH), quartz SiO_2, brookite TiO_2, rutile (TiO_2) and calcite $CaCO_3$, with a cementitious porous structure, mainly composed of structure-aggregate, structure-granule, and structure-agglomerate. The Bayer process red mud is dried at 105 °C to constant weight, ground through a 0.075 mm square hole sieve, and its mineral composition is tested by XRD, as shown in Fig. 2.1. It can be seen from the figure that there are significant differences in the content of main chemical components such as SiO_2, Al_2O_3, CaO, and Fe_2O_3 in sintered and Bayer process red mud. Since Bayer process red mud is produced by strong alkali leaching of high-alumina, high-iron, monohydrate and trihydrate bauxite, there is no high-temperature treatment during the production process, the content of Al_2O_3, Fe_2O_3, SiO_2 is high, and the content of CaO is relatively low, so the active components are relatively few, but the content of Fe_2O_3 is high, so Bayer process red mud generally appears dark red or bright red.

The Shanxi and Weiqiao red muds were dried at 105 °C after grinding, and XRF analysis was performed, the results are shown in Table 2.1. It can be seen from Table 2.1 that the composition elements of the two red muds are not much different, all are sodium, magnesium, aluminum, silicon, phosphorus, sulfur, chlorine, potassium, calcium, titanium, vanadium, manganese, iron, etc., but the content of each oxide is

Y. Cheng and X. Zhu, *Utilization and Environmental Impact Assessment of Red Mud Resources*, https://doi.org/10.1007/978-981-96-7411-4_2

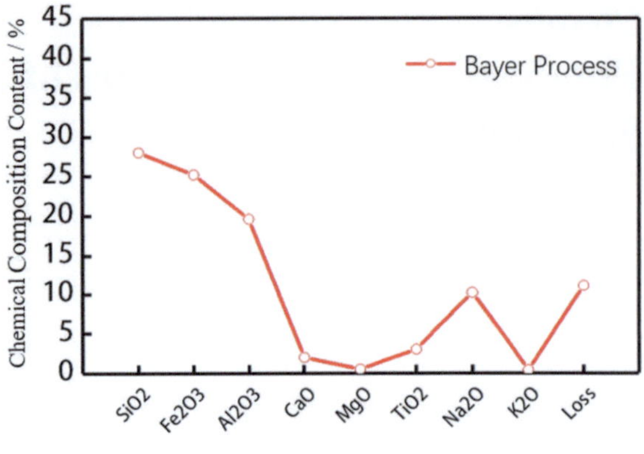

Fig. 2.1 Chemical composition of Bayer process red mud

Table 2.1 Chemical composition of Bayer process red mud used in the experiment

Red mud source	Na_2O	MgO	Al_2O_3	SiO_2	P_2O_5	SO_3	K_2O	TeO_2
Shandong aluminium	6.21	0.31	17.37	14.03	0.22	0.45	0.15	0.09
WeiQiao	5.84	0.44	22.25	10.78	0.19	0.91	0.10	0.07
Red mud source	CaO	TiO_2	V_2O_5	MnO	Fe_2O_3	SrO	ZrO_2	PdO
Shandong aluminium	9.56	4.87	0.66	0.04	24.87	0.01	0.08	0.22
WeiQiao	2.90	4.91	0.22	0.06	33.16	0.02	0.01	0.08

different. The content of silicon and calcium in Shanxi red mud is higher, while the content of iron and aluminum compounds in Weiqiao red mud is higher, with iron content 33.3% higher than that of Shanxi.

2.1.1 *The Impact of Production Process on the Composition of Red Mud*

The sintering method of alumina production mainly uses high-silica bauxite mixed with soda and lime for calcination (high temperature of 1200 ~ 1300 °C), and then alkali dissolution for solid–liquid separation and refining. Therefore, the minerals in the red mud from the sintering method mainly come from the insoluble minerals formed by the high-temperature reaction of the clinker and the derivatives, hydrates, and newly formed minerals formed by the hydration and hydrolysis during the leaching process. The production process of Bayer alumina is to dissolve the alumina hydrate in bauxite by crushing and wet grinding, and then using NaOH solution under high temperature and high pressure conditions to make sodium aluminate solution. The reaction process is

$$Al_2O_3 \cdot xH_2O + 2NaOH + aq \rightarrow 2NaAl(OH)_4 + aq$$

The silicon material in the bauxite reacts with the alkali to form insoluble hydrated sodium aluminate, which is commonly referred to as sodium silicate slag. The iron minerals in the bauxite eventually become hematite in the caustic alkali solution. Hematite is relatively stable and does not react or dissolve with alkali at 300 °C. Therefore, it is washed out with other insoluble substances to form Bayer red mud.

The X-ray diffraction diagram of the mineral composition of the red mud from the sintering method (Fig. 2.2) reveals that the main mineral components are β-dicalcium silicate, augite, calcite, α-hydrated alumina, and tricalcium aluminate, with β-dicalcium silicate being the main component, followed by calcium oxide and silicon dioxide. Augite or calcite is the result of adding quicklime and introducing CO_2 during the alumina production process. In addition, some mineral phases in the X-ray diffraction analysis did not show their characteristic spectra, such as $Al_2O_3 \cdot nH_2O$ and $SiO_2 \cdot nH_2O$, $Ca(OH)_2$, FeS(or FeS_x), $Al_2O_3 \cdot CaCO_3 \cdot nH_2O$, $Na[Fe_2S_4(OH)_2] \cdot 4H_2O$, etc. This is because some have too little content, some are colloidal minerals, and some are difficult to distinguish due to overlapping spectral lines. The XRD image of Bayer red mud is shown in Fig. 2.2b. The main mineral phases of Bayer red mud are hydrogarnet, mica, hematite, chlorite, and diaspore, etc. Research shows that the mineral composition of red mud varies with different stacking times. In fresh sintered red mud, goethite and colloidal $Fe(OH)_3$ coexist, while FeO exists in the form of siderite ($FeCO_3$). As the storage time increases, the characteristic spectrum of β-$2CaO \cdot SiO_2$ becomes less and less obvious, and the spectrum of diopside has an increasing trend, while other mineral spectra do not change significantly. The mineral composition of Bayer red mud stored at different times (Table 2.2) does not change much, only hematite slightly increases. Table 2.2 analysis of the main mineral composition of Bayer red mud of different storage years. The elements such as Si, Al, and Fe in red mud are high in content, mainly appearing in the form of stable oxides such as SiO_2, Al_2O_3, Fe_2O_3, and also contain trace amounts of heavy metals such as Mn, Zn, Cu, Cr, Pb. The phase composition of red mud is extremely complex. The general mineral composition analyzed by X-ray is as follows: β-$2CaO \cdot SiO_2$, β-$2CaO \cdot SiO_2 \cdot nH_2O$, $3CaO \cdot Al_2O_3 \cdot xSiO_2 \cdot yH_2O$, $xCaO \cdot ySiO_2 \cdot nH_2O$, α-$Al_2O_3 \cdot 3H_2O$, $Al_2O_3 \cdot nH_2O$ (gel state), $CaCO_3$ (calcite), $CaCO_3$ (aragonite), $Fe_2O_3 \cdot nH_2O$, $SiO_2 \cdot nH_2O$, β-$2CaO \cdot SiO_2$, $3CaO \cdot Al_2O_3 \cdot 6H_2O$ and $CaCO_3$, accounting for 45% ~ 65%, 5% ~ 10%, 10% ~ 15% respectively.

The main chemical elements of red mud produced by sintering method are Ca, Si, Al, Fe, Na, Ti and small amounts of Mg, K, S, with other elements in trace amounts. Due to different raw materials and sintering processes, sintered red mud is divided into ordinary red mud and ore-rich red mud, the basic chemical composition is shown in Table 2.3. Sintered red mud contains a large amount of valuable metal compounds such as TiO_2, Al_2O_3, Fe_2O_3 and Na_2O, especially the ore-rich red mud contains a higher content of TiO_2 (12.44%), which is the chemical composition of ilmenite and rutile contained in bauxite, and has strong chemical inertness. The content of Fe_2O_3 in ore-rich red mud is also higher than that in ordinary red mud, reaching 21.02%. This is due to the colloidal $Fe(OH)_3$ precipitate formed by the oxidation and hydrolysis of pyrite (FeS) associated with natural bauxite. The $Fe(OH)_3$ colloid is unstable under strong alkaline and heating conditions and tends to transform into

Table 2.2 Different storage years of bayer process red mud main mineral composition analysis

Sample	Hydrated Garnet	Calcitic feldspar	Calcium Titanite	Hematite	Monohydric Boehmite	Trihydrate Bauxite	Calcite
1-year red mud	33.2	8.9	12.5	28.3	7.8	-	4.2
8-year red mud	30.7	10.7	12.1	33.0	5.6	1.8	2.0

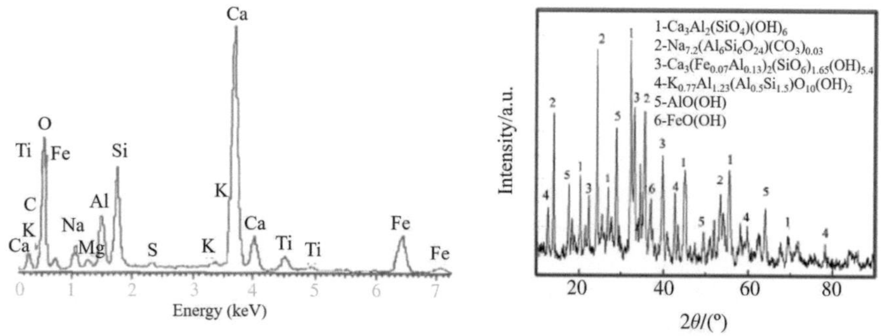

(a) Sintering method red mud (b) Bayer method red mud

Fig. 2.2 XRD images of red mud from different processes

goethite (FeOOH). The contents of CaO and SiO_2 in ordinary sintered red mud are the highest, accounting for about 60%. The precipitation and crystallization of $CaCO_3$ and SiO_2 in ordinary sintered red mud not only constitute the coarse particles of red mud, but also produce cementing effect due to their crystallization. This is because the SiO_2 liberated after the chemical decomposition of coexisting kaolinite in the alumina production process forms opal ($SiO_2 \cdot nH_2O$) and sodium silicate (Na_2SiO_3) in a strong alkaline environment, which produce irreversible cementing effect in the aging and hardening process of red mud. Al_2O_3 is 9.6%, which is the residual part after the production of $Al(OH)_3$ from bauxite. It exists in the form of $NaAlO_2$ and $Al(OH)_3$ under strong alkaline conditions. Na_2O is 3.63%, of which about 60% is Na_2CO_3, mainly in the form of Na^+ existing in the pore solution. Under dry conditions of red mud, in addition to the form of NaOH, it also forms Na_2CO_3, $NaHCO_3$, Na_2SiO_3 and $NaAlO_2$ and other salt precipitates or colloids. These soluble salts exist in the form of pore solution in the early stage of stacking, and appear on the surface as solid salts due to the dehydration, water loss and especially drying of red mud. The chemical composition of red mud will also change with the continuation of the stacking process, mainly due to the loss of water-soluble substances and reactions with water and air.

Table 2.3 Comparison of the composition of ordinary red mud and ore-rich red mud produced by sintering method

Category	Chemical composition (%)							
	SiO_2	Fe_2O_3	Al_2O_3	CaO	Na_2O	K_2O	TiO_2	Impurities
Ordinary red mud	18.87	13.25	9.6	41.1	3.63	0.25	3.71	8.45
Rich ore red mud	13.39	21.02	8.89	16.79	5.94	0.26	12.44	10

Note The remaining impurities are V, RE, Th, Er, Nb, among which Sc is 170 mg/L in rich ore red mud

Bayer process red mud is a special waste with high iron and low silicon. Due to the influence of the external environment, the red mud that has been piled up for a long time undergoes an oxidation reaction, and the content of ferric oxide and calcium oxide in the chemical composition changes, while other chemical components basically do not change. The difference in ferric oxide content is as much as 4.72%, which is also reflected in the color of these two soil samples. It is obvious that the 8-year-old red mud turns red, while the 1-year-old red mud appears more brown.

The main components and properties of red mud vary with the source of bauxite and the different alumina production processes, and also change with the different storage time. From its mineral composition, it is similar to general clay, so red mud can be used as a substitute for general clay.

2.1.2 *Influence of Ore Source on Red Mud Composition*

(1) Indonesian Conventional Ore Red Mud

The regular bauxite red mud has a particle size of 0.4 ~ 12 μm. The median particle size is 11.16 μm. The main minerals in the sample are hematite (a-Fe_2O_3), goethite (a-$Fe(OH)_3$), siderite Fe_2CO_3, hydrated sodium aluminosilicate ($Na_2O \cdot Al_2O_3 \cdot 1.68SiO_2 \cdot 1.73H_2O$), monohydrate gibbsite, followed by kaolinite and calcite, and a small amount of natural alkali, water glass, caustic soda, etc. (Table 2.4).

(2) Composition of low-alumina low-grade bauxite red mud

The main phases of low-alumina low-grade bauxite red mud are aluminum goethite, hematite, and α-$Al(OH)_3$. The red mud particle size is 0.3 ~ 350 μm, and the median particle size is 3.5 μm (Table 2.5).

(3) Australian bauxite red mud

The above are the particle size distributions of two Australian bauxite red muds. As can be seen from the figure, the particle size of Australian bauxite red mud is between 0.2 μm ~ 1000 μm, with an average particle size $d_{50} = 11.65$ μm, boundary particle

Table 2.4 Chemical composition of Indonesian conventional bauxite Bayer process red mud (%)

SiO_2	Fe_2O_3	Al_2O_3	CaO	TiO_2	Na_2O	K_2O	Loss on Ignition
30.40	34.24	13.82	3.40	1.95	4.90	0.21	10.1

Table 2.5 Chemical composition of low-alumina low-silicon bauxite red mud Bayer process red mud (%)

SiO_2	Fe_2O_3	Al_2O_3	CaO	TiO_2	Na_2O	K_2O	Loss on ignition
17.40	57.96	10.81	0.11	0.97	2.20	0.05	9.98

size $D_{10} = 0.635$ μm, $D_{90} = 224.361$ μm, the dispersion of particle size distribution is large, $(D_{90} - D_{10})/D_{50} = 8.12$, at the same time, the particle size distribution chart has "small double peaks" and hump signs, particle sizes of 0.5 and 30 μm occupy a higher proportion. From the results, the main mineral components of Australian bauxite red mud are: hematite (Fe_2O_3), sodium silicate slag ($Na_2O \cdot Al_2O_3 \cdot 2SiO_2 \cdot H_2O$), quartz, rutile, monohydrate soft alumina, trihydrate alumina, etc. There are also some vein stone phases and iron phases, with low content, which are not fully presented in the diffraction pattern (Table 2.6).

(4) Composition of Ghanaian bauxite red mud

As can be seen from the figure, the main minerals of Ghanaian red mud are hematite, sodium silicate slag, goethite, and quartz, among which the iron oxide content is relatively high at 68.16%, and the quartz content is low (Table 2.7).

(5) Malaysian bauxite red mud

From the figure, it can be seen that the main minerals of the Malaysian bauxite red mud are hematite, sodium silicate slag, goethite, quartz, and rutile, among which the titanium dioxide content is relatively high at 9.84%, and it is in the form of rutile, with a low quartz content (Table 2.8).

The particle size of Malaysian bauxite red mud is between 0.3 and 120 μm. The overall particle size of the red mud is fine. It has a bimodal structure. The content of goethite in the red mud is relatively high.

Table 2.6 Chemical composition of Australian Gove bauxite red mud (%)

SiO_2	Fe_2O_3	Al_2O_3	CaO	TiO_2	Na_2O	K_2O	Loss on ignition
11.66	40.11	22.49	0.14	5.97	5.74	0.11	8.43
13.6	37.38	21.2	0.92	7.65	7.24	0.49	8.82

Table 2.7 Chemical composition of Ghanaian bauxite red mud (%)

SiO_2	Fe_2O_3	Al_2O_3	Na_2O	Loss on ignition
3.04	68.16	10.91	0.84	8.86
3.58	65.66	12.70	1.24	8.66
3.78	65.74	12.42	1.59	8.49

Table 2.8 Chemical composition of Malaysian bauxite red mud (%)

SiO_2	Fe_2O_3	Al_2O_3	TiO_2	Na_2O	Loss on ignition
8.38	53.29	13.37	9.84	4.82	8.89
8.91	52.61	13.47	9.66	5.17	8.68

Table 2.9 Chemical composition of different bauxite red mud (%)

Bauxite source	SiO_2	Fe_2O_3	Al_2O_3	TiO_2	Na_2O	Loss on ignition
Brazilian ore	16.98	42.84	17.19	4.59	8.80	
Indian ore	7.38	42.96	18.68	15.60		9.84
Indian ore	18.76	10.47	21.33	9.44	10.73	11.94
Australian GOVE ore	14.12	37.14	23.30		7.50	8.09
Guyana ore	57.36	4.26	23.30	5.17	6.88	5.56
Venezuelan ore	22.05	47.60	11.84	4.22	3.39	9.14
Tanzanian ore	4.53	55.10	14.98	4.22	2.56	11.91
Dominican ore	6.46	44.84	25.32	5.78	3.84	9.20

(6) Other bauxite red mud components

In addition to the above-mentioned red mud, the study also measured the composition of other bauxite red mud provided by overseas centers, and the results are as follows (Table 2.9).

2.1.3 *Typical Enterprise Red Mud Component Analysis*

The chemical composition of red mud in Shandong Province is shown in the Table 2.10. The content of calcium oxide and silicon dioxide in sintered red mud is relatively high, reaching 65%, and the mineral composition is mainly β-dicalcium silicate (Table 2.11).

The chemical composition of red mud from the sintering process and the Bayer process can be seen in Table 2.12.

The main chemical components of red mud from the sintering method are about 50% SiO_2 and CaO, about 15% iron and aluminum oxides, a small amount of TiO_2, MgO, Na_2O, K_2O and dozens of trace elements (Table 2.13).

Table 2.10 Chemical composition of red mud from different production processes of Shandong aluminium (%)

Production process	SiO_2	Al_2O_3	Fe_2O_3	Na_2O	CaO	MgO	TiO_2	K_2O	Deduction
Sintering method	22.67	7.68	10.97	2.93	40.78	1.77	3.26	0.38	11.77
Combined method	20.56	8.10	12.10	2.77	44.86	2.02	5.09	0.35	8.18
Bayer method	12.04	17.21	23.07	4.00	14.90	1.36	5.46	1.0	6.35

Table 2.11 Chemical composition of red mud from various stockyards of Shandong aluminium (%)

Number	SiO_2	Fe_2O_3	Al_2O_3	CaO	MgO	TiO_2	Na_2O	K_2O	Loss on ignition	SO_3
1#	28.09	25.54	19.81	1.71		2.15	10.17	0.31	11.43	
2#	19.34	11.93	8.32	39.67	0.72	3.56	4.60	0.30	10.60	0.29
3#	16.81	40.61	17.63	0.75		4.79	10.02		8.55	0.32
4#	14.19	7.83	5.42	33.35	0.86	2.04	5.91	0.96	28.89	

1#: Bayer process red mud from alumina plant; 2#: sintering process red mud; 3#: secondary aluminum south line; 4#: one pile

Table 2.12 Chemical composition of red mud (measured in 2005)

	SiO_2	Fe_2O_3	Al_2O_3	CaO	MgO	TiO_2	Na_2O	K_2O	LOI	ST
Sintering process red mud	16.11	7.38	8.30	33.20	0.43	2.00	2.92	0.84	26.50	0.36
Bayer process red mud	17.87	32.66	20.30	3.18	–	2.07	8.06	0.08	11.59	–

Table 2.13 Chemical composition of red mud (measured in March 2014)

	SiO_2	Fe_2O_3	Al_2O_3	K_2O	Na_2O	TiO_2	CaO	MgO	SO_3	SC	Ignition loss
New red mud from sintering method	19.34	11.93	8.32	0.30	4.60	3.56	39.67	0.72	0.29	58	10.60
A pile of dry red mud	14.19	7.83	5.42	0.96	5.91	2.04	33.35	0.86	–	69	28.89
A pile of dam	15.63	6.77	6.65	–	3.66	–	34.8	–	–	–	28.47

The main chemical components of red mud from the Bayer process are about 20% SiO_2 and CaO, about 50% iron and aluminum oxides, a small amount of TiO_2, MgO, Na_2O, K_2O and dozens of trace elements. Compared with red mud from the sintering method, Bayer red mud has the following characteristics: it contains much more sodium, has a very low calcium oxide content, and contains a higher amount of iron.

2.2 Basic Physical Properties of Red Mud

From the perspective of geotechnical engineering alone, the physical and mechanical properties of the red mud from the Ji Qing Expressway demonstration section are similar to those of clay, mainly composed of extremely fine particles, with a particle group content of 60% or more in the range of 1 ~ 100 μm, as shown in Table 2.14. Its porosity is about 0.78, as shown in Tables 2.15 and 2.16.

2.2.1 *Red Mud Particle Characteristics*

Due to the differences in the origin of bauxite, mineral composition, and aging time of red mud, different batches of red mud samples may have differences in physical and chemical properties.

Red mud samples I come from the second stockyard of Chalco Shandong Branch (referred to as Shandong Aluminum Red Mud) and Shandong WeiQiao Aluminum Company (referred to as WeiQiao Red Mud), both with half a year of dry storage time, see Fig. 2.3. Shandong Aluminum Red Mud is brick-red and blocky, easy to crush into powder-like solids with small particle size; WeiQiao Red Mud is reddish-brown and muddy, sticky, with particles clumping together, no large particles appear. Red mud will clump when it loses water at rest. Before the test, it needs to be crushed and ground, and it can be tested after passing through a 4.75 mm square hole sieve.

Table 2.14 Particle size composition of Bayer red mud

Source of red mud	Particle size composition/%			
	< 1 μm	1 ~ 10 μm	10 ~ 100 μm	> 100 μm
Chalco Shandong branch	4.81	28.94	64.18	2.07
WeiQiao aluminum power	1.81	31.94	44.18	22.07

Table 2.15 Engineering mechanics index of Bayer process red mud I

Moisture Content %	Natural bulk density g/cm^2	Dry bulk density g/cm^2	Porosity ratio	Liquid limit %	Plastic limit %
33.3	2.03	1.52	0.780	58.5	42.1

Table 2.16 Engineering mechanics index of Bayer process red mud II

Compression coefficient MPa^{-1}	Compression modulus MPa	Cohesion kPa	Internal friction angle/degree	Saturated bulk density g/cm^2
0.19	9.37	40.0	25.5	1.96

Fig. 2.3 Morphology of two kinds of red mud **a** Shandong Aluminum Red Mud, **b** WeiQiao Red Mud

The moisture content of the stockyard red mud is mostly between 35 and 55%, and some of the stock materials even reach more than 70%, with strong water retention. The natural moisture content and liquid plastic limit of the two kinds of red mud are shown in Table 2.17.

The typical distribution of red mud particles is shown in Fig. 2.4. The content of red mud particles less than 75 μm exceeds 80%, and the content less than 10 μm is about 10%. Most particles are between 20 and 50 μm; The uniformity coefficient C_u is 29.0, and the curvature coefficient C_c is 0.31. The C_u value is large, the particle size distribution of red mud particles is wide, the gradation is good, and the particles are easy to compact, but the C_c value is small, and the material lacks large particles. Shan Aluminum Red Mud belongs to high liquid limit silt, Wei Bridge red mud belongs to low liquid limit clay.

This project obtained red mud samples from the Bayer red mud dump in Shandong Aluminum Industry (Zibo), with a natural moisture content between 42 and 47%, which is significantly higher than that of general soils such as cohesive soil. A total of 5 representative samples (codes A–E) were taken for the limit moisture content test, and the results are shown in Table 2.18.

According to Table 2.18, the liquid limit of Bayer red mud w_L and the plasticity index I_p are plotted into the plasticity chart, with Ip as the vertical coordinate and w_L as the horizontal coordinate for the soil quality classification of Bayer red mud, as shown in the following Fig. 2.5. As can be seen from the figure, Bayer red mud

Table 2.17 Natural moisture content and liquid plastic limit of red mud

Type of red mud	Natural moisture content/%	Liquid limit/%	Plastic limit/%	Plasticity index
Shandong Aluminum Red Mud	41.4	50.9	44.0	6.9
Wei Bridge red mud	40.73	48.0	33.4	14.6

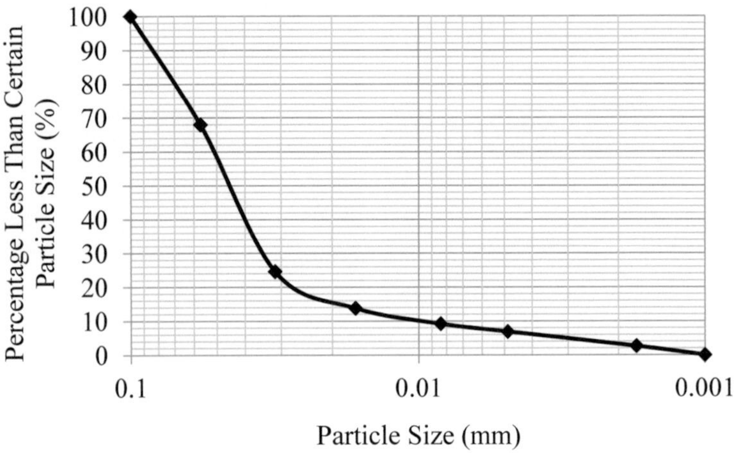

Fig. 2.4 Composition of red mud particles

Table 2.18 Results of Bayer red mud limit moisture content test

Index code	Liquid limit	Plastic limit	Plasticity index
Code A	44.12	31.56	12.56
Code B	45.99	31.62	14.38
Code C	45.17	31.61	13.57
Code D	45.91	30.64	15.26
Code E	44.56	31.47	13.09

w_L, I_p are mainly distributed below line A and to the left of line B of the plasticity chart. According to the classification method of soil according to the plasticity chart in the "Highway Soil Test Procedure", Bayer red mud has the relevant properties of low plasticity silty clay.

The particle size ratio and particle curve of the red mud sample are shown in Figs. 2.6 and 2.7 respectively. The particle sieving results show that the particle size of Bayer red mud is less than 0.074 mm, accounting for 94% of the total, up to 98%, indicating that the particle size of red mud is fine (Table 2.19).

The composition analysis of red mud particles at different times is given. The 8-year red mud is mainly composed of silt and clay particles in the fine particle group, with more silt particles than clay particles. The silt particles account for 72%, and the clay particles account for about 19.9%. The coefficient of non-uniformity $C_u = 13.0$, and the curvature coefficient $C_c = 0.692$. In the 1-year red mud, the silt content is as high as 86.5%, and the clay is only 4.3%. The particle size distribution is more concentrated, with $C_u = 5.0$ and $C_c = 0.80$. Both types of red mud are highly dispersed. According to the particle composition analysis results, red mud belongs to fine-grained soil and is classified as low liquid limit silt in engineering classification.

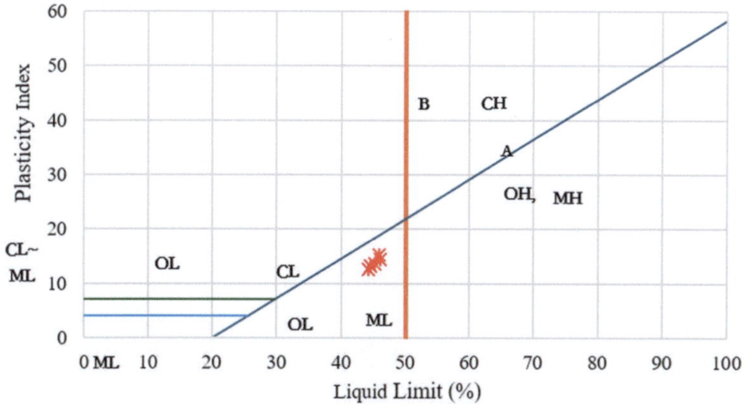

Fig. 2.5 Classification of Bayer red mud plasticity chart

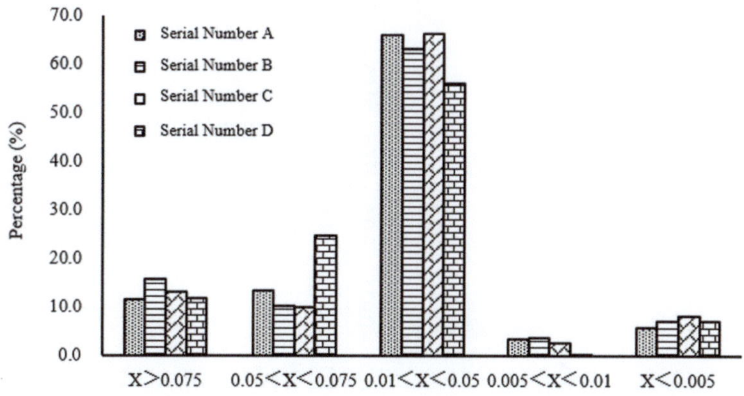

Fig. 2.6 Particle size ratio chart

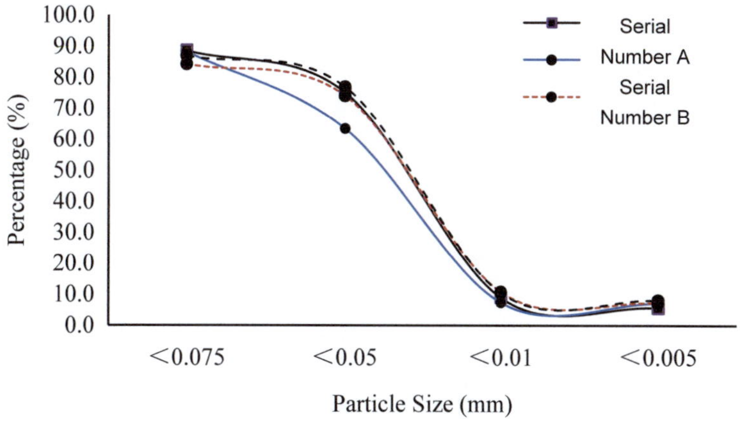

Fig. 2.7 Particle curve of red mud

Table 2.19 Composition analysis of red mud particles at different times

Sample	Coarse soil ($\geq$ 0.074 mm)/%				Fine soil (< 0.074 mm)/%				
	Gravel	Coarse sand	Medium sand	Fine sand	Silt particles				Clay particles
	20 ~ 2	2 ~ 0.5	0.5 ~ 0.25	0.25 ~ 0.074	0.074 ~ 0.05	0.05 ~ 0.01	0.01 ~ 0.005	0.005 ~ 0.002	< 0.002
8-year red mud	0.7	0.9	0.3	6.1	10.2	26.6	13.9	21.3	19.9
1-year red mud	1.2	1.8	0.4	5.9	8.8	21.6	20.2	35.9	4.3

2.2.2 Hydraulic Properties of Red Mud

The hydraulic properties of red mud, mainly including permeability, swelling and liquefaction, are constrained by the material composition of red mud and its evolution after stacking.

1. High water retention

Due to the addition of excess NaOH, CaO and CO_2 in the process of refining alumina, red mud becomes a strong alkaline material with a pH value above 11.0. Therefore, red mud contains a large amount of NaOH and $NaHCO_3$ electrolytes (dispersants), which determines that red mud is in a highly dispersed colloidal state in its natural state. Its high water content and low density are related to this. The dry weight of red mud is only 7.9 ~ 8.1 kN m^{-2}, the porosity is as high as 2.45 ~ 2.61, and the water content is as high as 36 ~ 80%, this is extremely rare for silt, but the composition and genesis of red mud are completely different from silt, so the study of its characteristics genesis helps to evaluate and modify the engineering properties of red mud. The results of mineral composition determination show that the water content of kaolinite and calcite, which constitute the skeleton minerals of red mud, is not large. The real hydrophilic substances in red mud are free SiO_2, Al_2O_3 and Fe_2O_3, which form a lot of adsorbed water due to their high content. However, the water content of red mud is as high as 36 ~ 90%, especially under the action of vibration, a large amount of gravitational water is precipitated, which is related to the special microstructure of red mud, that is, high porosity. Although the results of full dispersion particle analysis show that 0.001 ~ 0.01 mm particles account for 60 ~ 75% of the total amount of red mud, the results of scanning electron microscope observation show that most of the particle units constituting red mud are highly dispersed materials of < 1 μm. Due to the cementation of free oxides, aggregates with a diameter of 0.01 ~ 0.02 mm and round or elliptical granules (primary structures) with a diameter of 0.1 ~ 0.5 mm are formed, and a pore system between granules, aggregates and particles is formed accordingly, such as interparticle pores (d < 2 μm), interaggregate pores (d = 2 ~ 10 μm), intergranular pores (d = 100 ~ 500 μm). The high porosity of red mud is mainly related to the interparticle pores with a pore size of d < 2 μm, while the large and medium-sized pores with d > 10 μm control the size of permeability, but their proportion in the total pores is not high. The results of mercury intrusion porosimetry show that 68.27% of the total pore volume is small pores and micropores with a pore size of d < 2 μm (especially small pores with a pore size of 0.2 ~ 2.0 μm account for about 40%), which is related to the extremely high water content of red mud. The precipitation of water in red mud under vibration conditions is mainly related to the precipitation of contact water (capillary water) in pores with d > 20 μm, and most of the water is preserved on the surface of fine particles and in interparticle pores due to the huge suction force in micropores with d < 2 μm.

2. Permeability

Red mud is a weakly permeable material, its permeability coefficient is generally $2.57 \sim 3.62 \times 10^{-5}$ cm/s, which is similar to the permeability coefficient of loess. Under the action of seepage water, the permeability coefficient of red mud tends to increase gradually, but the increase is not large, which is the result of the leaching of soluble salts such as $NaHCO_3$ and the flocculation of Ca^{2+}. Under the action of HCO^{3-}–Ca^{2+} type seepage water, the soluble salts in red mud decrease rapidly, and the salt content of seepage water decreases by about 10 times in less than 6 h, that is, from 8 ~ 10 g/L to 0.87 ~ 0.96 g/L, which is mainly caused by the loss of Na salts and alkali leaching. It can be seen that the alkalization of groundwater and soil by red mud after stacking is inevitable.

The water content and porosity of Bayer red mud are relatively high, the permeability coefficient of natural red mud is relatively large, but the permeability coefficient decreases greatly after compaction. The test results show that the average permeability coefficient of saturated water content field natural red mud, when its water content is 36% ~ 42%, is 4.5×10^{-6} cm/s. However, after compaction of red mud, when the dry density reaches 1.70 g/cm^3, the saturated permeability coefficients of 8-year red mud and 1-year red mud of China Aluminum Shandong Co., Ltd. are 1.38×10^{-7} cm/s and 1.63×10^{-7} cm/s respectively, and the permeability is extremely small. Therefore, the permeability of red mud can be reduced by rolling.

3. Swelling and shrinkage characteristics

In general clay, there are clay colloid particles, because the silicon-aluminum ratio (SiO_2/Al_2O_3) is large, its surface can form a double electric layer in water, thus causing the property of clay swelling when encountering water and shrinking when losing water. Although red mud is high in water content, it does not shrink after drying and losing water, indicating that high water content is not the result of the existence of hydrophilic minerals. At the same time, red mud has no obvious expansibility. With the increase of drying degree, obvious hardening occurs, and there are a lot of white precipitates and cementation on the surface. The shrinkage test shows that the soil sample is air-dried for 45 days, although the water content is reduced, there is basically no shrinkage deformation. Bayer red mud does not contain clay minerals, mainly Fe_2O_3. Minerals with good stability in large amounts of water, therefore, the expansibility and shrinkability of red mud are very small. The free expansion rate of red mud stored for 1 year is measured to be – 1%; the average expansion rate of red mud CBR specimens soaked in water after 8 years of storage is only 0.44%, while the expansion rate of general clay can reach 2 ~ 3%, and that of expansive soil can reach as high as 7 ~ 8%. There is no volume shrinkage phenomenon in the process of red mud drying and losing water.

(4) Liquidity

Red mud has a high water content and strong water retention characteristics. The newly deposited red mud, due to its high water content, has a value that is mostly greater than the liquid limit, and the powder and sand particles are hydrophobic calcite and dolomite, therefore, the newly deposited red mud has the possibility of liquefaction under vibration. Compaction tests on red mud with a water content of 26 ~ 28% show that the surface of the specimen is soft and elastic and water is analyzed, so high water content red mud has the possibility of liquefaction.

2.3 Basic Engineering Characteristics of Red Mud

The mechanical properties of Bayer red mud, such as its shear resistance, are evaluated through unconfined compressive strength tests and triaxial shear tests, and its compressibility is evaluated through consolidation tests.

2.3.1 Compaction Performance

Compaction tests use mechanical energy to reduce the pore volume between soil particles and make the contact more compact. Compacted soil can improve its strength and resistance to deformation. Taking the red mud of Weiqiao Aluminum and Electricity Company and China Aluminum Shandong Co., Ltd. as examples, the test selects Bayer red mud samples that have been dried and crushed and passed through a 5 mm sieve. According to the relevant regulations in JTG E40-2007 "Highway Geotechnical Test Procedures", standard heavy compaction tests are carried out on red mud samples prepared with a water content 2% higher than the plastic limit. The test results are shown in Figs. 2.8 and 2.9. The maximum dry density of red mud from China Aluminum Shandong Branch is 1.34 g/cm^3, and the optimum water content is 38.3%. The maximum dry density of Weiqiao Aluminum and Electricity red mud is 1.65 g/cm^3, and the optimum water content is 27.3%.

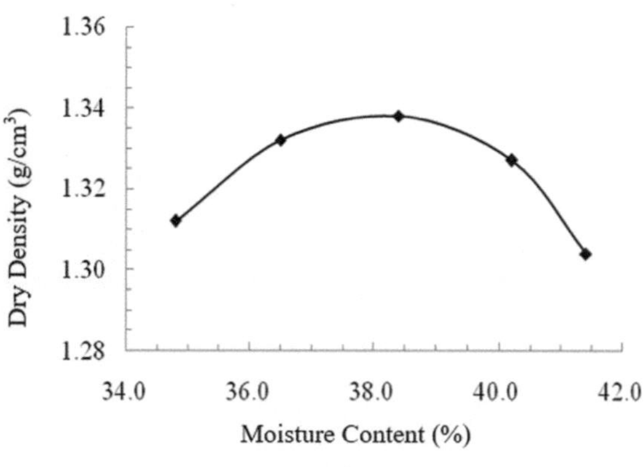

Fig. 2.8 Compaction curve of Shandong Aluminum Red Mud

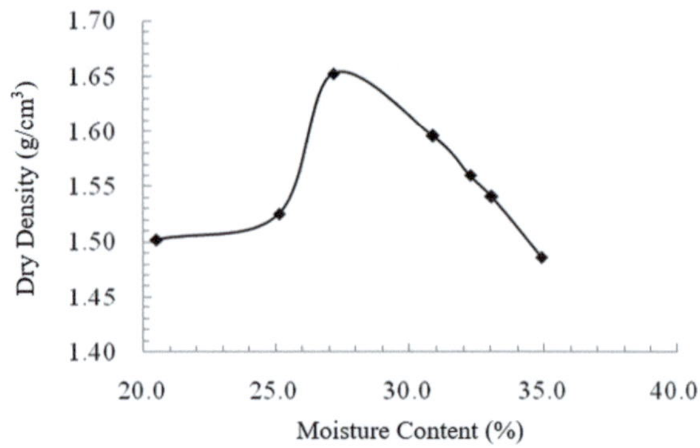

Fig. 2.9 Compaction curve of Weiqiao Red Mud

2.3.2 Unconfined Compressive Strength

Unconfined compressive strength is the limit strength index of the specimen to resist axial pressure under the condition of no lateral pressure, which is the simplest and most direct index reflecting the strength characteristics of the material. In order to study the feasibility of low-cost and large-scale utilization of red mud, the test is carried out according to the "Highway Geotechnical Test Procedures" (JTG E40-2007), using a lime soil strain unconfined compressive strength tester. After the specimen is demolded, it is sealed in a bag and placed in a constant temperature and humidity curing room for curing. The age is controlled at 1, 3, 7, and 28 days for unconfined compressive strength tests (Fig. 2.10).

According to the test results of the unconfined compressive strength of red mud statically pressed under a compaction degree of 96% (Fig. 2.11), it shows that the unconfined compressive strength of pure Bayer red mud is about 0.4 ~ 0.6 MPa, and the strength does not change significantly with age. This indicates that its own

Fig. 2.10 Unconfined compressive strength test of red mud

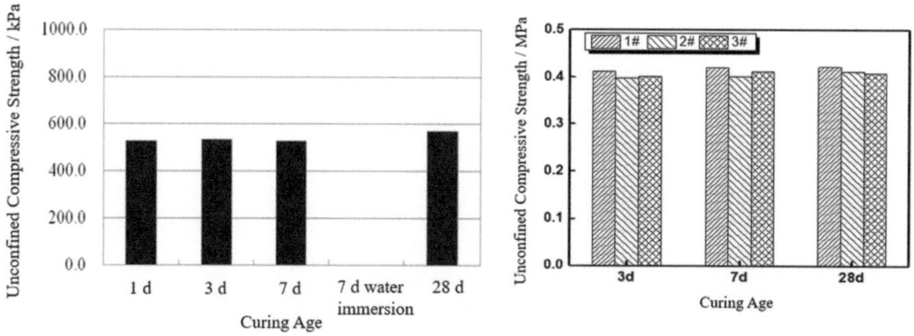

Fig. 2.11 Comparison of unconfined compressive strength of Bayer red mud at different curing ages

properties are relatively stable, and there is no hydration reaction activity, and no physical and chemical reactions occur under curing conditions. After soaking in water for 7 days, the specimen completely disintegrated in a short time.

2.3.3 Shear and Compression Deformation of Red Mud

Through the triaxial (UU) test, the shear performance of Bayer red mud is analyzed and evaluated. According to the actual engineering load, confining pressure is added to carry out the shear test and record the test data. Figure 2.12 shows the shear strength envelope of red mud. As can be seen from the figure, the internal friction angle of Bayer red mud is 21.73°, and the cohesion is 42.3 kPa, which indicates that the cohesion and internal friction angle of Bayer red mud are both small, and the shear resistance is poor.

Through the consolidation test, the compression coefficient index of Bayer red mud is measured, so as to evaluate the compressibility of Bayer red mud. A total

Fig. 2.12 Shear strength envelope of Bayer red mud

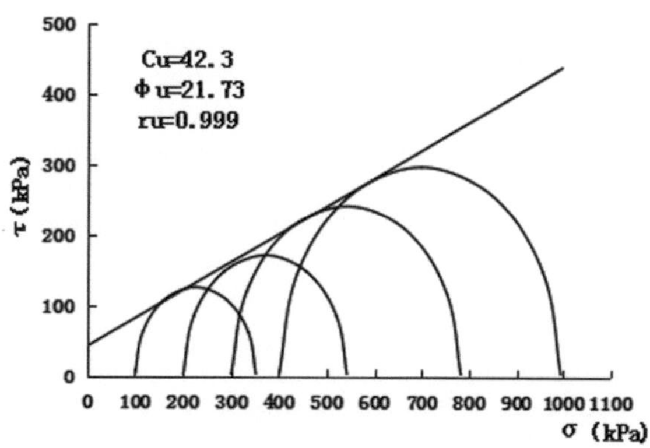

Table 2.20 Consolidation compression test results of red mud

Sample number	Compression coefficient/ MPa^{-1}	Compression modulus/ kPa	Volume compression coefficient/MPa^{-1}
1#	0.14	11,928.6	0.084
2#	0.31	5387.1	0.186
3#	0.28	5964.3	0.168
4#	0.39	4282.1	0.234
5#	0.18	9277.8	0.108

of 5 red mud samples were selected for the test, and the compression coefficient, compression modulus, and volume compression coefficient of red mud are shown in Table 2.20. From the results, it can be seen that the compression coefficient of Bayer red mud is between 0.1 and 0.4 MPa^{-1} Between them, it belongs to medium compressible soil with a small compression modulus, and there is a large compression deformation under the action of load. If Bayer method red mud is directly used as roadbed filling material, its compressibility cannot be guaranteed, and it needs to be treated to improve its compressibility and improve its engineering performance.

2.3.4 *Load Bearing Test*

The CBR value of the roadbed filler is a reflection of the soil's resistance to local shear, and is an important indicator for evaluating the strength and load-bearing capacity of the roadbed. The load-bearing capacity refers to the ability of the roadbed top to resist deformation under the action of the upper load (road surface structure, vehicle load). As the main bearer of the road surface structure and the upper vehicle load, the stress–strain characteristics of the roadbed under the action of the upper load have a significant impact on the overall strength and stability of the road surface structure. Insufficient load-bearing capacity is one of the reasons for the damage to the road surface structure.

This project uses the California Bearing Ratio (CBR) as the strength index of the Bayer method red mud roadbed. To characterize the degree of influence of immersion conditions on Bayer method red mud, consider the change in penetration value index under adverse conditions (immersion) and normal environment (non-immersion), and then analyze the load-bearing capacity of Bayer method red mud under immersion conditions, and evaluate the degree of influence of immersion conditions on Bayer method red mud.

(1) CBR penetration of red mud

This test conducted three parallel tests, numbered 1#, 2#, and 3#. Figures 2.13 and 2.14 show the change in penetration of immersed red mud and non-immersed red mud with unit penetration pressure. It can be seen from the figure that the penetration of the red mud sample gradually increases with the increase of the penetration pressure. Under the same unit penetration pressure, the penetration of the immersed red mud sample is significantly larger than that of the non-immersed red mud sample.

The CBR value test results of red mud under immersion and non-immersion conditions are shown in Table 2.21, Figs. 2.13 and 2.14. It was found from the results that the immersion conditions have a significant impact on the CBR value of Bayer method red mud. Compared with the non-immersion environment, the CBR value of immersed Bayer method red mud is significantly reduced. After immersion, the shear resistance of Bayer method red mud weakens, the strength decreases, and the load-bearing capacity also decreases. This shows that under the influence of adverse immersion conditions, the road performance of Bayer method red mud is significantly reduced (Figs. 2.15 and 2.16).

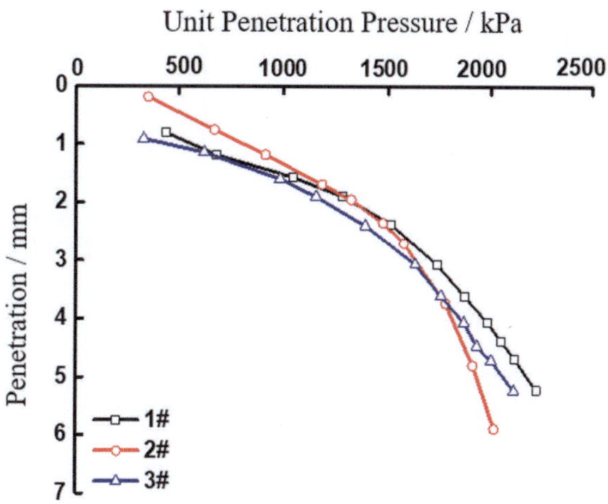

Fig. 2.13 p–l relationship curve of immersed red mud sample

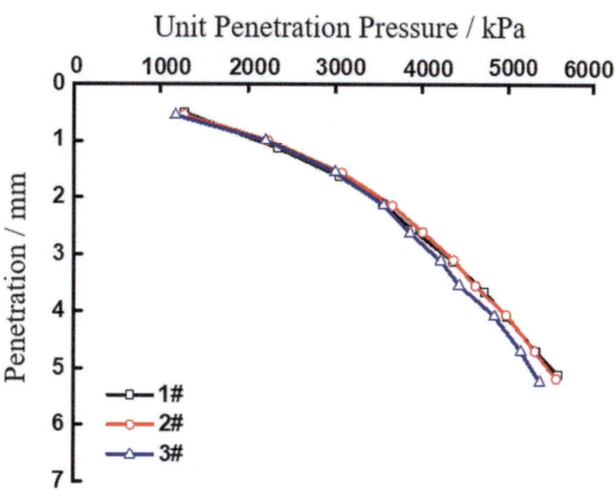

Fig. 2.14 p–l relationship curve of non-immersed red mud sample

Table 2.21 CBR values of Bayer method red mud under different conditions

Number condition	1#		2#		3#	
	$CBR_{2.5}$/%	CBR_5/%	$CBR_{2.5}$/%	CBR_5/%	$CBR_{2.5}$/%	CBR_5/%
Soaked	15.1	14.5	18.7	16.9	17.3	16.3
Unsoaked	45.1	42.3	47.9	45.8	41.2	49.1

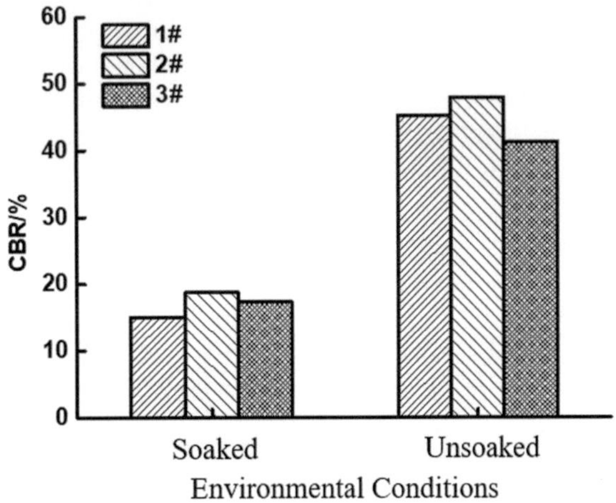

Fig. 2.15 $CBR_{2.5}$ values of Bayer process red mud under different conditions

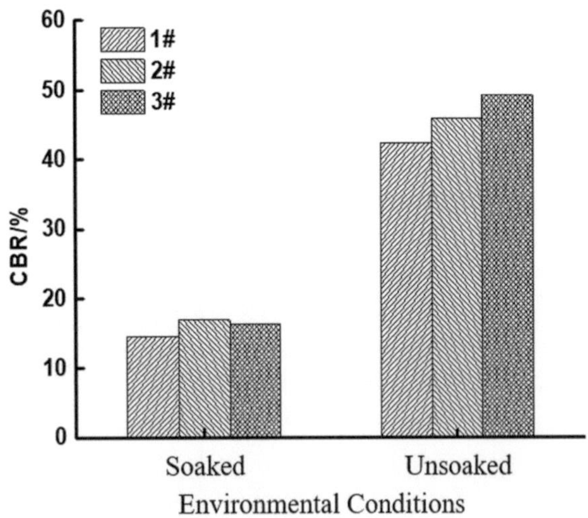

Fig. 2.16 CBR_5 values of Bayer process red mud under different conditions

(2) Swelling amount of red mud after soaking

The swelling amount of Bayer process red mud after soaking refers to the ratio of the change in the height of the specimen after soaking for 4 days and nights to the original height of the specimen (120 mm). The change in the height of the sample over time is shown in Fig. 2.17. As can be seen from the figure, the swelling amount of the red mud after soaking can reach a stable state at about 30 h, and the swelling amount hardly changes after stabilization.

Table 2.22 and Fig. 2.18 show the swelling amount of red mud in soaked and unsoaked environments. The results show that the swelling amount of Bayer process red mud changes significantly under different environmental conditions. Compared with unsoaked, the swelling amount of Bayer process red mud increases under soaked conditions. This indicates that soaking conditions, while reducing the strength and bearing capacity of Bayer process red mud, increase its swelling amount, thereby greatly reducing its engineering properties.

After the red mud sample is soaked, the water on the top of the specimen is poured out and the water on the wall of the cylinder is wiped dry, and the weight can obtain the water absorption of the red mud, as shown in Table 2.23. The specimen has a certain water absorption during the soaking process, and according to the above analysis of the physical properties of Bayer process red mud, this is related to its high plasticity index and fine particles.

In conclusion, Bayer process red mud is sensitive to water. After encountering water, its strength decreases and its expansibility increases, greatly increasing the engineering risk. At the same time, a large amount of metal ions and radioactive elements seep into the ground during the soaking process, causing pollution to the environment while reducing its engineering properties. Therefore, when designing the foundation of Bayer process red mud, it is necessary to control its seepage and drainage facilities to ensure the safety of the project.

2.4 Summary of This Chapter

Based on the above experimental research on the mineral chemical characteristics, basic physical and mechanical properties, and road use characteristics of Bayer process red mud, Bayer process red mud is a special soil with high liquid limit, high water content, low weight, and high alkalinity. According to geotechnical engineering specifications, Bayer process red mud should be classified as high liquid limit clay, which can be used in highway subgrade, foundation, backfill dam and other engineering constructions. In order to better solve the engineering performance and environmental safety needs of red mud, it can also be further improved by adding a suitable amount of solidification modification materials to improve the foundation bearing capacity, reduce the pollution of red mud, and achieve good engineering characteristics.

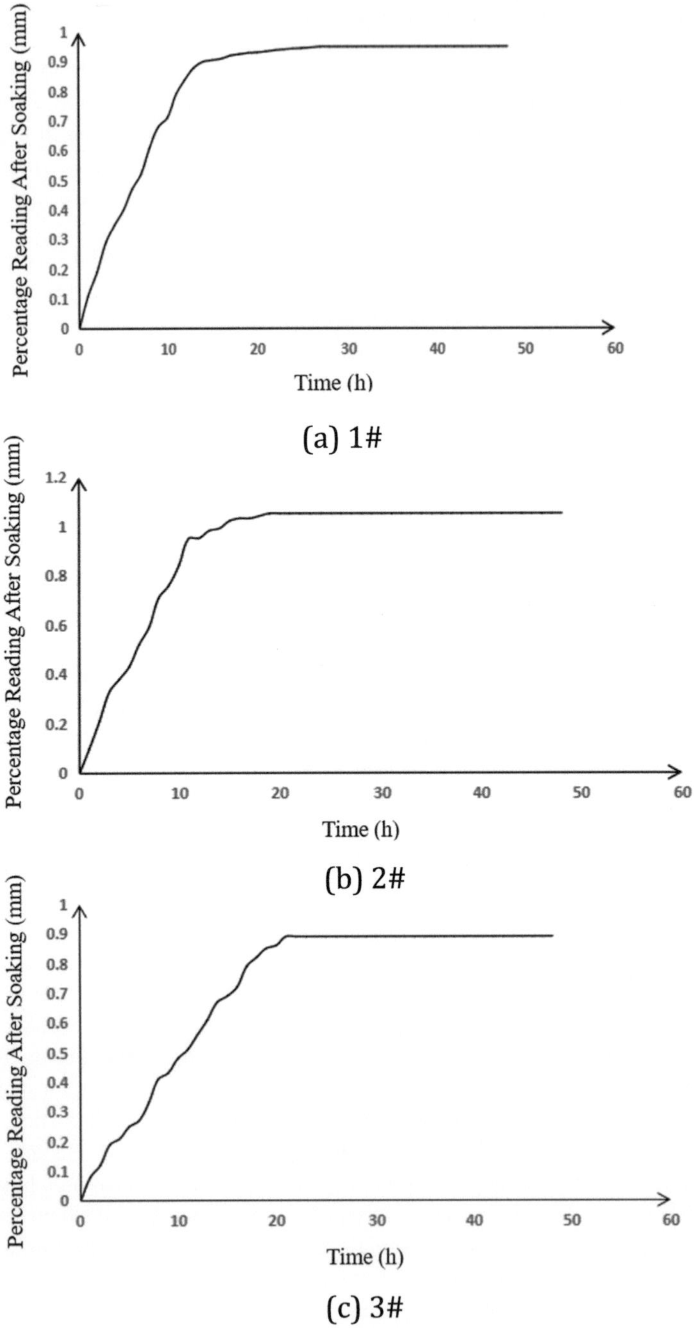

Fig. 2.17 Change in the swelling amount of red mud specimens over time

Table 2.22 Swelling amount of Bayer process red mud after soaking

Test number	Swelling amount	
	Soaked	Unsoaked
1#	0.79	0.12
2#	0.88	0.20
3#	0.74	0.09

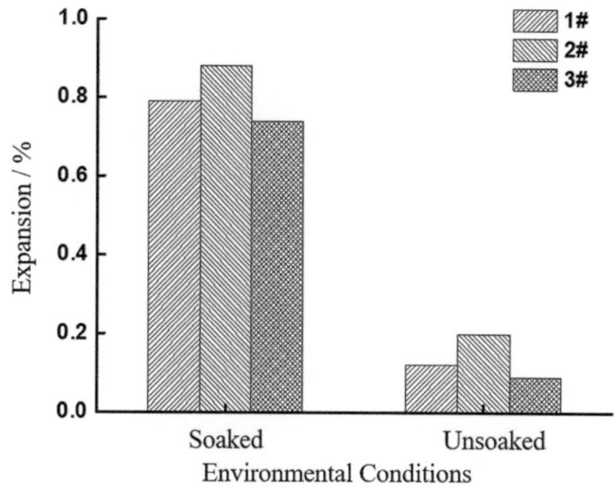

Fig. 2.18 Swelling amount of Bayer process red mud under different environments

Table 2.23 Water absorption of red mud

Number	Cylinder mass/g	Cylinder + test piece mass/g	Cylinder + test piece mass after soaking/g	Water absorption/g	Average water absorption/g
1#	4591.2	9085.3	9285.3	200.0	205.6
2#	4596.6	9102.2	9290.1	187.9	
3#	4592.5	8997.6	9226.4	228.8	

Chapter 3
Environmental Protection Characteristics of Red Mud Road Engineering

3.1 Overview

Due to the addition of high-concentration sodium hydroxide solution in the process of alumina production, the complex chemical reaction between bauxite and sodium hydroxide results in the emission of red mud containing a certain amount of alkali, fluorides, heavy metals and other pollutants. Red mud and its filtrate, in the process of entering the groundwater body, not only bring their own pollutants, but also bring a large amount of sulfate and calcium ions into the groundwater body through the solute filtration effect, causing secondary pollution. Other ions in red mud also have a significant impact on water quality. Due to the harmful substances contained in red mud, with the washing of rainwater, red mud filtrate seeps into the soil water system and rivers, causing serious pollution to groundwater, surface water and the surrounding environment. At the same time, after the surface of the red mud storage is dried, it will produce dust in windy days, causing pollution to the atmosphere and the surrounding environment.

3.1.1 Impact on the Water Environment

The main pollutants in red mud are alkali, fluorides and heavy metals, which seep into groundwater through various channels. Generally, alkali does not directly harm the human body. On the one hand, high-alkaline wastewater seeps into groundwater or enters surface water, causing the pH of the water body to rise, exceeding the national minimum standard, and causing water pollution; on the other hand, the toxicity of compounds in water is often affected by pH. It is generally believed that the suitable range of alkali content for public water sources is 30 ~ 400 mg/L ($CaCO_3$), while the alkalinity of red mud washing water is high, and a large amount of alkaline liquid enters the water body, causing great pollution to the water body.

Y. Cheng and X. Zhu, *Utilization and Environmental Impact Assessment of Red Mud Resources*, https://doi.org/10.1007/978-981-96-7411-4_3

Red mud is transported to the red mud dump after washing and concentration, usually still containing 2.5 ~ 5.0% of Na_2O, and some Na The oxygen content even exceeds 7%, and the waste liquid carried in the red mud will gradually seep into the ground, causing the pH value of the groundwater to significantly increase, and the content of aluminum salts and other harmful impurities to exceed the standard, causing serious pollution to the water bodies around the dump site. Some countries dump red mud into rivers, lakes or oceans due to lack of dumping sites, causing direct pollution to the water quality, seriously harming the living conditions of aquatic organisms, and affecting the full use of water resources. In addition, a large amount of red mud in the red mud dump site is washed and soaked by rainwater, decomposed by the red mud itself, and transformed and migrated by leachate and other harmful substances, which will cause serious pollution to the rivers and groundwater around the dump site. Dumping solid red mud into the water body will also reduce the effective area of the river and lake surface, leading to a decrease in its flood discharge and irrigation capacity.

3.1.2 Impact on the Atmospheric Environment

Due to the extremely fine particle size and alkalinity of red mud, the dry part in the red mud dump site joins the atmosphere in the form of dust under the wind, increasing the dust content in the air, thereby causing pollution to the atmospheric environment. Studies have shown that when a wind force of level four or above occurs, powder with a particle size of 1 ~ 1.5 cm or more on the surface of the red mud dump will peel off, and its floating height can reach 20 ~ 50 m or more. The harmful components such as alkali in the red mud can inhibit the growth and development of plants, and in areas lacking vegetation, the surface of the soil layer will be peeled off due to erosion.

3.1.3 Impact on the Soil Environment

An alumina plant produces as many as millions or even tens of millions of tons of red mud slurry each year. The huge volume and corrosiveness of red mud is a problem for red mud disposal. The storage of red mud will occupy a large amount of land resources. Red mud and its attached liquid and leachate contain a large amount of harmful substances, which will change the properties and structure of the surrounding soil, cause large-scale soil marshification and salinization, and have harmful effects on the microorganisms in the soil. The existence of pollutants in red mud not only causes the roots of plants to grow and develop slowly, but also accumulates in the organic body of plants, and reaches the human body through the food chain, endangering human health.

3.2 Identification and Qualification of Red Mud

According to the definition in the "General Principles for Identification of Hazardous Waste" (GB 5085.7-2007), solid waste refers to items and substances that have lost their original use value or have been discarded or abandoned in the form of solid, semi-solid, and gaseous in containers in production, life and other activities, as well as items and substances that are included in solid waste management according to laws and administrative regulations. Hazardous waste refers to those listed in the national hazardous waste list or identified according to the national hazardous waste identification standards and identification methods, which have one or more dangerous characteristics such as corrosiveness, toxicity, flammability, reactivity, and infectivity, as well as solid waste that does not exclude the above dangerous characteristics.

Corrosiveness and leaching toxicity are the main dangerous characteristics for identifying red mud as hazardous waste. According to the provisions in the "Hazardous Waste Identification Standard Corrosiveness Identification" (GB 5085.1-2007) and "Hazardous Waste Identification Standard Leaching Toxicity Identification" (GB 5085.3-2007), the corrosiveness determination range of hazardous waste is the pH value of the leachate $\geq$ 12.5, or $\leq$ 2.0. After a large number of sampling tests, the pH value of red mud is in the range of 10.4 ~ 11.9, and the leaching toxicity concentration of each component is lower than the specification limit. Therefore, red mud should be classified as solid waste.

In the "General Industrial Solid Waste Storage and Disposal Site Pollution Control Standard" (GB18599-2001), industrial solid waste without dangerous characteristics is further classified. Class I general industrial solid waste refers to general industrial solid waste in which the concentration of any pollutant in the leachate obtained by the leaching test according to GB 5086 does not exceed the maximum allowable discharge concentration of GB 8979, and the pH value is within the range of 6 ~ 9. Class II general industrial solid waste refers to general industrial solid waste in which one or more pollutant concentrations in the leachate obtained by the leaching test according to GB 5086 exceed the maximum allowable discharge concentration of GB 8978, or the pH value is outside the range of 6 ~ 9. If classified according to GB18599-2001, red mud should belong to Class II general industrial solid waste.

According to the "General Solid Waste Classification and Code" (GB/T 39198-2020), red mud is not listed in the "National Hazardous Waste List", and according to the GB 5058.7 identification standard and GB 50586.1, HJ557 and GB/T 15555.1, GB/T 15555.3, GB/T 15555.4, GB/T 15555.5, GB/T 15555.7, GB/T 15555.8, GB/T 15555.10, GB/T 15555.11, GB/T 15555.12, HJ 751, HJ 786 identification methods determine that it does not have hazardous characteristics and belongs to general solid waste.

The pollutants in red mud and its attached liquid mainly include fluorides, alkalis, chlorides, etc. The pH of the red mud is 10.0 ~ 12.0, and the fluoride content is 4.89 ~ 8.6 mg/L; the pH of the leachate is 12.1 ~ 13.0, and the fluoride content is 11.5 ~ 26.7 mg/L. According to the "Non-ferrous Metal Industry Solid Waste Pollution

Control Standard" (Gb5058-85), because the pH of the red mud is less than 12.5 and the fluoride content is less than 50 mg/L, the red mud is considered general solid waste. However, the pH of the attached liquid of the red mud is greater than 12.5, and the fluoride content is less than 50 mg/L, the comprehensive discharge of sewage is classified as exceeding the standard wastewater. Therefore, red mud (including attached liquid) is considered hazardous waste (strongly alkaline soil).

3.3 Red Mud Radioactivity

During the long formation process of bauxite, it is affected by frequent geological structures. Many unstable atomic nuclei in the earth's crust, rocks, and soil undergo decay, releasing α particles, β particles, and γ photons, and then become other atomic nuclei. Due to their different half-lives, atomic nuclei with short half-lives have disappeared, but several atomic nuclei with long half-lives, such as uranium-238, radium-226, thorium-232, and potassium-40, still exist (Table 3.1), and they exist in the ore. Iron tailings come from ore. During the mining and ore selection process, the radioactive elements in the ore may enter the red mud with the waste slag. The radioactive hazards of solid waste such as red mud mainly have two aspects, namely external radiation hazards and internal radiation hazards.

(1) External radiation hazards

External radiation hazards refer to the ionization and destruction of human cell molecular structures caused by radiation from radioactive substances from the outside. The radiation emitted by radioactive substances mainly includes three types: α rays, β rays, and γ rays. However, γ rays ionize less in the air, have a short range, can radiate from building materials, and can cause serious harm to the human body from the outside. Therefore, the main cause of external radiation hazards is γ rays.

(2) Internal radiation hazards

Internal radiation hazards refer to the destruction of human cell molecular structures caused by radioactive substances entering the human body. Internal radiation hazards are mainly caused by radon-222. Radon-222 is produced by the decay of radium-226 and is inhaled into the human body as an inert gas. The atomic nucleus of radon-222

Table 3.1 Main radiation characteristics of three radioactive nuclides

Nuclide name	Internal and external radiation relationship	γ Ray energy (kev)
Radium-226	Internal and external radiation	352.5
Thorium-232	External radiation	238
Potassium-40	External radiation	14,602

emits α particles. Although α particles are difficult to harm the human body from the outside, α particles that enter the human body can cause ionization and destruction of atoms, damage cell structure molecules, and cause greater damage to human cells.

3.3.1 Radioactivity Evaluation Standard

Excessive radioactivity can cause serious radiation damage to the human body and the surrounding environment, but natural elements in the soil and rocks contain certain radioactive elements, so both traditional building materials and building materials produced using industrial waste have a certain level of radioactivity. Generally, the content of natural radioactive nuclides in industrial waste is usually higher than that in natural soil and rocks. Therefore, the country has strict requirements for the radioactivity indicators of building materials, and building materials with excessive radioactivity cannot be used. According to the standard "Limit of Radioactive Nuclides in Building Materials" (GB 6566-2010), the radioactivity evaluation of the main building materials mainly includes the following two evaluation indicators:

(1) Internal radiation index

Definition: The radioactivity ratio of the natural radioactive nuclide radium-226 in building materials, divided by the limit specified in this standard, results in a product. The calculation formula is as follows:

$$I_{\mathrm{Ra}} = \frac{C_{\mathrm{Ra}}}{200}$$

where

I_{Ra} Internal irradiation index;
C_{Ra} The radioactivity ratio of the natural radioactive nuclide radium-226 in building materials, in becquerels per kilogram (Bq kg^{-1});
200 Considering only the internal irradiation situation, the radioactivity ratio limit of the radioactive nuclide radium-226 in building materials specified in this standard, in becquerels per kilogram (Bq kg^{-1}).

(2) External irradiation index

Definition: The radioactivity ratios of the natural radioactive nuclides radium-226, thorium-232, and potassium-40 in building materials are divided by their respective limits when they exist separately as specified in this standard, and the sum of the products is obtained. The calculation formula is as follows:

$$I_{\gamma} = \frac{C_{\mathrm{Ra}}}{370} + \frac{C_{\mathrm{Th}}}{260} + \frac{C_{\mathrm{K}}}{4200}$$

Table 3.2 "National regulations on the limits of radioactive nuclides in building materials" (GB6566-2010)

Category	Control Index		Scope of use
	Internal irradiation I_{Ra}	External irradiation I_r	
Class A	≤ 1.0	≤ 1.3	The production, sale, and use are not restricted
Class B	≤ 1.3	≤ 1.9	Can be used for structures (such as walls, road surfaces, etc.)
Class C	–	≤ 2.8	Can be used for exterior decoration of buildings and outdoors
Others	Can only be used for roadbeds, culverts, deep underground pipes, and other places far from people's lives		

where

I_γ External irradiation index;
C_{Ra}, C_{Th}, C_K Are the radioactivity ratios of the natural radioactive nuclides radium-226, thorium-232, and potassium-40 in building materials, in becquerels per kilogram (Bq kg^{-1});
370, 260, 4200 Are the limits of the natural radioactive nuclides radium-226, thorium-232, and potassium-40 in building materials when they exist separately, considering only the external irradiation situation, as specified in this standard, in becquerels per kilogram (Bq kg^{-1}).

The "Limits of Radioactive Nuclides in Building Materials" GB 6566-2010 divides the radioactive control of building materials into control of main materials and control of decoration materials. Their limits are listed in Table 3.2.

The third article of the "Limits of Radioactive Nuclides in Building Materials" GB 6566-2010 has the following mandatory requirements for radioactive indicators: When the radioactivity ratios of the radioactive nuclides radium-226, thorium-232, and potassium-40 in the main building materials simultaneously satisfy $I_{Ra} \leq 1.0$ and $I_\gamma \leq 1.0$, their production, sale, and use are not restricted.

3.3.2 *Radioactivity Index of Red Mud*

According to the radioactive detection and evaluation of red mud from Bayer process of Chinalco Shandong Co., Ltd., as shown in Table 3.3.

The red mud samples were tested for the radioactive Table 3.4 activity concentration of 226Ra, 232Th, and 40K in the samples by the Shandong Academy of Medical Sciences, as shown in Tables 3.5 and 3.6.

Table 3.3 Radioactivity evaluation results of Chinalco red mud

Sample	Radioactive element test results Bq/kg			Internal exposure value	External exposure value	Total evaluation
	226Ra	232Th	40K			
Bayer process red mud	97.9	218.6	69	0.4895	1.121	Class A
Sintering process red mud	103.2	143.3	36	0.516	0.839	Class A

Table 3.4 Radioactive detection and evaluation of Chinalco red mud stacking field

Sample	Radioactive element test results Bq/kg			Internal exposure value	External exposure value	Total evaluation
	226Ra	232Th	40K			
Top layer red mud	64.2	105.5	10	0.321	0.581	Class A
Middle layer red mud	64.3	90.6	11	0.322	0.525	Class A
Bottom layer red mud	99.2	120.3	59	0.496	0.745	Class A
Fresh red mud	103.2	143.3	36	0.516	0.839	Class A

Table 3.5 Radioactive detection results of sintering process red mud

Test nuclide name	Unit of measurement	Test result	Remark
226Ra	Bq kg^{-1}	222.92 ± 2.33	Confidence level is 68.3%
232Th	Bq kg^{-1}	372.91 ± 1.90	Confidence level is 68.3%
40K	Bq kg^{-1}	Not detected	The detection limit of this method is approximately 1×10^1
IRa	–	1.11	–
Ir	–	2.04	–

Table 3.6 Results of radioactivity detection in Bayer process red mud

Test nuclide name	Unit of measurement	Test result	Remarks
226Ra	Bq kg^{-1}	59.17 ± 1.66	Confidence level is 68.3%
232Th	Bq kg^{-1}	291.45 ± 2.36	Confidence level is 68.3%
40K	Bq kg^{-1}	Not detected	The detection limit of this method is approximately 1×10^1
IRa	–	0.30	–
Ir	–	1.27	–

Table 3.7 Comparison of radioactivity indices of typical solid industrial waste

Sample type	Fly ash	Coal gangue	Phosphogypsum	Red mud
Internal exposure index	0.52–0.73	0.3–1.2	0.3–1.2	0.14–1.31
External exposure index	0.76–0.93	0.5–2.58	0.53–1.5	0.17–0.76

Judging from the determination and calculation results of radioactive elements (Table 3.7).

(1) The 232Th in Bayer process red mud is 1.53 times that of sintering process red mud, 40K is 1.91 times that of sintering process red mud, and 226Ra is basically the same.
(2) Both the internal and external irradiation indicators of Bayer process red mud have reached the national A-level building material standards, and its production, sales, and usage are not restricted.
(3) Compared with the radioactivity index of other common typical industrial waste, the radioactivity of red mud is lower than that of fly ash, coal gangue, and phosphogypsum.

3.3.3 *Radioactive Shielding Mechanism of Red Mud*

Some encouraging research results have been achieved regarding the resource characteristics of red mud, the enrichment and removal mechanism of alkali in red mud, the natural radioactivity of red mud, the active characteristics of red mud, and the safe application of red mud in the field of building materials. However, research on the enrichment mechanism of radioactive nuclides in red mud and the shielding methods of radioactivity is still in its early stages. Therefore, it is necessary to further systematically study the changes in the environment of radioactive nuclides during the transformation process from bauxite to red mud, to clarify their migration and enrichment mechanisms. It is necessary to study the retention ability of the mineral composition and structural characteristics of the red mud solidified body on radioactive nuclides, to clarify the shielding and sealing mechanism of red mud. It is necessary to study the radioactive characteristics and absorption ion existence state of radioactive nuclides under long-term environmental conditions after the addition of external shielding materials, to achieve the safe and harmless application of red mud. To achieve large-scale resource utilization of red mud, it is necessary to systematically and deeply study the above key issues, fundamentally solve the application difficulties brought by the radioactivity of red mud, and promote the comprehensive, efficient, and harmless utilization of red mud.

The radiation protection of building materials mainly considers the shielding of *y* rays and neutron rays. *y* rays have strong penetration power, and their energy is weakened when passing through high-density building materials. When the density and thickness of the building materials reach a certain level, their energy can be completely absorbed. Neutron rays are composed of uncharged nuclear particles and have high penetration power, so neutron shielding materials not only need to contain heavy elements but also must contain a suitable amount of light elements. The *y* rays emitted by building materials are usually absorbed by cement concrete, mineral admixtures, steel slag, barite powder, and zeolite powder, etc., through physical solidification and chemical absorption mechanisms to reduce the internal and external irradiation index of radioactive nuclides. There is relatively little research on the radioactive shielding technology of red mud. The literature research mainly focuses on the separation of radioactive minerals, cement solidification, or reference to the research results of radiation protection cement, and the radioactivity of red mud is reduced by adding shielding materials (barite, silica fume, steel slag).

Solidification is the process of converting radioactive waste into a solidified body, the main purpose of which is to convert the waste into an appropriate solid form to reduce the migration or dispersion of radioactive nuclides that may be caused by natural processes during storage, transportation, and processing. Physical encapsulation is to use the solidification and sealing effect of cement and other solidifiers to form a dense solidified body to block the penetration of rays. By improving the pore structure of the solidified body, the diffusion and leaching of nuclide ions can be controlled. The hydrated calcium silicate (CSH) gel of cement hydration products has a large specific surface area and has a certain ion adsorption and exchange capacity. Its adsorption and exchange capacity and the mechanical properties of cement solidified bodies increase with the decrease of the calcium-silicon ratio (C/S) in CSH. Yang Jiujun and others studied the regulatory effect of cement hydration solidification on radioactivity. When the internal and external irradiation indexes are 2.05 and 3.27 respectively, and the content of sintered red mud is not more than 20% (mass fraction), the mechanical properties of red mud composite silicate cement meet the requirements of 42.5 general cement, and both the internal irradiation index and the external irradiation index are less than 1, and it can be used as a building material without restriction. Based on the testing and evaluation of the natural radioactivity level of Bayer process red mud and sintered red mud from Guizhou Aluminum Plant, the formula calculated that the mixing range of red mud for use in the field of building materials is 28 ~ 44% (mass fraction).

3.4 Leaching Toxicity of Red Mud

Considering that the heavy metal elements that may be contained in red mud will diffuse and cause harm to the environment under the long-term soaking action of rainwater and groundwater. On the basis of radioactive evaluation, further research is needed on the leaching toxicity of metal elements in red mud. The main research is on the ability of heavy metal elements in red mud to migrate and diffuse to the surrounding environment under the soaking and scouring action of rainwater and groundwater, so as to judge whether the heavy metal elements contained in red mud will enter the surrounding environment and whether they will cause potential harm to the environment.

Leaching is the process by which soluble components dissolve into the leachate from a solid through methods such as permeation or diffusion. When solid waste and liquid (including infiltrating rainwater, surface water, groundwater, or moisture contained in waste materials) come into contact, the components in the solid phase will dissolve into the liquid phase to form leachate. The degree of component dissolution depends on the points of contact between the liquid and solid phases, the characteristics of the material (chemical, physical, biological factors), and the duration of contact. The leaching process can be conceptually described as: water exposes the landfill material to a damp environment, leading to the leaching of the landfill material, and then the leachate infiltrates groundwater or surface water through diffusion, permeation, or sedimentation.

When solid waste is soaked in water, the harmful substances that are leached out migrate and transform, polluting the environment. This harmful characteristic is known as leaching toxicity. The basic principles of the release of toxic substances from solid waste into the environment include volatilization and dissolution leaching, which can be explained as: solid waste is leached by water, and the harmful substances in it dissolve into the liquid phase. Leaching toxicity is used to evaluate the indirect or potential toxic hazards of harmful substances in solid waste leaching into the liquid phase, producing leachate, and polluting the environment. Leaching toxicity simulates the toxic hazards of hazardous waste under certain disposal conditions, and leaching toxicity is directly related to waste characteristics, disposal methods, and the environment.

3.4.1 Leaching Evaluation Standards

The modified solidified red mud road material filler must comply with the current environmental protection regulations of the state and localities. The material supplier needs to carry out environmental impact detection and evaluation work to assess its impact on water bodies, soil, etc. It is strictly forbidden to use red mud materials that do not meet the standard for harmful substance content as road embankment fillers. Specific detection items include but are not limited to the following items, and the pretreatment and detection methods must meet the requirements of Table 3.8.

At present, China has not yet established specific regulations for the management or remediation of heavy metal pollution sites, but for the safe use of heavy metal-containing materials, corresponding environmental quality safety evaluation standards have been established, including the "Soil Environmental Quality Standard" GB 15618-1995 and the "Surface Water Environmental Quality Standard" GB 3838-2002. At the same time, according to the characteristics of highway engineering, the "Environmental Impact Assessment Specification for Highway Construction Projects" JTG B03-2006 was formulated. The "Environmental Impact Assessment Specification for Highway Construction Projects" is consistent with the "Surface

Table 3.8 Red mud test detection indicators

Characteristics	Detection items	Pretreatment method	Detection method
Physical and chemical properties	PH	GBT 15555.12-1995	
Content	Inorganic fluorides	GB5085.3 Appendix F	GB5085.3 Appendix F or GBT 15555.11-1995
	Cyanides (as CN–)	GB5085.3 Appendix G	
	Total mercury, lead, cadmium, total chromium, copper, zinc, beryllium, barium, nickel, arsenic, cobalt, selenium, molybdenum, silver, antimony, vanadium, manganese	GB5085.3 Appendix S	GB5085.3 Appendix B or GBT 15555
	Hexavalent chromium	GB5085.3 Appendix T	GBT 15555.4-1995
Leaching	Total mercury, lead, cadmium, total chromium, copper, zinc, beryllium, barium, nickel, arsenic, cobalt, selenium, molybdenum, silver, antimony, vanadium, manganese	HJ/T 299	GB5085.3 Appendix B or GBT 15555
	Hexavalent chromium	HJ/T 299	GBT 15555.4-1995

Water Environmental Quality Standard" for the analysis method and evaluation standard of surface water samples in highway engineering. Considering that the heavy metals in red mud will eventually enter the soil and water, the "Soil Environmental Quality Standard" and "Surface Water Environmental Quality Standard" are used to evaluate the leaching toxicity of red mud. China divides the soil environmental quality standard into three levels, as shown in Table 3.9, and the surface water environmental quality standard is divided into five categories, as shown in Table 3.10.

It should be noted that these standard values for heavy metals are merely benchmark values and cannot be used as the sole criterion for evaluating actual soil samples. The mobility and activity of all heavy metals are also influenced by soil properties, such as the pH value of the soil, clay content, humus content, etc. For potential pollution sites, specific on-site investigations, classifications, and evaluations need to be carried out.

Leaching tests are field or laboratory simulations of leaching behavior in the environment. Leaching tests are generally divided into two categories, one is field leaching tests; the other is laboratory leaching tests.

Field tests usually use a controlled landfill unit or field seepage well, collect on-site leachate, and analyze the content of specific elements in the leachate.

Table 3.9 China soil environmental quality standard values (mg/kg)

Items	Level 1	Level 2			Level 3
	Natural background	pH < 6.5	pH 6.5 ~ 7.5	pH > 7.5	pH > 6.5
Cadmium	0.2	0.3	0.30	0.60	1.0
Mercury	0.15	0.3	0.50	1.0	1.5
Arsenic (paddy field)	15	30	25	20	30
Arsenic (dry land)	15	40	30	25	40
Copper (farmland, etc.)	35	50	100	100	400
Copper (orchard)	–	150	200	200	400
Lead	35	250	300	350	500
Chromium (paddy field)	90	250	300	350	400
Chromium (dry land)	90	150	200	250	300
Zinc	100	200	250	300	500
Nickel	40	40	50	60	200

Note Level 1 standard: The limit value for maintaining the soil environmental quality of the natural ecological protection area
Level 2 standard: The limit value for ensuring agricultural production and maintaining human health
Level 3 standard: The critical value of soil for ensuring agricultural and forestry production and normal plant growth

Table 3.10 Surface water environmental quality standard values (mg/kg)

Test item	I	II	III	IV	V
Copper (Cu)	≤ 0.01	≤ 1.0	≤ 1.0	≤ 1.0	≤ 1.0
Zinc (Zn)	≤ 0.05	≤ 1.0	≤ 1.0	≤ 2.0	≤ 2.0
Chromium (Cr)	≤ 0.01	≤ 0.05	≤ 0.05	≤ 0.05	≤ 0.1
Cadmium (Cd)	≤ 0.001	≤ 0.005	≤ 0.005	≤ 0.005	≤ 0.01
Lead (Pb)	≤ 0.01	≤ 0.01	≤ 0.05	≤ 0.05	≤ 0.1
Arsenic (As)	≤ 0.05	≤ 0.05	≤ 0.05	≤ 0.1	≤ 0.1
Fluoride	≤ 1.0	≤ 1.0	≤ 1.0	≤ 1.5	≤ 1.5

Note Category I: Mainly applicable to source water, national nature reserves
Category II: Mainly applicable to the primary protection area of surface water sources for centralized domestic drinking water
Category III: Mainly applicable to the secondary protection area of surface water sources for centralized domestic drinking water
Category IV: Mainly applicable to general industrial water areas and recreational water areas not in direct contact with the human body
Category V: Mainly applicable to agricultural water areas and general landscape requirement water bodies

Laboratory leaching tests study one or more samples of the material to determine which components can be leached or dissolved by the liquid and the possibility of being released into the environment in the liquid phase. Laboratory leaching tests are divided into two categories: single-stage extraction/batch tests (sometimes referred to as static extraction tests); multi-stage extraction tests/circulation and direct flow leaching tests (sometimes referred to as dynamic tests).

The single-stage extraction experiment is a test in which a certain amount of leaching solution and a certain amount of waste are mixed and extracted within a certain period of time, without updating the leaching solution. In the single-stage extraction experiment, it is crucial for the system to reach equilibrium when predicting long-term leaching behavior. If the experiment is conducted under unbalanced conditions, the leaching amount does not reach the actual leaching capacity, and the prediction of long-term leaching behavior will be too high or too low.

The multi-stage extraction experiment maintains the trend of leaching by continuously or intermittently updating the leaching solution. In the multi-stage extraction experiment, the sample particles and the leaching solution are mixed and oscillated for a certain period of time at a certain liquid–solid ratio, then the leaching solution and the solid are separated, and new leaching solution is added to repeat the extraction until the predetermined number of leaching times. The multi-stage extraction experiment provides kinetic information on the migration of pollutants.

The content of elements such as Si, Al, Fe in red mud is relatively high, mainly appearing in the form of stable oxides such as SiO_2, Al_2O_3, Fe_2O_3. It also contains heavy metals with significant biological toxicity such as mercury, cadmium, lead, and metalloid arsenic, as well as trace amounts of Mn, Zn, Cu, Cr heavy metals. The risk of heavy metal substances leaching and releasing from modified red mud is referred to as the leaching characteristics of heavy metal substances.

(1) TCLP Leaching Test

The TCLP leaching test (Toxic Characteristic Leaching Procedure), also known as the toxicity leaching test, is currently the most widely used leaching test procedure and is the only routine test procedure for hazardous waste in the United States and Canada. It can simulate the most unfavorable leaching conditions. The TCLP method is a set of hazardous waste toxicity leaching procedures established by the United States Resource Conservation and Recovery Act for the management of hazardous waste and solid waste in 1984. It is used to determine the migration of 40 toxicity indicators (TC) in liquid, solid, and municipal waste. If any TC component in the TCLP leachate is equal to or greater than the concentration limit specified in the RCRA clause 40CFR16 1.24, it is considered that the waste contains this TC component and is hazardous waste.

The TCLP test accelerates the leaching process of the sample by using crushed test material (increasing the solid/liquid contact area) and acidified leaching solution. During the test, the test material and the leaching solution are placed in a sealed polyethylene bottle, and continuous shaking extraction promotes the dissolution and diffusion of pollutants in the liquid phase in the sample. Pollutants dissolve in the leaching solution, and the solution at this time is called the leachate, as shown in

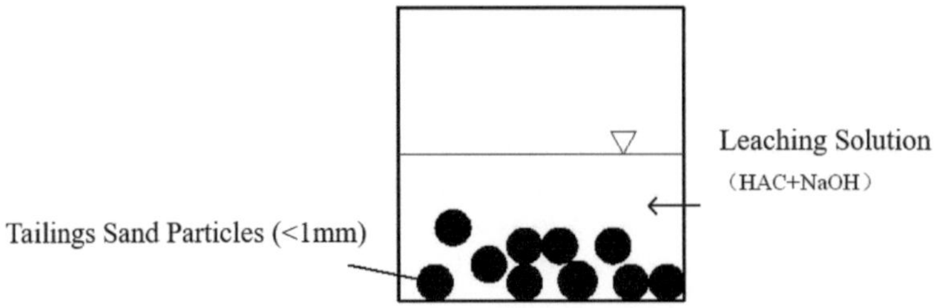

Fig. 3.1 Schematic diagram of TCLP leaching test

Fig. 3.1. The composition of the leachate and its potential impact on water quality are important bases for determining the type of waste, and are also key factors in evaluating the disposal technology suitable for this type of waste.

Because the TCLP test uses crushed test materials, it can be considered to use the TCLP test to simulate the leaching characteristics of modified red mud road embankment under the most unfavorable conditions, that is, the road embankment is completely destroyed, and the modified red mud is in a completely loose state. Considering the difference between the indoor and construction site red mud modification effects, this topic selected different doses of indoor modified red mud and modified red mud from the construction site of the test section for testing.

According to the TCLP standard, the experimental steps include:

(1) Crush the modified red mud sample that has been standard cured for 7 days through a 1 mm sieve and collect it for later use.
(2) Determine the leaching solution used in the experiment according to the method provided by the TCLP standard. The specific process is: weigh 5 g of the sample from step (1) in a beaker, add 96.5 mL of distilled water, stir with a magnetic stirrer for 5 min, and then test the pH of the solution. Heat the mixture to 50 °C and maintain this temperature for 10 min, then cool the test solution to room temperature and test the pH.
(3) Take 5.7 mL of glacial acetic acid (HAC) and add it to 500 distilled water, then add 64.3 mL of 1 mol/L sodium hydroxide (NaOH), dilute the solution to 1 L to get the leaching solution, its pH range is around 4.93 ± 0.05.
(4) Weigh 10 g of the soil sample from step (1) and mix it with 200 mL of the leaching solution from step (2) (1:20 solid–liquid mass ratio), place it in a polyethylene bottle, and oscillate it at a speed of 180mot/min for 18 h at room temperature of 25 °C;
(5) After the oscillation is completed, test the pH value of the leachate, then filter it with a 0.45 μm filter membrane, and collect the clear leachate sample;
(6) Acidify the collected clear sample to pH < 2 with nitric acid, and use a plasma atomic emission spectrometer to test the concentrations of Mn^{2+}, Zn^{2+}, Cu^{2+}, Cr^{3+}.

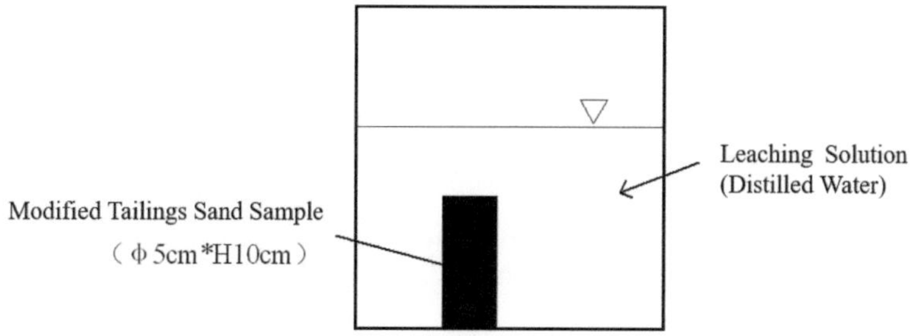

Fig. 3.2 Schematic diagram of the tank leaching test

(2) Tank Leaching Test

The Tank Leaching Test, also known as the water tank leaching test, commonly uses the European EA NEN 7375:2004 and the American ANSI/ANS-16.1-2003 standards. This test is used to evaluate the environmental risks of waste in landfills. It determines the leaching characteristics of non-organic components in the whole sample through diffusion tests. The test can also derive parameters such as the degree of surface clarification and the effective diffusion coefficient, which can be used to estimate the long-term leaching characteristics of the sample. The principle of the Tank Leaching Test is shown in Fig. 3.2. The whole test sample is completely submerged in the leaching solution, left stationary, and the pollutants are released into the leaching solution through dissolution and diffusion. During the test, the trend of pollutant leaching is maintained by intermittently updating the leaching agent, thereby providing kinetic information on pollutant migration. The amount of leaching solution is controlled by the mass ratio, volume ratio, or surface area/volume ratio of the sample to the leaching solution. This repeated leaching process can determine the highest concentration of leachable components in the waste in actual landfills.

In this project, the Tank test is used to simulate the leaching characteristics of modified red mud road embankments under normal use conditions. Different doses of modified red mud mixed in the laboratory or on-site are selected. Considering that the modified red mud is used in the subgrade and sub-base layers of the road, corresponding to 94 and 96% compaction states, the leaching characteristics of the modified red mud under 94 and 96% compaction states were studied. The specific test steps are:

(1) Acid-wash the polyethylene water tank, then rinse with distilled water, immerse the sample cured for 7 days in distilled water (the mass ratio of the sample to distilled water is 1:10), and seal the water tank to prevent absorption of carbon dioxide from the air;
(2) Collect the leaching solution at intervals according to the test time, and repeat step (1) to put the sample into a new leaching solution;
(3) Test the concentrations of Mn^{2+}, Zn^{2+}, Cu^{2+}, and Cr^{6+} in the leachate.

3.4.2 Leaching Test Results

Sampling points and methods are carried out according to HJ/T298. The pretreatment methods of inorganic elements and their compounds (except hexavalent chromium, inorganic fluoride, and cyanide) refer to Appendix S of GB 5085.3-2007: The pretreatment methods of hexavalent chromium and its compounds refer to Appendix T. The test standards and results are shown in Tables 3.11, 3.12 and 3.13.

3.5 Analysis of Alkali Migration in Red Mud

The Bayer process red mud contains a large amount of alkali. In practical applications, it can cause alkali-aggregate reaction, product efflorescence, and corrosion of steel parts. Therefore, reducing the alkali content in red mud is very important for the utilization of red mud. There are two forms of alkali in red mud, one is the soluble alkali solution brought in during the production of alumina, such as NaOH, sodium aluminate, $NaCO_3$ and so on. The other is sodium aluminosilicate hydrate in the form of cancrinite, which is an insoluble alkali. Among them, the content of soluble alkali accounts for 54% of the total alkali, and the content of non-soluble alkali accounts for 46% of the total alkali. Only the soluble alkali in red mud is harmful alkali. Therefore, distinguishing the soluble alkali from the non-soluble alkali in red mud is very critical for the utilization of red mud, and how to reduce the soluble alkali in red mud is also crucial for the large-scale utilization of red mud.

(1) Non-soluble alkali

The Bayer process is to mix bauxite with a high A/S ratio with NaOH solution. Under high temperature and pressure, the alumina in bauxite reacts with NaOH to form a sodium aluminate solution. The silicate minerals in bauxite are decomposed by NaOH to form sodium silicate into the solution, and react with the sodium aluminate solution

Table 3.11 Red mud digestion test indicators

	Be	Cr	Mn	Co	Ni	Cu	As	Se
Red mud digestion	0.9650	241.0350	275.0200	4.8750	10.9950	14.6500	20.6000	1.7300
Red mud digestion	1.1489	231.5927	154.4083	4.5118	9.7337	12.4753	13.4000	2.0513
Detection limit	0.0004	1.4780	0.0748	0.0089	0.1491	0.3152	0.1442	0.2045

	Ag	Cd	Sb	Ba	Hg	Tl	Pb
Red mud digestion	1.2850	1.8000	1.7900	64.7150	0.6574	0.0250	16.1850
Red mud digestion	1.3905	1.8491	4.5020	103.4172	4.6647	0.0247	20.5227
Detection limit	0.0844	0.0543	0.0100	0.2136	0.6574	0.0004	0.1607

Table 3.12 Detection of leachate from red mud and modified red mud

Test item	Standard basis	Test results	
		Red mud	Modified red mud (CPS-12)
Total mercury	GB 5085.3-2007	< 0.00005	< 0.00005
Total chromium	GB 5085.3-2007	1.66	0.15
Lead	GB 5085.3-2007	< 0.0002	< 0.0002
Cadmium	GB 5085.3-2007	< 0.0005	< 0.0005
Copper	GB 5085.3-2007	0.012	0.011
Zinc	GB 5085.3-2007	0.003	< 0.001
Beryllium	GB 5085.3-2007	< 0.0001	< 0.0001
Barium	GB 5085.3-2007	< 0.001	< 0.001
Nickel	GB 5085.3-2007	0.005	< 0.001
Arsenic	GB 5085.3-2007	0.0058	0.002
Cobalt	GB 5085.3-2007	< 0.001	< 0.001
Selenium	GB 5085.3-2007	0.0061	< 0.0005
Molybdenum	GB 5085.3-2007	0.12	0.12
Silver	GB 5085.3-2007	< 0.0001	< 0.0001
Vanadium	GB 5085.3-2007	0.79	0.34
Antimony	GB 5085.3-2007	< 0.001	< 0.001
Manganese	GB 5085.3-2007	< 0.001	< 0.001
Hexavalent chromium	GB/T 15555.4-1995	1.56	0.051

to form sodium aluminosilicate hydrate (sodium silicate slag). Taking kaolinite as an example to represent silicate minerals, the reaction equation of sodium silicate slag is as follows:

$$Al_2O_3 \cdot 2SiO_2 \cdot 2H_2O + 6NaOH + aq \rightarrow 2NaAl(OH)_4 + 2Na_2[H_2SiO_4] + aq$$

$$\begin{aligned} & xNa_2[H_2SiO_4] + 2NaAl(OH)_4 + aq \\ & \quad \rightarrow Na_2O \cdot Al_2O_3 \cdot xSiO_2 \cdot n_2O + 2xNaOH + aq \end{aligned}$$

The sodium silicate slag and the iron, titanium, and excess silicon minerals in the bauxite are discharged in the form of precipitation, and after repeated washing, they are sent to the yard to form red mud. As can be seen from the above process, the insoluble alkali in the red mud is sodium aluminate. Its chemical formula can generally be expressed as $Na_2O{\cdot}Al_2O_3 \cdot xSiO_2{\cdot}nH_2O$. Under different reaction conditions, the composition and structure of the sodium silicate slag produced also have certain differences, because in the sodium aluminate solution, there are $Al(OH)^{4-}$, $SiO_2\ (OH)_2{}^{2-}$ and other complex aluminosilicate anions. Under different production

Table 3.13 Original sample detection of red mud and modified red mud

Test item	Standard basis	Test results	
		Red mud	Modified red mud (CPS-12)
Inorganic fluorides	GB/T 22104-2008	751	642
Cyanides	CJ/T 221-2005	< 0.20	< 0.20
pH	CJ/T 221-2005	10.63	10.99
Total mercury	GB/T 22105.1-2008	0.086	0.11
Total chromium	GB 5085.3-2007 Appendix S	629	518
Lead	GB 5085.3-2007 Appendix S	25	20.6
Cadmium	GB 5085.3-2007 Appendix S	0.386	0.264
Copper	GB 5085.3-2007 Appendix S	48	43
Zinc	GB 5085.3-2007 Appendix S	72	63
Beryllium	GB 5085.3-2007 Appendix S	0.7	0.6
Barium	GB 5085.3-2007 Appendix S	73	98
Nickel	GB 5085.3-2007 Appendix S	47	40
Arsenic	GB/T 22105.2-2008	16.3	14.3
Cobalt	GB 5085.3-2007 Appendix S	13	9.96
Selenium	NY/T 1104-2006	1.12	1.59
Molybdenum	GB 5085.3-2007 Appendix S	6.67	5.54
Silver	GB 5085.3-2007 Appendix S	0.37	0.276
Vanadium	GB 5085.3-2007 Appendix S	640	516
Antimony	GB 5085.3-2007 Appendix S	0.52	0.52
Manganese	GB 5085.3-2007 Appendix S	409	362
Hexavalent chromium	GB 5085.3-2007	18.6	4.71

conditions, the above ions will transform each other and undergo condensation reactions to form small structural units and gradually form nuclei. As the nuclei grow, different types of sodium silicate slag are formed. The common sodium silicate slags produced in the production process of alumina include: cancrinite, sodalite, A-type zeolite, yugawaralite, and amorphous hydrated aluminosilicate. And under suitable conditions, different types of sodium silicate slag will transform, and the sodium silicate slag formed at lower temperatures will transform into cancrinite at high temperatures. And the stability of sodium silicate slag in different forms in sodium aluminate solution is ranked as: cancrinite > yugawaralite > sodalite > A-type zeolite > amorphous.

(2) Soluble alkali

In the Bayer process of extracting alumina, a high concentration of NaOH solution is used to leach the aluminum in the bauxite. NaOH reacts with the aluminum-containing substances in the bauxite to form sodium aluminate, sodium silicate and other soluble substances, and there is a part of NaOH that does not participate in the reaction and remains. This part of NaOH and soluble sodium aluminate and other solutions are discharged with red mud, and NaOH reacts with CO_2 in the air during storage to form Na_2C_3, the longer the storage time, the greater the content of Na_2CO_3. Therefore, the soluble alkali in the red mud is mainly NaOH, Na_2CO_3, sodium aluminate, etc.

3.5.1 Red Mud pH Value

The so-called pH value, sometimes also called the hydrogen ion index, because the concentration of hydrogen ions is often very small, it is very inconvenient to use, so the concept of pH value is used as the judgment indicator of the acidity and alkalinity of the aqueous solution. Moreover, the negative logarithm of the activity of hydrogen ions can represent the order of magnitude of the change in acidity and alkalinity, which is very convenient to use, and thus obtained (under the condition of 25 °C):

Neutral aqueous solution, $pH = 7$.

Acidic aqueous solution, $pH < 7$, the smaller the pH value, the stronger the acidity;

Alkaline aqueous solution, $pH > 7$, the larger the pH value, the stronger the alkalinity.

The pH value is one of the most important physicochemical parameters of the aqueous solution. It needs to be measured in the fields of industry, agriculture, medicine, environmental protection, and scientific research. Considering that the heavy metal elements and PH value in the red mud may diffuse under the long-term soaking action of rainwater and groundwater, causing harm to the environment. On the basis of the radioactivity evaluation, further research is needed on the leaching toxicity of metal elements in red mud. The main research is on the ability of heavy metal elements in red mud to migrate and diffuse into the surrounding environment under the action of rainwater and groundwater infiltration and erosion, in order to judge whether the heavy metal elements contained in the red mud will enter the surrounding environment and whether they will pose potential harm to the environment.

Red mud is a kind of industrial solid waste discharged during the production of alumina, with complex physical composition, small particle size, large discharge volume, high alkalinity and radioactivity. The comprehensive utilization and resource utilization of red mud is a global problem. The open-air dam storage of red mud not only requires a large amount of land, but also brings serious environmental pollution and ecological destruction problems. The high alkalinity (such as sodium, potassium and other impurities) and radioactivity of red mud are both related to environmental issues and are important factors restricting the comprehensive utilization of red mud. With the development of the aluminum industry and the decline in ore grades, the amount of red mud will increase. Therefore, the problem of alkali in red mud cannot be ignored.

Leaching experiments are a method of environmental geochemistry to simulate atmospheric precipitation conditions to study the form and migration rules of elements in soil and surface rocks. According to the leaching experiments conducted on Bayer red mud and sintered red mud, different pH leaching solutions are used to simulate rainwater for leaching experiments, and then the release and migration rules of Na and K in red mud are discussed. This provides necessary basic data and scientific basis for the environmental effects of red mud and the study of alkali removal.

Micro-area analysis shows that the content of Na and K in Bayer red mud and sintered red mud shows a positive correlation at each micro-area point. Na and K show a significant positive correlation with Si, and a significant negative correlation with Ca; there is no obvious correlation with Al, Fe, Ti, Mn, and Mg at each point.

The leaching action, including rainwater, can filter out part of Na and K, which may have a certain impact on the environment. Through the uniform leaching experiment of red mud with different leaching solutions with pH of 2, 4, and 8, it can be seen that in the three systems, the sodium filtered out in the first 10 samples is 17.6, 16.8, and 19.6% of Na in Bayer red mud; while the Na filtered out in sintered red mud is 36.7, 53.1, and 54.1%. The absolute and relative amounts of K filtered out by the three different PH leaching solutions from sintered red mud are greater than those from Bayer red mud. In the three systems with PH of 2, 4, and 8, the K filtered out in the first 10 samples is 10.5, 10.1, and 11.4% of K in Bayer red mud; while the K filtered out in sintered red mud is 40.0, 56.8, and 57.5% of K in red mud.

At the beginning of the leaching experiment, the first sample was taken every 3 h. Starting from the second time, samples were taken every 24 h, and a total of 11 samples were taken. Each sampling also measured the pH of the filtrate within 24 h. After 14d, the pH of the filtrate in each group basically reached equilibrium, as shown in Table 3.14.

The trend of the change in the pH of the static solution is shown in Fig. 3.3. As can be seen from Fig. 3.4, the pH range of the static solution of red mud is between 10.50 and 12.00, and the pH of the field-modified red mud is between 10.0 and 11.0.

Table 3.14 Changes in filtrate pH over time

Test material	Time interval/h									
	3	24	48	96	120	144	170	193	218	316
Original red mud	11.90	11.85	11.70	11.35	11.05	10.95	10.90	10.80	10.75	10.40
Field modified red mud	10.60	10.85	11.15	11.05	10.90	10.90	10.55	10.50	10.25	9.95
Indoor modified red mud	11.95	11.45	11.85	11.40	11.10	11.05	10.65	10.70	10.70	10.05
5% lime soil	10.45	11.10	11.10	11.25	10.75	10.75	10.30	10.15	10.35	9.80
Cement test block	12.55	13.00	12.85	12.95	12.95	12.90	12.85	12.90	12.75	12.65
Jihe high-speed water stable	12.00	12.30	12.15	12.05	11.80	11.90	11.65	11.65	11.6	11.60

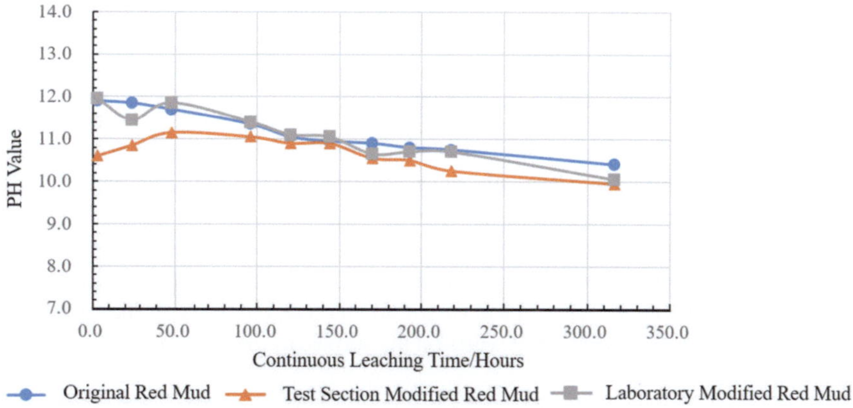

Fig. 3.3 Changes in pH of red mud and modified red mud over time

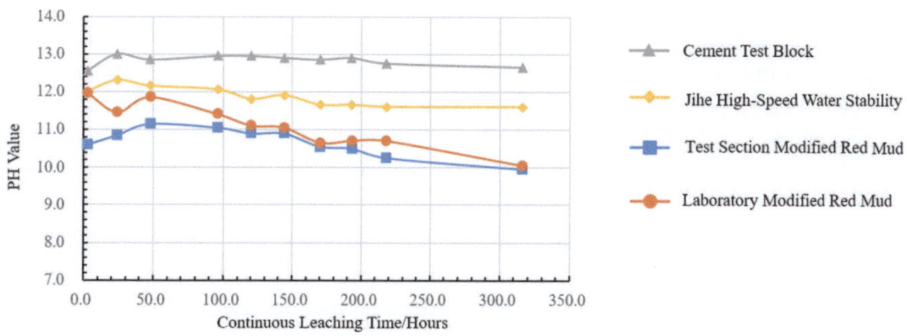

Fig. 3.4 Changes in pH of different materials over time

As the soaking time increases, the pH of the red mud of each year shows a gradually decreasing trend. Compared with other materials, the pH of red mud is generally lower than that of water-stable materials (Fig. 3.4). This also indirectly reflects that the soluble alkaline substances in the red mud can continuously dissolve into the water, preparing material conditions for the "alkalization" of red mud.

In the figure, the pH of the water-stable material is significantly higher than that of other material samples, and the pH of the modified red mud, especially the field-modified red mud, is significantly lower than the predicted pH of the red mud. Red mud is subjected to long-term accumulation and rainwater erosion, a certain amount of soluble alkaline substances infiltrate with rainwater, and as the storage time increases, the mineralization of red mud becomes obvious (as can be seen from scanning electron microscope photos), some soluble alkaline substances are transformed into non-minerals, these reasons together cause the pH of multi-year red mud to be lower than that of other years of red mud.

3.5.2 Conditions and Process of Red Mud Alkalization

(1) Conditions for Red Mud "Alkalization"

Combining the process of red mud "alkalization", influencing factors, and the determination experiment of alkaline substances, the mechanism and conditions of red mud "frosting" are summarized as follows:

The first type of Na, K soluble salt substances: ① There are a certain amount of Na, K metal ions and some other non-metal ions in the red mud. ② The red mud contains water, which serves as a carrier for alkali metals to move to the surface of the red mud. ③ The evaporation of water on the surface of the red mud forms sodium and potassium salts.

The second type of $Ca(OH)_2$, $CaCO_3$ slightly soluble or insoluble substances "alkalization" conditions: ① The entire reaction process must be carried out in an alkaline environment, in an acidic environment, ${CO_3}^{2-}$ is very easy to react with H^+ to generate CO_2 and H_2O, the precipitation of calcium carbonate cannot form. ② There is a certain amount of CaO in red mud, which can form $Ca(OH)_2$ with water. ③Red mud contains water, which serves as a carrier for alkali metals to move to the surface of red mud. ④ There is CO_2 gas on the surface of red mud, which reacts with the gray slurry of $Ca(OH)_2$ precipitated on the surface of red mud to form $CaCO_3$.

(2) The "alkalization" process of red mud

According to the X-ray diffraction analysis results and chemical composition of insoluble "alkaline" substances, the insoluble alkaline substances also contain a certain amount of soluble alkaline substances, indicating that the two different types of alkaline substances are formed at the same time, there is no order of precedence, but the formation process of insoluble alkaline substances is more complex than that of soluble alkaline substances, and can stay on the surface of red mud for a long time.

1) Soluble alkaline substance "alkalization" process

Most of the sodium and potassium elements in red mud exist in the form of free Na^+ and K^+. The red mud washing water contains a large amount of free ${CO_3}^{2-}$, ${HCO_3}^{2-}$, Cl^-, NO^{3-}, etc. Under the condition of water, Na^+, K^+ combine with non-metal ions such as ${CO_3}^{2-}$, ${HCO_3}^{2-}$, Cl^-, $NO^{3-}$to form soluble sodium and potassium salts such as Na_2CO_3, $KHCO_3$, NaCl, KCl, $NaNO_3$. With the evaporation of water, soluble salts continuously crystallize and accumulate on the surface of red mud, forming a large area of thin layer and film-like soluble "alkaline" substances on the red mud dam body.

2) Insoluble alkaline substance "alkalization" process

The formation process of insoluble "alkaline" substances is relatively complex, which includes $CaCO_3$, $MgCO_3$, CaF_2 and other insoluble substances. In the production process, sodium salt and potassium salt act as reactants to participate in the reaction process of insoluble alkaline substances. Both of them have the same effect. Taking sodium salt as an example, the process of sodium salt and calcium salt forming insoluble alkaline substances is explained.

First, calcium oxide reacts with water to form calcium hydroxide slurry, sodium hydroxide reacts with carbon dioxide and water to form sodium bicarbonate, and sodium bicarbonate reacts with sodium hydroxide to form sodium carbonate. Reaction (3.1) provides free Ca^{2+} for the subsequent formation of calcium carbonate precipitation. Reactions (3.2) and (3.3) provide HCO^{3-}, $CO_3{}^{2-}$ for the solution.

$$H_2O + CaO = Ca(OH)_2 \tag{3.1}$$

$$NaOH + 2CO_2\,(\text{excess}) + H_2O = 2NaHCO_3 \downarrow \tag{3.2}$$

or $2NaOH + CO_2\,(\text{small amount}) = Na_2CO_3 + H_2O$

$$NaOH + NaHCO_3 = Na_2CO_3 + H_2O \tag{3.3}$$

Next, calcium hydroxide reacts with carbon dioxide, sodium carbonate, and sodium bicarbonate in water to form calcium carbonate precipitate. At the same time, it also reacts with F^- in the red mud wash water to form insoluble CaF_2.

$$Ca(OH)_2 + CO_2 = CaCO_3 \downarrow + H_2O \tag{3.4}$$

$$Ca(OH)_2 + 2NaHCO_3 = CaCO_3 \downarrow + Na_2CO_3 + 2H_2O \tag{3.5}$$

$$Ca(OH)_2 + Na_2CO_3 = CaCO_3 \downarrow + 2NaOH \tag{3.6}$$

$$Ca^{2+} + 2F^- = CaF_2 \downarrow \tag{3.7}$$

From the above chemical reaction equations, it can be seen that during the formation of insoluble alkaline substances in red mud, soluble alkaline substances are also generated, as shown in (3.5) and (3.6). As calcium carbonate is formed, the moisture in the red mud continuously evaporates, and the Na^+ in the solution also flows to the surface of the red mud, finally forming crystalline powders such as NaCl, $NaNO_3$ with some non-metallic elements (such as Cl^-, $NO_3{}^-$, NO_3, etc.).

(3) Factors affecting the "alkalization" process of red mud

A special "alkalization" simulation experiment was conducted to study the "alkalization" process. The water-containing red mud obtained from the red mud stacking yard was initially dark green and contained a certain amount of water. In the first 11 h, the moisture on the surface of the red mud slowly evaporated, the surface began to dry, and the color of the red mud gradually turned red. In the next 20 h, the color of the red mud completely turned to reddish-brown, and white alkaline substances began to appear on the surface, and the moisture in the red mud became less and less. After 48 h, the white substance that the red mud exuded covered the surface of the red mud. The entire "alkalization" process lasted until about 72 h, when the white substance exuded from the red mud reached its maximum, and the moisture in the red mud also evaporated nearly dry. After that, even if the red mud was left for a long time, no white substance was produced.

1) The impact of moisture on the "alkalization" of red mud

In order to study whether moisture has an impact on the "alkalization" process, new red mud with a high moisture content was put into a thermostat, quickly dried at 105 °C, and no white alkaline substance was formed. Then the dried red mud was left at room temperature, and no white substance was formed after 72 h.

2) The impact of air on the "alkalization" of red mud

By observing the "alkalization" process of red mud, it was found that white alkaline substances only formed on the surface of the red mud, suggesting that some gaseous components in the air reacted with substances in the red mud to form white alkaline substances. Therefore, to study the impact of air factors on the "alkalization" of red mud, the experiment was designed to place the new water-containing red mud in a sealed plastic container for observation.

After 72 h of placement, only a very thin layer of red mud was reddish-brown, and most of it was still dark green, with no white substance formed. Moreover, the wall of the plastic container was indented, indicating that the pressure inside the container was lower than the outside, reflecting that some components in the red mud reacted with the air to form solid substances, causing the container pressure to decrease. From the color change, it can be inferred that due to the lack of sufficient O_2 in the container, the Fe in the red mud cannot be fully oxidized, causing only the skin to turn red. In addition, according to the alumina production process, the lime put into the bauxite is calcined at high temperature to become CaO, so that a large amount of CaO is contained in the newly produced red mud, which has not undergone chemical reaction to form $CaCO_3$. The X-ray diffraction result of the "alkalization" substance of red mud shows that "alkalization" contains a large amount of calcium and magnesium carbonates. And to convert CaO into $CaCO_3$ The main substance is CO_2, which can react with CaO and MgO under certain environmental conditions to ultimately form $CaCO_3$ or $MgCO_3$. However, there is not enough CO_2 in the container for the reaction to proceed, so it is difficult to generate white alkaline substances. The experiment shows that the elements such as Ca and Mg in red mud must react with CO_2 in the air to form "pan-alkali" compounds such as $CaCO_3$ and $MgCO_3$.

3.5.3 Analysis of Alkaline Substances in Red Mud

Red mud contains a large amount of alkaline substances (mainly CaO, MgO, Na_2O, K_2O) that can cause "pan-alkali". These alkaline substances mainly come from the use of high-concentration alkaline mother liquor soaking or the addition of a large amount of Na_2CO_3 to extract alumina from bauxite in the alumina production industry. However, due to economic and technical constraints, some alkaline substances are not completely recovered and remain in the red mud. The pH of the leaching solution of the sintered red mud from the Zhongzhou Aluminum Plant is 12.19 ~ 12.87, the R_2O (Na_2O, K_2O) content is 1.72 ~ 4.61%, the CaO content is 26.54 ~ 29.90%, and the MgO content is 1.17 ~ 2.22%. These alkaline substances continuously emerge from the red mud during the storage process, forming a large "pan-alkali" area, as shown in Figs. 3.5 and 3.6. Figure 3.5 is a photo of the "pan-alkali" on the surface of the one-year sintered red mud dam. From the picture, you can see a large amount of white alkaline substances covering the surface of the red mud dam, showing the characteristic of high alkali content in red mud. Figure 3.6 is a photo of the surface of the multi-year red mud dam. Although the red mud has been stored for many years, the "pan-alkali" phenomenon still exists, and plants cannot grow on the surface of the "pan-alkali" dam, which also reflects from the side that although the sintered red mud has experienced many years of rain washing, it still contains a large amount of alkaline substances.

The mineral composition and chemical composition of the soluble and insoluble white "pan-alkali" substances in red mud were determined by atomic absorption, X-ray diffraction, and scanning electron microscopy, providing a theoretical basis for the study of the "pan-alkali" process.

The insoluble alkaline substances were dried, crushed, passed through a 0.4 mm sieve, and 0.2 g was taken. 10 mL of HCl, 1 mL of HNO_3, and 1 mL of $HClO_4$ were added in order. After the reaction was complete, it was placed in an MDS-2002A microwave digestion instrument for digestion. After digestion, it was made up to

Fig. 3.5 One-year red mud dam pan-alkali

Fig. 3.6 Multi-year red mud dam surface pan-alkali

100 mL in a volumetric flask and left to stand. Then, the concentrations of K, Na, Ca, and Mg were measured using a ContrAA300 atomic absorption spectrophotometer.

The experimental data are shown in Table 3.15. The table shows that most of the alkaline substances are calcium compounds, but they also contain some potassium, sodium, and magnesium compounds.

As can be seen from Fig. 3.7, the "pan-alkali" substances are mostly rod-shaped, radiated, stalactite-shaped, bean-shaped calcite (composition $CaCO_3$), forming cluster or radiated aggregates, which are easy to transform into calcite. Figure 3.8 shows that there are also some magnesite (composition $MgCO_3$) in the insoluble alkaline substances, which mainly appear as granular aggregates or porcelain-like dense blocks (with shell-like fractures), and the cleavage is parallel to the rhombohedron.

(1) Soluble Alkaline Substances

The soluble salts mainly include sodium and potassium salts. The main components are NaCl, KCl, Na_2CO_3, $KHCO_3$, $NaNO_3$ and other soluble sodium and potassium salts. Most of these substances are bright white in color, but in some places, these substances are slightly brownish-yellow due to the presence of other soluble iron metal ions. These alkaline substances are mostly thin and flaky, and their texture is loose. They can be easily peeled off by hand and dissolve immediately when they come into contact with water. Figure 3.9 is a photo of soluble alkaline substances accumulated on the surface of the red mud dam.

Figure 3.9a and b are comparative photos of soluble alkaline substances at the same location before and after rain wash. In Fig. 3.9a and b, you can see that a layer of white alkaline substance is accumulated on the entire surface of the dam body in the second section of the dam. During the previous investigation, no such

Table 3.15 Elemental content of insoluble alkaline substances

Element type	K_2O	Na_2O	CaO	MgO
Content (%)	0.53	1.70	31.27	1.49

Fig. 3.7 Insoluble alkaline substance magnified 2000 times

Fig. 3.8 Insoluble alkaline substance magnified 10,000 times

large-scale "alkalization" phenomenon was found. It was found that this layer of the dam body did not set up a red mud washing water diversion channel. The red mud washing water discharged from the first layer of the dam body could not be drained in time during the downward seepage process, A large amount of it gathered on the surface of the second layer of the dam body. At that time, it was the hot summer, and the washing water was constantly evaporating and crystallizing under the sun, forming a large area of white "alkalization" substance like snow. If you dig a small pit 30 cm deep on the surface of this dam body, you will immediately find that water is constantly seeping into the pit wall. The water is clear and bright, And it has a certain viscosity. Figure 3.9d is a photo taken at the same location 4 months later. This place had experienced several large-scale rainfalls before, and at this time, there was little left of the large-scale soluble white "alkalization" substance on the surface

Fig. 3.9 Soluble alkaline substance photo

of the dam body, which further confirmed that this type of alkaline substance is a soluble salt.

The "alkalization" process of this type of alkaline substance mainly involves: the accumulated washing water or seepage washing water constantly evaporates and crystallizes, forming a large area of crystalline salt, similar to the natural evaporation and crystallization process of making table salt in a salt factory by the sea.

The soluble white "alkalization" substance is put into a thermostatic box to dry and crystallize. It is sent to the Key Laboratory of Henan Polytechnic University for X-ray diffraction analysis of the crystals. The X-ray diffraction analysis found that the composition of the crystals was Na_2CO_3, $KHCO_3$, $K_3Na(SO_4)$, $NaNO_3$. The results of the X-ray diffraction spectrum analysis are shown in Fig. 3.10.

From the X-ray diffraction results, it can be seen that the red mud contains a large amount of soluble sodium salts and potassium salts, which are the main substances causing the production of soluble salts in the red mud.

(2) Insoluble Carbonate Substances

When the soluble alkaline substances are washed away by washing water or rainwater, what remains are insoluble alkaline substances, as shown in Fig. 3.11a. Figure 3.11b shows the lime slurry formed by CaO in red mud after soaking in water, seeping out from the red mud and gathering on the surface of the lower dam body, and curling into flakes after drying. At this time, the "alkalization" substance on the surface of the red mud is mostly gray-white block or granular, tightly wrapped on the surface of the red mud particles, not easy to peel off by hand, and the gap between the particles is larger. Figure 3.11c and d are photos of the cross-section of the red mud dam body.

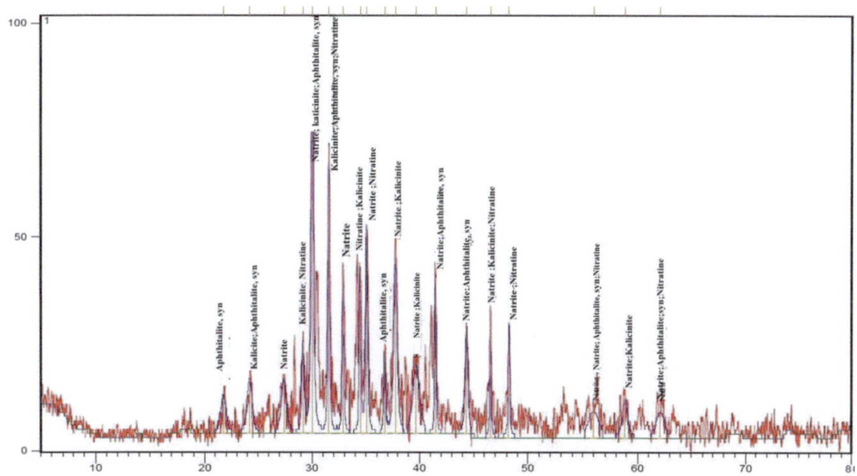

Fig. 3.10 X-ray diffraction diagram of soluble alkaline substances in red mud

From the figure, it can be seen that the insoluble white alkaline substance and the red mud are layered and alternately accumulated, which is related to the intermittent discharge of red mud washing water. In the alternating layers, the alkaline substance is granular and has more gaps, while the red mud is dense and hard, which is obviously contrasted with the alkaline substance.

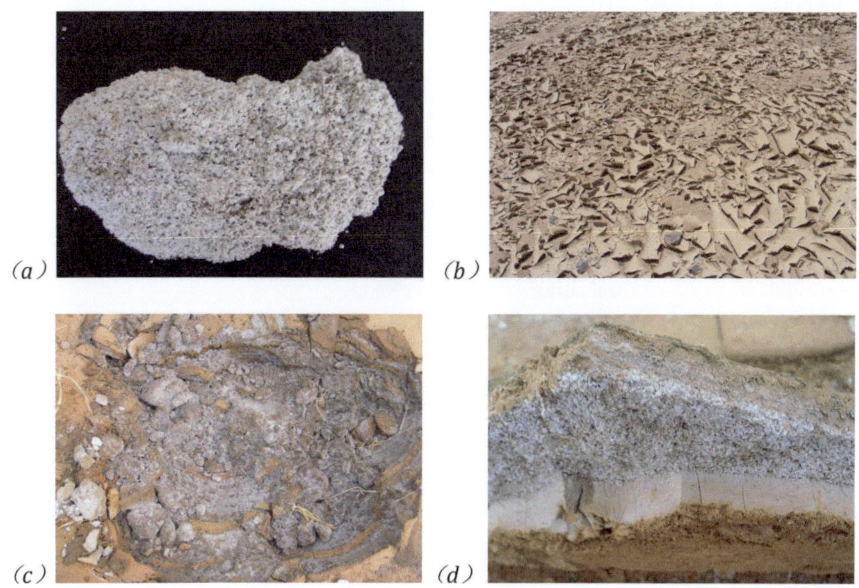

Fig. 3.11 Insoluble alkaline substance photo

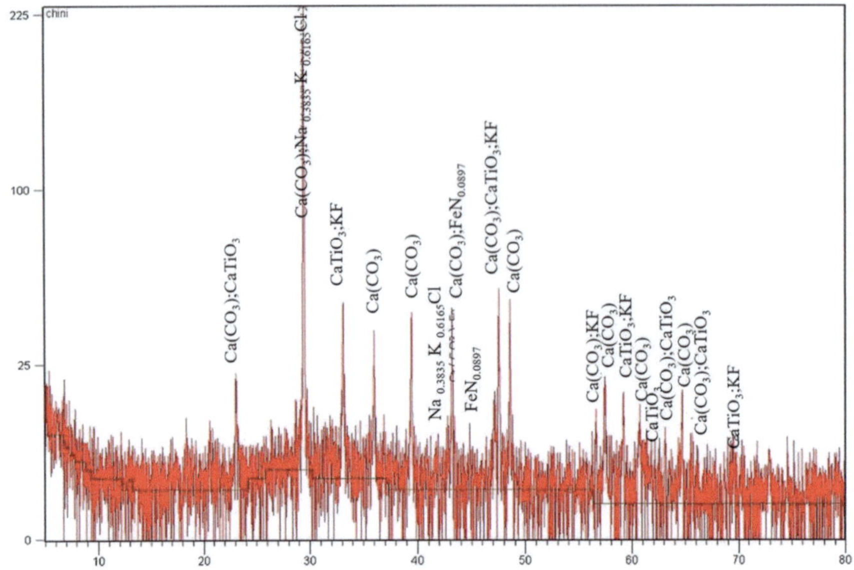

X-ray diffraction analysis of the insoluble white "pan-alkali" substances in red mud after drying, grinding, and sieving. The crystalline components obtained from the X-ray diffraction analysis are insoluble calcite ($CaCO_3$), perovskite ($CaTiO_3$), and iron nitride (FeN). In addition, there are some soluble sodium potassium salt compounds (NaCl, KCl), and potassium fluoride (KF) mixed in the insoluble alkaline substances. See Table 3.16.

Table 3.16 X-ray diffraction results of insoluble alkaline substances

Chinese name	English name	Chemical formula	Relative content (%)
Calcite	Calcite	$CaCO_3$	48
Potassium sodium salt compound	Sylvine sodian, syn	$Na_{(0.3835)}K_{(0.6165)}Cl$	27
Perovskite	Calcium titanate	$CaTiO_3$	14
Iron nitride	Iron nitride	$FeN_{(0.0897)}$	5
Potassium-rich stone	Carobbiite, syn	KF	6

3.6 Improvement and Solidification of Red Mud's Environmental Control Mechanism

3.6.1 Red Mud Alkali Control

The soluble alkali in red mud can cause alkali-aggregate reaction, pan-alkali, corrosion of steel parts, etc. How to solidify the soluble alkali in red mud and turn it from soluble alkali to insoluble alkali has become the object of many scholars' research.

At present, there are mainly three ways to solidify soluble alkali:

(1) Add various active mixed materials to hydrate and generate C-S-H to adsorb Na^+, K^+ and other alkali ions. Different types of C-S-H have different absorption capacities for alkali ions. When Al^{3+} is more, Al^{3+} will replace Si^{4+} in the silicate tetrahedron, causing an imbalance of valence, thereby adsorbing Na^+, K^+ ions to balance the valence, which has the effect of solidifying Na^+, K^+. In addition, the generation of low-calcium-silicon-ratio C-S-H is also prone to Al^{3+} replacing Si^{4+}, thereby adsorbing Na^+, K^+.

(2) Add active mixed materials to generate zeolite or zeolite-like structures, that is, generate sodium silicate, potassium silicate, etc., so that Na^+, K^+ are solidified in the zeolite structure and cannot be dissolved out, thereby reducing the content of soluble alkali.

(3) Add a certain amount of Li compounds. Because the ionic radius of Li^+ is smaller than that of Na^+, K^+, and the charge density is also high, it binds with the active silicon in the aggregate before Na^+, K^+, quickly generating insoluble substances to prevent Na^+, K^+ from contacting the active silicon, thereby inhibiting the occurrence of alkali-aggregate.

Under different activation methods, the dissolution of Na^+ indirectly reflects the active SiO_2, Al_2O_3 in red mud. The leaching reflects the impact of different activation methods on the soluble alkali content and activity of red mud, providing some theoretical support for the modification of red mud. Distinguishing soluble alkali from insoluble alkali in red mud, it is clear that only soluble alkali will bring harmful effects. Using the principle that alkali-activated fly ash can form hydrated sodium aluminosilicate, red mud is mixed with fly ash, and the soluble alkali in red mud is combined with fly ash at a temperature of 80 °C to form low-activity, insoluble hydrated sodium aluminosilicate, thereby transforming harmful soluble alkali into insoluble alkali, proposing a new idea for the alkali removal problem of red mud. Fly ash, as a common industrial waste, is rich in substances such as SiO_2, Al_2O_3, etc., and can form insoluble hydrated silicoaluminate after being activated by alkali, thus playing a role in solidifying soluble alkali. Literature shows that the dosage, fineness, and curing age of fly ash have an important impact on the synthesis of zeolite aluminosilicate, and the study of different factors on the adsorption of soluble alkali in red mud by fly ash plays an important role in solidifying harmful alkali. The use of Bayer red mud is mainly limited by two factors, namely low activity and high alkali content. Therefore, studying the impact of different treatment methods on the activity of red mud and reducing the content of harmful alkali in red mud has important guiding significance for the large-scale use of red mud. When using fly ash to adsorb soluble alkali in red mud, as the amount of fly ash increases, the soluble alkali content in red mud gradually decreases, and the corrosion degree of quartz glass also decreases. When the ratio of red mud to fly ash reaches 1:2.5, the alkali leaching rate of the sample reaches the lowest, decreasing from 54% without treatment to 33.5%. After the fineness of fly ash increases, the soluble alkali content in red mud significantly decreases, especially when the fineness of fly ash reaches 763m2/kg, the alkali leaching rate of the sample drops to 22.5%. Through the corrosion situation of quartz glass, it can be clearly seen that after the fineness of fly ash increases, the corrosion degree of quartz glass significantly weakens. As the curing age increases, the reaction degree of soluble alkali in red mud with fly ash gradually deepens. The corrosion condition of quartz glass at 14d is not much different from that at 7d, indicating that as time goes on, fly ash has adsorbed the soluble alkali in red mud, reducing the corrosion of quartz glass. The experimental results show that after fly ash is coupled with red mud, it can solidify a certain amount of soluble alkali, which plays an important role in reducing the soluble alkali content in red mud.

3.6.2 Red Mud Heavy Metal Ion Inhibition

The solidification effect is mainly that the cementing material encapsulates heavy metal ions or precipitates through physical wrapping and adsorption, while the stabilization effect is to incorporate heavy metal ions into the matrix structure through chemical bonds.

In the reaction process of stabilizing red mud, the amorphous silicon-aluminum phase dissolves quickly under the action of alkali activator, showing a gelation reaction phenomenon, forming a low-polymerized gel, and gradually forming a network structure with the aggregation effect. During this period, heavy metal ions are sealed in the gel body by the encapsulation of low-polymerized gel. Cr^{3+}, Cd^{2+}, and Pb^{2+} are not only encapsulated in ion form, but Cr^{3+}, Cd^{2+} form $Cr(OH)_3$, $Cd(OH)_2$ under alkaline conditions, and Pb^{2+} reacts with the dissolved silicon phase to form Pb_3SiO_5, forming a precipitate sealed by the gel body.

Chemical adsorption and solidification are manifested as some harmful ions entering the lattice of the hydrate during the hydration process of stable red mud through chemical reactions. The hydrate combines with harmful ions through chemical bonds, and harmful metal ions can play a role in charge balance like alkali metal ions inside the polymer network. Pb^{2+} is mainly solidified by chemical action. Research shows that it may be solidified in the newly formed amorphous phase in the form of Pb_3SiO_5. Harmful element As is partially combined with the hydration product by chemical bonds to form a whole, and Sr, Cs, Mo, etc. are also solidified in the solidified body structure due to chemical action.

In the structure of stabilized red mud mixture, Al binds with the surrounding 4 oxygen in a + 3 valence state, so the aluminum tetrahedron as a whole shows electronegativity. To balance the charge, cations are needed for charge balance, at which point heavy metal ions directly replace alkali metal ions to participate in charge balance. The cations that solidify and stabilize heavy metals in stabilized red mud mainly include Na^+, Ca^{2+}, etc., they can all enter the structure of stabilized red mud as charge balancing ions, adsorbed around the aluminum components, and compete with heavy metal ions. In addition, there are differences in the ionic radii of different heavy metals. For heavy metal ions with ionic radii close to Na^+ and Ca^{2+}, they are more likely to replace Na^+ and Ca^+ in the framework of stabilized red mud. Ettringite can solidify and stabilize lead, zinc, etc. Attempts to synthesize ettringite using dissolved Pb(II), lime, and aluminum-containing solutions have shown that in the absence of $SO_4{}^{2-}$, $Pb(OH)^{3-}$ cannot replace $SO_4{}^{2-}$ to bind with Ca^{2+}, AlO^{2-} to form ettringite; but when $SO_4{}^{2-}$ is present in the solution, some Pb(II), Zn(II), Cd(II), etc. can be captured by ettringite and enter the ettringite lattice. Ettringite has a strong chemical adsorption of dissolved Pb(II) under alkaline conditions, and Pb(II) may enter the ettringite lattice through adsorption.

3.7 Chapter Summary

(1) The alkali content of red mud washing water is high, and it contains a lot of pollutants such as heavy metals. These substances seep into the water body, causing the pH of the water body to be high, the content of harmful impurities exceeds the standard, causing great pollution to the water body, endangering the survival of aquatic organisms, and affecting the flood discharge and irrigation capacity of rivers and lakes. Because the particle size of red mud is extremely fine, when red mud enters the atmosphere in the form of dust, it pollutes the atmosphere. When the wind reaches level 4 or above, the powder with a particle size of 1 ~ 1.5 cm or more on the surface of the red mud pile will be peeled off, and its floating height reaches 20 ~ 50 m or more, inhibiting the growth and development of plants. The annual output of red mud is quite large, and the harmful substances it contains during storage will cause the surrounding soil to become marshy and salinized, affecting plant growth and ultimately harming the human body.

(2) The radioactive hazards of solid wastes such as red mud are mainly manifested in external irradiation hazards and internal irradiation hazards. The main cause of external irradiation hazards is γ-rays, and the calculation formula for the external irradiation index is; the internal irradiation hazards are mainly caused by radon-222, and the calculation formula for the internal irradiation index is. The radon-222 nucleus emits α particles, which can damage cell structures when inhaled into the human body. The radioactivity of Bayer red mud is lower than that of fly ash, coal gangue, and phosphogypsum, and the indicators meet the national A-level building material standards.

(3) Using the "Soil Environmental Quality Standard" and the "Surface Water Environmental Quality Standard" to evaluate the leaching toxicity of red mud, the leaching toxicity of modified red mud obtained through TCLP leaching test and Tank leaching test is lower than the leaching toxicity of red mud.
(4) Due to the high alkali content in Bayer red mud, excessive alkali can cause problems such as corrosion of steel parts and product efflorescence, so it is necessary to reduce the alkali content in red mud. Through leaching experiments, it was found that the pH value of red mud is generally lower than that of water-stable materials.
(5) Modified red mud will produce efflorescence and precipitation of heavy metal ions during actual use, and certain measures need to be taken to control or suppress it. The measures to control alkali mainly include adding various active mixed materials, adding a certain amount of Li compounds, and adsorbing Na^+, K^+ to achieve the purpose of solidifying soluble alkali. In addition, experimental results show that after fly ash is coupled with red mud, it can solidify a certain amount of soluble alkali, which plays an important role in reducing the soluble alkali content of red mud. The measures to suppress metal ions are mainly through solidification and stabilization, binding with harmful ions through chemical bonds, or being sealed in the gel by the encapsulation of low-polymer gels, thereby solidifying heavy metal ions in the solidified body structure.

Chapter 4
Modification and Solidification Methods and Mechanisms of Red Mud

Due to the special process of alumina production, the resulting red mud has characteristics such as high water content, high plasticity, low strength, and low bearing capacity. Red mud is not suitable for direct use as road fill material and must undergo modification and solidification treatment. From the perspective of modifying and solidifying road base fill materials, research is conducted on the selection of modified materials, methods, mixing ratios, and control of road use conditions. Through a large number of indoor experiments, research is conducted on the modification and solidification treatment methods, action mechanisms, and effect evaluation of red mud, and an optimized red mud modification and solidification system is proposed; the solidification mechanism is revealed through microscopic structure analysis.

4.1 Red Mud Modification and Solidification Scheme

4.1.1 Selection of Basic Materials for Modification and Solidification

(1) Lime

The experimental basic material uses the effective calcium oxide and magnesium oxide detection method (T813-1994) to test the sample, determines the water content of the lime by the drying method, and uses the lime fineness test method (T814-2009) to detect the fineness of the lime. The test results are shown in Table 4.1.

(2) Fly Ash

Fly ash is a volcanic ash mixed material formed under high temperature conditions in a blast furnace, and is a type of cementitious material. The effective components of fly ash include active SiO_2 and Al_2O_3, these effective components are active

Y. Cheng and X. Zhu, *Utilization and Environmental Impact Assessment of Red Mud Resources*, https://doi.org/10.1007/978-981-96-7411-4_4

Table 4.1 Test of quicklime raw material

Index	Measured value/%	Test method
Calcium magnesium content	55.23	T 813
Moisture content	29.34	T 801
Residue on 0.60 mm square hole sieve	18.9	T 814
Residue on 0.15 mm square hole sieve	17.6	T 814
Magnesium oxide content	1.1	T 812

materials, under the condition of water, they can react with $Ca(OH)_2$ in lime to form a water-hardening gel, giving the solidified body a certain mechanical strength, this process is called pozzolanic reaction. The fly ash used was tested for the content of the above substances by the silica, iron oxide and alumina content determination method (T816-2009); the loss on ignition of the sample was determined by the fly ash loss on ignition determination method (T817-2009); the fineness of the sample was determined by the fly ash fineness test method (T818-2009), and the moisture content was determined by the drying method. The test results are shown in Table 4.2, and all indicators meet the performance parameter requirements for fly ash in JTG/T F20-2015 "Technical Specifications for Highway Pavement Base Construction".

(3) Phosphogypsum

In the experiment, phosphogypsum was used as an additive. Phosphogypsum is an acidic solid waste produced and discharged during the wet production of phosphate fertilizers. Each ton of phosphoric acid produced will discharge about 4.5–5 tons of phosphogypsum. Depending on the amount of bound water, phosphogypsum can be divided into dihydrate phosphogypsum ($CaSO_4 \cdot 2H_2O$) and hemihydrate phosphogypsum ($CaSO_4 \cdot 1/2H_2O$). In addition to the main components, phosphogypsum also contains a small amount of phosphoric acid, so it is acidic. The pH value of the phosphogypsum sample is 2.4, and the purity is 86%.

(4) New Polymer Composite Material

A certain proportion of cement, phosphogypsum and polymer are mixed to form a composite stabilization treatment material, fully utilizing the mechanism of each

Table 4.2 Test of fly ash raw material

Index	Measured value/%	Test method
Total content of SiO_2, Al_2O_3, and Fe_2O_3	80.20	T 816
Loss on ignition	7.90	T 817
0.3 mm sieve pass rate	98.30	T 818
0.075 mm sieve pass rate	81.59	T 818
Moisture content	0.75	T 801

component material on the Bayer process red mud, studying the comprehensive stabilization effect of the composite material on the Bayer process red mud, comparing the physical and mechanical properties of the stabilized red mud, and observing its impact on environmental indicators such as metal ions, which is also the focus of this topic analysis.

4.1.2 Research Experiment Scheme

The particle size of red mud particles is mainly distributed between 3 and 8 μm, and the liquid limit is higher than 40, which is a high liquid limit clay. For the orthogonal test of lime-modified red mud, it is necessary to complete the heavy compaction test and the unconfined saturated compressive strength test. The main instruments are:

(1) Heavy Compaction Test

The main equipment of this test are: TLD-III multifunctional electric compactor, compaction cylinder with an inner diameter of 100 mm and a height of 127 mm, electronic balance, square hole sieve, electric demoulding machine, scraper, leveling ruler, mixing tools, oven.

(2) Unconfined Compressive Strength Test

The main equipment of this test are: cast iron mold with diameter × height = φ50 mm × 50 mm, LD-127s type road material strength tester, electric demoulding machine, standard curing box, constant temperature water bath, electronic balance, square hole sieve, mixing tools, oven, etc.

(3) Thermal Shrinkage Characteristics Test

Molding equipment: cast iron mold, inner wall size 50 mm, cast iron mold, inner wall size, inner surface polished, with upper and lower pressure blocks of the same area as the specimen, about 5 cm thick, as shown in Fig. 4.1; LD-127s type road material strength tester, pad, standard curing box, constant temperature water bath, electronic balance, square hole sieve, mixing tools, oven, etc. Test equipment: GDJS-150L type high and low temperature alternating test box: can perform temperature change in the range of − 20 ~ 60 °C, and can change humidity in the range of 0 ~ 100%, can set the rate of temperature and humidity change, the test box is shown in Fig. 4.2; shrinkage meter, micrometer, glass rod, lubricating oil, etc.

This section aims to improve the comprehensive road performance of red mud and meet the requirements for use in highway subgrades. It has conducted experimental research on the stabilization of red mud using two stabilization methods: inorganic solidification and composite solidification. Inorganic solidification mainly involves adding lime, fly ash, clay, and phosphogypsum to red mud samples in different mixing ratios for stabilization and solidification, and studying and analyzing the influence of the two ash ratios, the total amount of two ashes, and the type of red mud on the performance of stabilized red mud; Composite solidification mainly

Fig. 4.1 Thermal shrinkage test mold

Fig. 4.2 High and low temperature alternating test box

involves mixing cement, lime, phosphogypsum, and polymer in a certain ratio and adding them to red mud samples for stabilization and solidification, mainly exploring the influence of the type and ratio of additives on the stabilization of red mud. Tables 4.3 and 4.4 respectively show the proportioning schemes for treating red mud with inorganic solidification technology and composite solidification technology, and the stabilization treatment effect of red mud is evaluated through indoor standard heavy compaction tests and unconfined compressive strength tests.

When inorganic binder stabilized materials are used in the base layer of the road surface, temperature stress is generated in winter due to temperature differences. The surface temperature is lower than the internal temperature, and there is tensile stress

Table 4.3 Inorganic solidification technology treatment red mud scheme

Serial number	Lime/%	Fly ash/%	Red mud/%	Soil/%	Lime-ash ratio (A)	Total lime-ash/% (B)	Type of red mud (C)
1	2.3	22.7	75.0	0.0	0.1:1	25	Shan aluminum
2	4.6	45.4	50.0	0.0	0.1:1	50	Wei bridge
3	6.8	68.2	17.5	7.5	0.1:1	75	Shan aluminum + soil
4	8.3	16.7	75.0	0.0	0.5:1	25	Wei bridge
5	16.7	33.3	35.0	15.0	0.5:1	50	Shan aluminum + soil
6	25.0	50.0	25.0	0.0	0.5:1	75	Shan aluminum
7	15.0	10.0	52.5	22.5	1.5:1	25	Shan aluminum + soil
8	30.0	20.0	50.0	0.0	1.5:1	50	Shan aluminum
9	45.0	30.0	25.0	0.0	1.5:1	75	Wei bridge

at the top of the base layer. Combined with the tensile stress generated by the vehicle load at the bottom of the base layer, it can cause structural layer cracking and other diseases. Thermal shrinkage refers to the performance of the size of inorganic binder stabilized materials shrinking after cooling when the environmental temperature drops. Therefore, this section also conducted thermal shrinkage tests on inorganic solidified red mud, and the test plan is shown in Table 4.5.

4.1.3 Test Plan

1. Orthogonal Test

(1) Test factors, levels, and indicators

The orthogonal test method is a scientific arrangement of multi-factor, multi-level tests, using mathematical statistical methods to analyze performance indicators. Compared with ordinary test methods, orthogonal tests can reduce the number of tests as much as possible while ensuring the comprehensiveness of the data obtained. The orthogonal test method has been widely used in the fields of chemistry, medicine, etc., and in recent years, the application of the orthogonal test design method to the field of road engineering has also appeared.

Table 4.4 Red mud treatment scheme with composite solidification technology

Stable component	Test number	Mixing ratio (%)	Stable component	Test number	Mixing ratio (%)	Stable component	Test number	Mixing ratio (%)	Stable component	Test number	Mixing ratio (%)
Cement	C1	4	Lime	L1	4	Phosphogypsum	P1	4	Composite Material	F1	4
	C2	6		L2	6		P2	6		F2	6
	C3	8		L3	8		P3	8		F3	8
	C4	10		L4	10		P4	10		F4	10
	C5	12		L5	12		P5	12		F5	12

Table 4.5 Thermal shrinkage test plan

Serial number	Lime/%	Fly ash/%	Red mud/%	Soil/%	Type of red mud
1	0.0	0.0	100.0	0.0	Shan aluminum
2	0.0	0.0	100.0	0.0	Wei bridge
3	30.0	20.0	35.0	15.0	Shan aluminum + clay
4	30.0	20.0	50.0	0.0	Shan aluminum

When optimizing the ratio of lime fly ash stabilized red mud structure layer, use L_9 (3^4) orthogonal table, considering the two ash ratio, the total amount of two ash and the type of red mud three factors (this test only has three factors, the fourth factor is a blank column for error analysis). The two ash ratio is the ratio of lime to fly ash in the two ash stabilized materials (lime: fly ash), which will affect the excitation ability of the stabilized materials and ultimately affect the curing strength of the materials. The levels in the test are 0.1:1, 0.5:1, and 1.5:1 respectively. The total amount of two ash is the percentage of the amount of lime and fly ash in the mixture to the total mass, which will affect the strength of the stabilized materials. As the total amount of two ash increases, the strength of the material will also increase. But it's not to say that the more two ash, the better. Too much two ash, the strength will decrease. In addition, too much two ash content will increase the cost of highway construction and affect its practical promotion. The levels in the test are 25%, 50%, and 75% respectively. Different types of red mud have different chemical compositions, and their curing potential is also different. The levels in the test are Shan Aluminum red mud, Wei Bridge red mud, and 70% Shan Aluminum red mud mixed with 30% clay. The orthogonal test plan of the test is shown in Table 4.6.

Table 4.6 Orthogonal test plan

Serial number	Lime/%	Fly ash/%	Red mud/%	Soil/%	Two ash ratio (A)	Total two ash/% (B)	Type of red mud (C)
1	2.3	22.7	75.0	0.0	0.1:1	25	Shan aluminum
2	4.6	45.4	50.0	0.0	0.1:1	50	Wei bridge
3	6.8	68.2	17.5	7.5	0.1:1	75	Shan aluminum + soil
4	8.3	16.7	75.0	0.0	0.5:1	25	Wei bridge
5	16.7	33.3	35.0	15.0	0.5:1	50	Bauxite + soil
6	25.0	50.0	25.0	0.0	0.5:1	75	Bauxite
7	15.0	10.0	52.5	22.5	1.5:1	25	Bauxite + soil
8	30.0	20.0	50.0	0.0	1.5:1	50	Bauxite
9	45.0	30.0	25.0	0.0	1.5:1	75	Wei bridge

The test indicators must be representative and must objectively evaluate the target properties of the test object. For the road performance of the pavement base, refer to JTG/T F20-2015 "Technical Details for Highway Pavement Base Construction", and choose the unconfined compressive strength as the performance indicator, which is representative for project acceptance and also convenient for analyzing test results. In order to study the strength of different curing ages at the same time, orthogonal tests are carried out with the unconfined compressive strength of 7, 28, and 90 days as technical indicators. Each test is divided into 9 ratios, and each group of tests only differs in curing time, other conditions are the same.

The test results are analyzed and processed using range analysis and variance analysis methods.

(2) Range Analysis

The orthogonal test has orthogonality and comprehensive comparability. Taking the three-factor three-level orthogonal test as an example, although the combinations of the three factors are different, the benefits of the remaining two levels can be approximately offset under a fixed level of one. In the L_9 (3^4) orthogonal table, for the *i* factor *j* level, its test indicator is X_{ij}. If T represents the sum of all test indicators, then T_{ij} represents the sum of the three indicators of the *i* factor *j* level in the orthogonal table, and its average T_{ij} is. The formula for the range R_i of factor *i* is:

$$R_{\mathrm{i}} = \max\left\{\overline{T}_{\mathrm{i}1}, \overline{T}_{\mathrm{i}2}, \overline{T}_{\mathrm{i}3}\ldots\right\} - \min\left\{\overline{T}_{\mathrm{i}1}, \overline{T}_{\mathrm{i}2}, \overline{T}_{\mathrm{i}3}\ldots\right\} \tag{4.1}$$

The range can characterize the dispersion of a set of data, and is used to describe the range and dispersion of variable distribution. The difference between the standard values of any two units in the population is less than or equal to the range. The range can depict the fluctuation range of a set of data, and the size of the range represents the influence level of a factor on the test results.

(3) Variance Analysis

In actual tests, there are subtle differences in the samples after completion, and the temperature and humidity conditions in different positions of the curing box cannot be exactly the same, so there will inevitably be some errors in the test results. In order to distinguish the impact of factor level and error on performance indicators, variance analysis needs to be completed. Variance analysis, also known as analysis of variance, is a method used for significance testing of differences between two or more sample means. During the analysis, the fluctuation of the performance indicator is divided into fluctuations caused by factor levels and fluctuations caused by test errors. For the L_9 (3^4) orthogonal table, the total variation of the measured data is composed of four parts: factor A, factor B, factor C, and error variation *e*, so in variance analysis, the sum of squares SS_T and degrees of freedom df_T can be expressed as:

$$SS_T = SS_A + SS_B + SS_C + SS_e \tag{4.2}$$

$$df_T = df_A + df_B + df_C + df_e \tag{4.3}$$

Then:

$$SS_T = \sum_{i=1}^{i}\sum_{j=1}^{j} X_{ij}^2 - \frac{T^2}{9} \tag{4.4}$$

The sum of squares of each factor is:

$$SS_i = \sum \frac{T_i^2}{3} - \frac{T^2}{9} \tag{4.5}$$

The sum of squares of the error SS e is:

$$SS_e = SS_T - \sum SS_i \tag{4.6}$$

The size of SS_i and SS_e is related to their degrees of freedom, and they need to be divided by their respective degrees of freedom before they can be compared. The method for determining the degrees of freedom is as follows:

Total degrees of freedom:

$$df_T = 9 - 1 \tag{4.7}$$

Degrees of freedom for each factor:

$$df_i = 3 - 1 \tag{4.8}$$

Error degrees of freedom:

$$df_e = df_T - \sum df_i \tag{4.9}$$

The total degrees of freedom in the experiment is 8, and the degrees of freedom for each factor and error is 2.

After obtaining A, B, C, the sum of squares of errors and degrees of freedom, the significance of the data is judged through the F test. The formula for the F test is:

$$F = \frac{SS_i/df_i}{SS_e/df_e} \tag{4.10}$$

The size of the F value reflects the degree of influence of changes in the levels of each factor on the experimental indicators. The larger the F value, the greater the impact of this factor on the experimental indicators. If the confidence level α is 0.05 in the experiment, and the F value of the experimental data is greater than 19, it is considered to pass the F test.

2. Standard Heavy Compaction Test

Similar to soil, the compaction effect of stabilized red mud is related to the moisture content of the raw material. The optimum moisture content is the moisture content that can make the fill reach the maximum dry density under a certain function of compaction. According to the optimum moisture content, the optimum moisture content of a certain mass of mixture can be calculated. Referring to this parameter, the mixture is compacted to achieve the maximum density. The dry gravity density in this state is called the maximum dry density. The optimum moisture content is closely related to the type of mixture. Different mixtures have different compaction efforts. The greater the compaction effort, the less the fine particle content of the material, the smaller the optimum moisture content, and the larger the maximum density. Therefore, the optimum moisture content refers to the moisture content corresponding to the maximum compact state under a certain compaction energy for a specific mixture.

The compaction test is to make specimens according to multiple different moisture contents (ω), measure their dry density (ρ_d), plot the $\rho_d - \omega$ curve and determine the peak value. The moisture content and dry density corresponding to this peak value are used as the optimum moisture content (ω_0) and maximum dry density (ρ_{dm}). The test steps are as follows:

(1) Determine the compaction cylinder specifications and the number of compactions per layer based on the particle size of the sample. The sample in this test belongs to fine-grained soil, so a compaction cylinder with a diameter × height = φ100 mm × 127 mm is used, with 27 compactions per layer and 5 layers of hammering for each specimen;
(2) Determine the compaction parameters of the compactor, the bottom diameter of the hammer is 50 mm, the mass is 4.5 kg, the total travel of the hammer in the guide tube is 45 cm, and the hammer marks are evenly distributed;
(3) Refer to previous literature to determine the approximate range of the optimum moisture content, and plan 6 ~ 8 groups of different moisture contents around it, each differing by 1.5%;
(4) Make samples according to the moisture content, mix the red mud, lime and other test materials together, evenly spray the specified mass of water, then put it into a sealed container and soak for 12 h;
(5) Compact each group of samples, install the compactor, inject the mixture into the compaction cylinder in 5 times, and compact 27 times after each loading, as shown in Fig. 4.3;
(6) Measure the height of the sample, weigh the total mass of the compaction cylinder and the specimen, and calculate the mass of the specimen, as shown in Fig. 4.4;
(7) Demold and measure the actual moisture content of the specimen, and plot the $\rho_d - \omega$ curve, and make each curve a camelback curve, ensuring that there are at least two points on both sides of the vertex. If the curve does not meet the requirements, additional experiments need to be conducted.

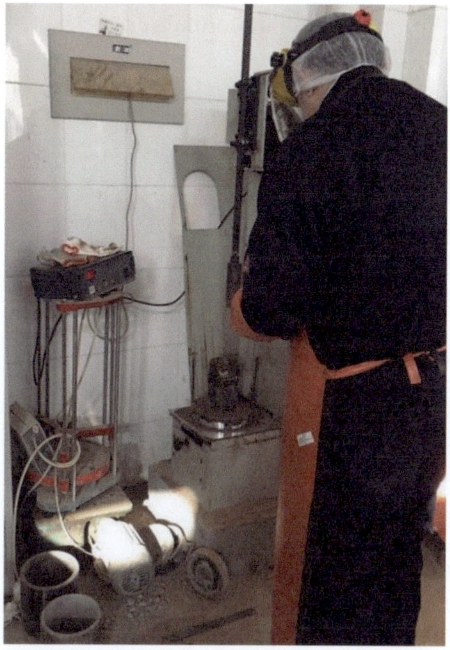

Fig. 4.3 Standard heavy compaction test

(a) (b)

Fig. 4.4 Compaction test specimens **a** before demolding, **b** after demolding

For the dry density ρ_d, its calculation formula can be expressed as:

$$\rho_d = \frac{m_1 - m_2}{(1 + 0.01\omega) \cdot V} \tag{4.11}$$

where

ρ_d the dry density of the sample (g/cm^3);
ω the actual moisture content of the sample (%);
m_1 the total mass of the compaction cylinder and the sample (g);
m_2 the mass of the compaction cylinder (g);
V the volume of the compaction cylinder (cm^3).

After determining the vertex of each ratio $\rho_d - \omega$ curve, draw the curve with the vertex and two points on the left and right respectively, and fit it with the quadratic curve method to obtain the $\rho_d - \omega$ curve of each ratio.

3. Unconfined compressive strength test

The unconfined compressive strength is the ultimate strength of the soil to resist axial pressure under unrestricted lateral conditions. According to the parameters determined by compaction, the unconfined compressive strength test mix design is formulated. Determine three curing ages for the specimens, which are 7, 28, and 90 days.

The test piece production and detection steps can be briefly described as follows:

(1) Calculate and determine the amount of red mud, lime, fly ash, clay and additional water required for each group of sample pieces;
(2) Mix the raw materials and mix them, while evenly sprinkling water, so that the moisture content of the mixture is close to the optimum moisture content, and is 3% smaller than the optimum moisture content;
(3) Soak the mixture in a sealed container for 12 h, and add the remaining 3% of water;
(4) Use a press to press the material into a mold in a φ50 mm × 50 mm cylinder mold, the loading rate is 1 mm/min, and maintain the pressure for 2 min;
(5) Unload, use a demolding device to demold the specimen. According to the specifications, at least six pieces should be made for each ratio. To reduce experimental errors and avoid accidental damage to the specimens, eight pieces are made for each ratio;
(6) The specimen is cured in a standard curing box, the curing temperature is 20 ± 2 °C, the relative humidity is 97 ± 2%, and the specimen does not directly contact the water flow during the curing process;
(7) Cure the specimens according to different ages, and soak in water for 24 h on the last day of curing, the water temperature is 20 ± 2 °C. The weight loss of the specimen during curing and soaking does not exceed 1 g;
(8) Use a pavement material strength tester to perform a compression test on the specimen, maintain a loading rate of 1 mm/min and measure the maximum

pressure when the specimen is broken. Take the average value of each group of results and ensure that the coefficient of variation meets the requirements, and use the Grubbs criterion to judge outliers. The curing and strength testing process of the pieces are shown in Figs. 4.5 and 4.6 respectively.

For the unconfined compressive strength of the specimen, refer to the following formula for calculation.

$$R_c = \frac{4P}{\pi D^2} \tag{4.12}$$

where

R_c the unconfined compressive strength of the specimen (MPa);
P the maximum pressure when the sample is broken (N);
D the diameter of the specimen (mm).

(a) (b)

Fig. 4.5 Unconfined compressive strength specimen curing process **a** specimen curing, **b** specimen soaking

(a) (b)

Fig. 4.6 Unconfined compressive strength testing process **a** testing conditions, **b** specimen failure process

In a set of measurement data, if individual data deviate far from the average value, they are called suspicious values. If the suspicious value is removed from the set of measurement data and does not participate in the calculation of the average value, the value can be regarded as an outlier. The Grubbs criterion method is used to eliminate outliers, and the judgment method is briefly described as follows:

(1) Assume there is a set of measurement data, for example, measured 10 times ($n = 10$), the data obtained are: 3.2, 4.2, 3.8, 6.5, 1.7, 2.8, 5.7, 3.9, 4.1, 3.3.
(2) Sort the above data in ascending order, to get 1.7, 2.8, 3.2, 3.3, 3.8, 3.9, 4.1, 4.2, 5.7, 6.5. The suspicious values could only be the minimum or maximum value.
(3) Calculate the average value of the 10 data points $\overline{x}$ and the standard deviation s: $\overline{x} = 3.92$; standard deviation $s = 1.376$.
(4) Calculate the difference between the average value and the extreme values: the difference between the average value and the minimum value is 3.92–1.7 = 2.22; the difference between the maximum value and the average value is 6.5–3.92 = 2.58.
(5) Comparing the data, the difference of the maximum value 2.58 is greater than the difference of the minimum value 2.22, so the maximum value 6.5 is a suspicious value.
(6) Calculate the G_i value, as shown in formula 4.13

$$G_i = \frac{x_i - \overline{x}}{s} \tag{4.13}$$

where

G_i the ratio of the residual to the standard deviation;
x_i the i th measurement data;
$\overline{x}$ the average of all measurement data;
S the standard deviation of all measurement data.

For the maximum value 6.5, since it is the 10th, substitute into the formula, G10 = (6.5–3.92)/1.376 = 1.875. Compare the G_i value with the critical value $G_{P(n)}$ listed in the Grubbs table. If the G_i value is greater than $G_{P(n)}$, the corresponding data is considered an outlier and needs to be removed. The critical value $G_{P(n)}$ is related to two parameters: the detection level α (related to the confidence probability P) and the number of measurements n (related to the degrees of freedom f).

(7) Set the detection level α: usually set $\alpha = 0.05$, $P = 0.95$.
(8) Determine the P value and the number of measurements n, look up the Grubbs table, and get the critical value $G_{95(10)} = 2.176$.
(9) Compare the calculated value G_i and the critical value $G_{95(10)}$: $G_i = 1.875$, $G_{95(10)} = 2.176$, $G_i < G_{95(10)}$
(10) Because $G_i < G_{95(10)}$, it can be judged that the measurement value 14.0 is not an outlier and can be retained.

4. Thermal shrinkage property test

The finished parts and test steps of the thermal shrinkage test are as shown in (1) and (2).

(1) Finished parts steps

1) Determine the optimal moisture content and maximum dry density;
2) For the same group of ratios, prepare 3 parallel specimens, calculate and determine the required amount of red mud, lime, fly ash, clay and water;
3) Mix the raw materials and stir, while evenly sprinkling water, so that the moisture content of the mixture is 97% of the optimal moisture content, reserving 3% of the moisture content;
4) Soak the mixture in a sealed container for 12 h, and add 3% water 1 h before the specimen is formed;

   - Initial length of the specimen (mm).

5) Place the lower pad of the mold at the bottom of the mold and expose 2 cm, weigh the specified mass of the mixture and place it in the mold in 2 ~ 3 times, and gently insert it with a tamping rod, then place the upper pad in the upper part of the mold and expose 2 cm, as shown in Fig. 4.7;
6) The material in the mold is pressed into shape by a press using the static method, the loading rate is 1 mm/min, until both pads are pressed into the mold, then maintain the pressure for 2 min, as shown in Fig. 4.8;
7) Unload, demold the specimen using the method of disassembling the mold after 3 h. Weigh the specimen and measure the cross-sectional dimensions and length with a vernier caliper, those that do not meet the molding standards are treated as waste, and re-molded, the molded specimens are shown in Fig. 4.9.
8) The specimen is cured in a standard curing box, the curing temperature is 20 ± 1 °C, the relative humidity is 97%, and the specimen does not directly contact with the water flow during the curing process, as shown in Fig. 4.10a;

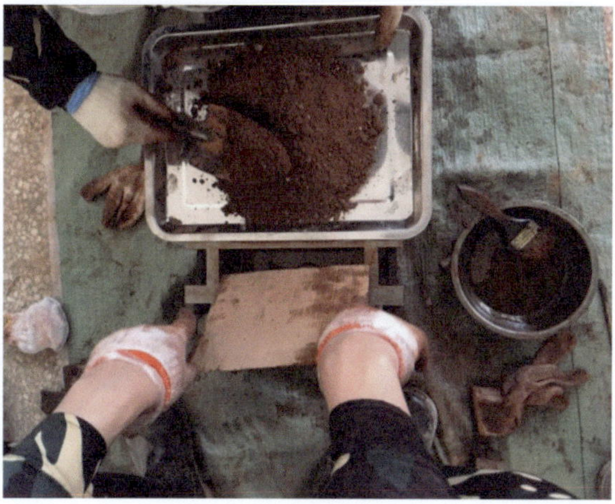

Fig. 4.7 Mixing and molding

(a) (b)

Fig. 4.8 Static compaction **a** operation process, **b** pressure maintenance process

Fig. 4.9 Thermal shrinkage test specimen

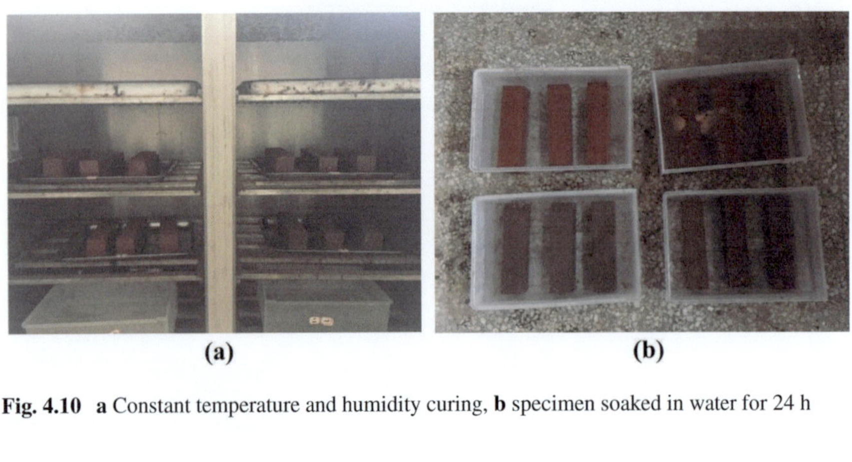

(a) (b)

Fig. 4.10 **a** Constant temperature and humidity curing, **b** specimen soaked in water for 24 h

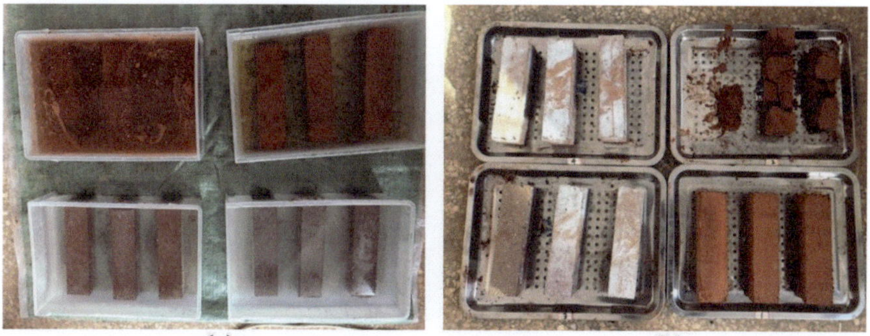

Fig. 4.11 **a** Specimen after soaking, **b** specimen after drying

9) According to different curing ages, the specimens are cured, and soaked in water for 24 h on the last day of the curing age, the water temperature is 20 ± 1 °C. The weight loss during curing and soaking should not exceed 1 g, as shown in Fig. 4.10b;

(2) Test steps

1) After soaking, dry in an oven at 105 °C until constant weight, then place in a dry and ventilated place until room temperature, the specimens after soaking and drying are shown in Fig. 4.11;
2) Use a vernier caliper to measure the initial length of the specimen, measure once at the edges and middle of both ends of the specimen, and take the average value;
3) Stick supports on both ends of the specimen, place three glass rods coated with lubricating oil on the bottom surface of the shrinkage meter;
4) Place the specimen with the smooth surface down on the glass rods, ensure that the shrinkage meter is level and the glass rods can roll, install the micrometer, and the micrometer head is against the center of the support;
5) Set the temperature and humidity control program of the high and low temperature alternating test box, the temperature is set to 40, 30, 20, 10, 0 °C, the cooling rate is 0.5 °C/min, the time for each temperature gradient is 3 h, the humidity of the test box is set to zero (dry);
6) Place the shrinkage meter and specimen in the test box, push the micrometer to a larger reading, and zero them together, the placement of the specimen is shown in Fig. 4.12;
7) According to the set temperature and humidity control program, gradually cool from high temperature to low temperature, the temperature difference between adjacent gradients is 10 °C, according to the set cooling rate, when it drops to the set temperature (about 20 min), keep it for 3 h. Read the micrometer in the last 5 min of heat preservation, the length of thermal shrinkage deformation is the sum of the extensions of the two micrometers on the left and right of a specimen during the cooling process;

(a) (b)

Fig. 4.12 **a** Placement of the specimen, **b** layout of the test box

8) Calculate the thermal shrinkage ε_i strain and the thermal shrinkage coefficient, α_t their calculation formulas are shown in (4.14) and (4.15) respectively:

$$\varepsilon_i = \frac{l_i - l_{i+1}}{L_0} \tag{4.14}$$

$$\alpha_t = \frac{\varepsilon_i}{t_i - t_{i+1}} \tag{4.15}$$

where

l_I Average value of the sum of the micrometer readings in the i-th temperature interval (mm);

t_I The i-th temperature interval set by the temperature control program (°C);

L_0 Initial length of the specimen (mm);

ε_i The average shrinkage strain (%) under the *I*th temperature;

α_t Thermal shrinkage coefficient, i.e., the linear shrinkage strain under unit temperature change.

4.2 Research on the Mechanism of Red Mud Solidification

4.2.1 Basic Principle of Modified Solidified Red Mud

This material is a new type of environmentally friendly composite polymer material based on Bayer process red mud. A certain amount of Haiyi Technology solidifier (CPS-12) is added to the Bayer process red mud, and after mixing and compaction, it can meet the required road performance indicators, while significantly reducing the strong alkalinity and pollution of red mud, making it meet environmental requirements. This modified solidifier product is stable to temperature changes, has the function of fixing metal ions, can prevent the negative effects of metal ions on the product, and is a surfactant with various special properties. Under normal temperature and pressure and weak alkaline environment, this high polymer composite material can achieve good adsorption effect in a short time, greatly reducing the leaching amount of 90% of Cr^{2+}, Hg^{2+} and other heavy metal ions.

(1) The qualitative description of the reaction of modified solidified red mud road material is

$$m\mathrm{Ca(OH)_2} + \mathrm{SiO_2} + (n-1)\mathrm{H_2O} = m\mathrm{CaO}\cdot\mathrm{SiO_2}\cdot n\mathrm{H_2O}$$

$$m\mathrm{Ca(OH)_2} + \mathrm{Al_2O_3} + (n-1)\mathrm{H_2O} = m\mathrm{CaO}\cdot\mathrm{SiO_2}\cdot\mathrm{Al_2O_3}\cdot n\mathrm{H_2O}$$

When a large amount of ${SO_2}^{4-}$ exists, the reaction product of active Al_2O_3 will be replaced by trisulfur type calcium aluminate (Aft):

$$\begin{aligned}&3\mathrm{Ca(OH)_2} + \mathrm{Al_2O_3} + 3\mathrm{CaSO_4}\cdot 2\mathrm{H_2O} + 26\mathrm{H_2O}\\&= 3\mathrm{CaO}\cdot\mathrm{Al_2O_3}\cdot 3\mathrm{CaSO_4}\cdot 32\mathrm{H_2O}\end{aligned}$$

Aft is needle-rod shaped, its formation and growth will produce a certain expansion, which is conducive to the formation of the structure connection inside the material. The presence of ${SO_2}^{4-}$ ions can stimulate the activity of red mud and accelerate the reaction process. Aft is needle-shaped, plate-shaped and other shapes, which increases the internal friction between particles in the consolidated material, which is beneficial to the formation of early strength. However, it may be due to the generation of too much calcium aluminate that restricts the degree of reaction, resulting in the slow growth of the later strength of red mud concrete (Figs. 4.13, 4.14, 4.15 and 4.16).

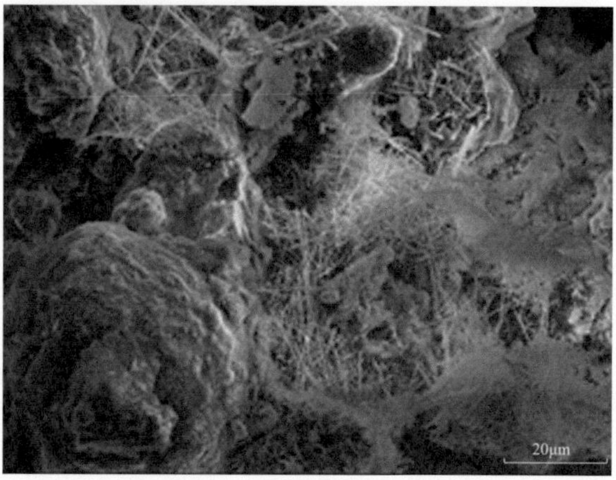

Fig. 4.13 SEM image of 1000 × sample cross section

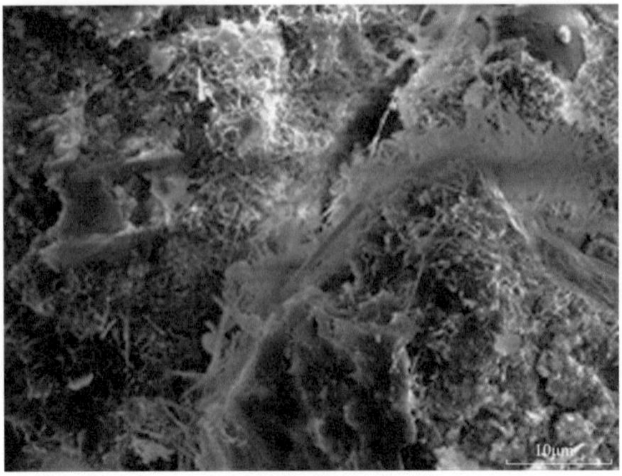

Fig. 4.14 SEM image of 2000 × sample cross section

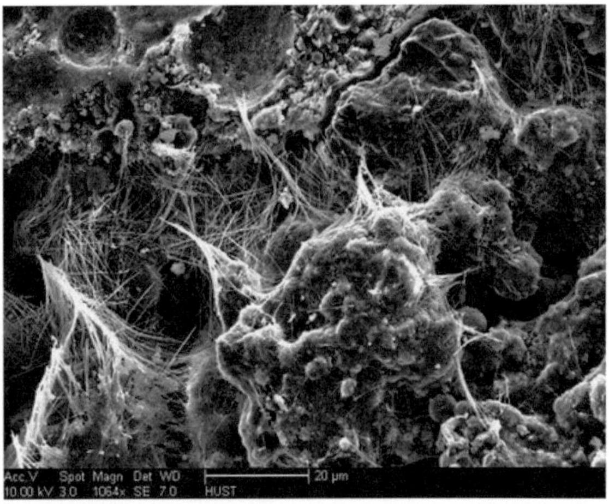

Fig. 4.15 SEM image of 3000 × sample cross section

From the figure, it can be clearly seen that the sample has produced a lot of hydrated cementitious substances and formed a space network structure of colloid plus crystal.

(2) Under alkaline conditions, the high polymer composite material can transform red mud into active SiO_2 and Al_2O_3. Increasing the alkalinity of the solution can accelerate the dissolution rate of active SiO_2 and Al_2O_3, and gradually react with $Ca(OH)_2$ to form calcium silicate, calcium aluminosilicate and other compounds. When the concentration of the product in the mixed system reaches a certain value, they will interlock with each other to form a network structure, and then form a gel, thereby improving the early strength. The addition of the polymer composite material is to improve the alkalinity of the solution to achieve

Fig. 4.16 SEM image of 3000 × sample cross section

early strength. As the alkalinity of the modified solidified red mud road material decreases, chemical bonds are formed, and silicon, aluminum and other oxygen-containing acid compounds are slowly generated. The newly formed crystals will gradually grow and develop, forming a network structure, and gradually dehydrate and dry to become a stable crystalline condensation structure, becoming a crystalline whole, and becoming a water-stable material with higher strength. Therefore, it cannot play the best skeleton network structure and excitation effect, and the later strength growth is slow.

(3) The formation of red mud has gone through the leaching and washing processes of alumina production, which makes the surface of the silicate minerals in it form a layer of hydrolyzed and decomposed colloidal compounds, which generally do not show the hydration activity of cementitious minerals, but show the potential water hardness. The addition of high polymer composite material can form CSH plate-like crystals on the surface of red mud particles, accelerate the destruction of the original colloidal compound film formed on the surface of red mud particles, cause its internal structure to dissociate, and release various cations and anion groups (such as $[SiO_4]^{4-}$ $[AlO_4]^{5-}$), and then various cations and anions form hydrated products through chemical reactions, ultimately leading to the system's condensation, hardening, and increased mechanical strength. The driving force is the excitation of alkaline substances, thereby achieving the purpose of exerting its cohesiveness.

(4) Through the analysis of the mechanism of strength formation of the red mud base layer, it can be seen that water plays a very important role in the process of strength formation. It is an indispensable participant, carrier, reaction condition, and physical and chemical environment in the reaction. The optimal moisture content during construction is just the condition that is most likely to reach the maximum dry density, not the best condition to reach the maximum strength. Therefore, it is recommended that in the future modification and solidification of red mud road materials, the moisture content of the mixture should be kept

near the optimal moisture content during the rolling process, and an appropriate amount of water should be added to the modified and solidified red mud road materials as soon as possible after rolling to meet the needs of the base layer strength formation reaction.

Because the surface of Bayer process red mud particles has a large residual valence force, molecular force, and hydrogen bond force, it can adsorb charged ions in the solution and can have a strong solvation effect with the medium, forming a certain thickness of diffusion layer, which hinders the interaction between particles, making it difficult for red mud particles to agglomerate into large particles. The mechanism of action of the new type of environmentally friendly composite polymer material is to effectively bind the red mud particles together through charge neutralization, adsorption bridging, and surface adsorption, gradually forming a large particle plate structure.

Charge Neutralization

Charge neutralization refers to the process where the potential of colloidal particles is reduced enough to overcome the energy barrier and cause flocculation precipitation. When the charge on the surface of colloidal particles is neutralized, the distance between colloidal particles is reduced. Under van der Waals forces, the interaction between colloidal particles can form a stable flocculation body. After the addition of red mud polymer modification materials, the surface charge of colloidal particles is neutralized. The surface charge of colloidal particles can not only be reduced to zero, but can also carry the opposite charge. Since the neutralization of charge is caused by adsorption, it leads to a change in the interface between colloidal particles and water, thereby changing the physicochemical properties, and the colloidal particles become unstable and flocculate. The neutralization of the ions in the diffusion layer on the surface of the colloid by the modified polymer material destroys its double-layer stability, causing the particles to agglomerate. Two particles with opposite charges agglomerate due to the compression of their diffusion layer by the neutralization of charge, reducing the point position between particles so that the two particles collide and agglomerate. The neutralization of charge reduces the electrostatic repulsion between charged particles, making it easier for them to aggregate with other colloidal particles.

Bridging Action

After the addition of red mud polymer modification materials to red mud, the long chains of polymers adsorbed on the surface of red mud particles may also adsorb other particle surfaces, connecting multiple particles together like a "bridge", thereby causing particle agglomeration. This bridging action mechanism can be represented by the model in Fig. 4.17.

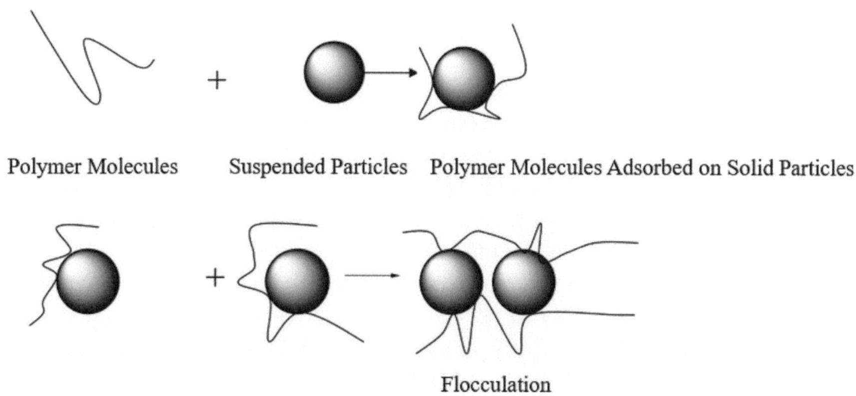

Fig. 4.17 Schematic diagram of the bridging action model of modified solidified red mud road materials

Net Catching Action

As a red mud polymer modification material, it can form a three-dimensional structure in red mud, which is suitable for capturing colloidal particles. As the volume of the polymer compound shrinks, precipitates and suspended matter sweep colloidal and suspended particles from the water like a porous net, forming flocculent precipitation.

Adsorption Action

Red mud polymer modification materials can effectively reduce high-valence heavy metal ions and fix them at multiple points on the solid–liquid interface of organic functional groups. This fixation can be achieved through physical adsorption, chemical adsorption, or both.

Physical adsorption includes electrostatic force, molecular force, and hydrophobic action. Electrostatic force is the adsorption force of polyelectrolytes on solid particles with opposite charges. Chemical adsorption includes covalent bonds, ionic bonds, coordination bonds, and hydrogen bonds. The functional groups of red mud modification materials can generally form insoluble compounds with metal ions on the particle surface through covalent bonds or ionic bonds. This adsorption is not only selective, but also has a large chemical bond energy. At the same time, the flocculant can form complexes or chelates on the surface of mineral particles through coordination bonds, thereby reducing the leaching concentration of more than 95% of heavy metal ions.

Mechanical Compaction

After the red mud-based road concrete is compacted and formed, the system consists of three phases: solid phase (red mud aggregate), liquid phase (aqueous solution), and gas phase (air). The interaction between these three phases results in the red mud-based road concrete having high strength and rigidity. Macroscopically, the factors that contribute to the strength of the modified cured red mud road material include two aspects: the internal friction and cohesion between the red mud polymer particles, and the "original cohesion" formed by the molecular attraction between the water film between the particles and the adjacent particles through mechanical compaction. Over time, a series of physical, physicochemical, and chemical actions occur between the solid and liquid phases in the concrete, and a series of substances with cementing effects are generated, which strengthen the connection between the particles in the mixture and form "cured cohesion", which is the main source of strength for the modified cured red mud road material.

The formation of the strength of the modified cured red mud road material mainly relies on a series of reactions that occur in the mixture, which tightly connect the red mud particles together. After the modified cured red mud road material is fully mixed, under certain weak alkaline conditions, through a series of chemical reactions and ion exchange effects through various pathways, the mineral components in the mixture form various hydrates with crystal structures around the micro particles. This colloid particle composed of micro particles and hydrates aggregates into a network structure under the action of molecular forces, hardens under the action of heat released during the chemical reaction process, external pressure (rolling), temperature (air temperature), and forms a crystalline network structure skeleton combined with chemical bonds to form strength, generating a dense and hard semi-rigid plate.

The calcium hydroxide in the aqueous solution of the liquid phase can undergo a reverse reaction of dissolution—crystallization under the supersaturated state caused by the evaporation of solution water, forming a water-containing lattice and gradually becoming a crystal from a colloid, these crystals connect with each other and combine with red mud to form a co-crystal, cementing the solid particles into a whole, thereby producing a certain structural strength. Compared with the amorphous $Ca(OH)_2$, the solubility of the crystal $Ca(OH)_2$ is almost halved, so the water stability of the crystalline structure formed by the crystal $Ca(OH)_2$ is better than the water stability of the agglomerate structure formed by the $Ca(OH)_2$ colloid, which improves the stability of the modified cured red mud road material.

The above reactions are not completed immediately, but gradually develop over time, and will end after a period of time. During this period, colloids and crystals continue to increase and grow, gradually contacting and intersecting each other, not only cementing other minerals in the red mud-based concrete that have not participated in the chemical reaction together, but also forming a space network structure of colloid plus crystal. This solid spatial network is the structural reason for the formation of the strength of the red mud base layer.

Therefore, a necessary condition for the formation of the strength of the red mud base layer is compaction. After the modified cured red mud road material is mixed, it should be rolled in time, so that the above chemical reactions occur in the compacted concrete. If not compacted, the chemical reaction will proceed as usual, but it will not form a network structure. At most, the reacted mixture becomes a loose aggregate with certain water stability, without strength, and cannot form a plate structure.

4.2.2 Analysis of Modified Curing Mechanism

1. Mechanism analysis techniques and experimental materials

(1) Experimental materials

The mechanism analysis mainly targets two raw materials, red mud and lime fly ash stabilized red mud solidification body. The ratio of red mud solidification body includes uncured, 7d age and 28d age samples. The mechanism analysis uses the sample ratio as shown in Table 4.7. The broken specimens in the strength test are collected according to the ratio, ground in a mortar, soaked in anhydrous ethanol for 48 h to terminate hydration, then dried at 40 °C and passed through an 800 mesh sieve, packed into sample bottles according to different ratios and ages, and tested within 2 working days. The samples are shown in Fig. 4.18.

(2) Mechanism analysis methods

Mechanism analysis methods include X-ray fluorescence spectroscopy (X-Ray fluorescence spectroscopy), X-ray diffraction (X-Ray diffraction) and scanning electron microscopy and energy-dispersive X-ray spectroscopy. The instrument parameters are briefly described as follows.

Table 4.7 Mechanism analysis ratio

Serial number	Lime/%	Fly ash/%	Red mud/%	Soil/%	Type of red mud
#1	0.0	0.0	100.0	0.0	Shandong aluminium
#2	0.0	0.0	100.0	0.0	WeiQiao
#3	30.0	20.0	35.0	15.0	Shandong aluminium + clay

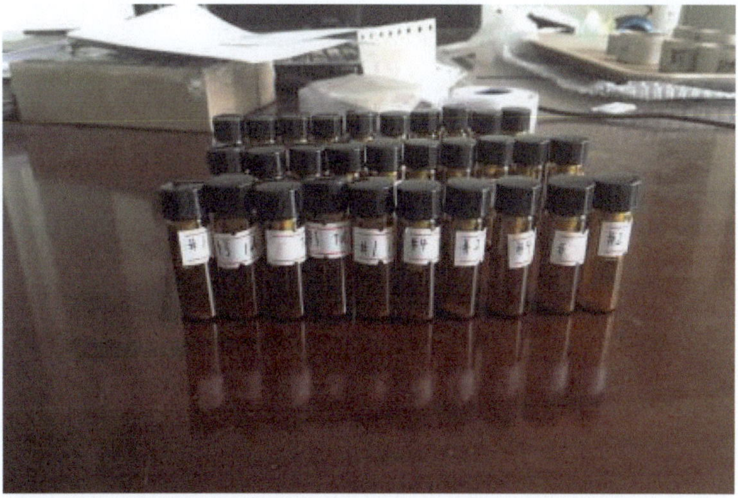

Fig. 4.18 Sample for mechanism analysis

(1) X-ray fluorescence spectroscopy analysis: Using the Japanese Rigaku ZSX Primus II automatic scanning X-ray fluorescence spectrometer, it can quickly qualitatively and quantitatively analyze elements from 6C to 92U. The measurement range of elements is 0.0001 to 100% (it can be reduced by 2–3 orders of magnitude under pretreatment);
(2) X-ray diffraction analysis: Using the Dutch PANalytical EMPYREAN X-ray diffractometer, the test operating conditions are voltage 60 kV, current 60 mA, Cu target, wavelength 0.15418 nm, scanning angle 2θ is 10 ~ 90°, step length is 0.0001°, using PIXcel^{3D} full energy matrix detector, the analysis database version is ICDD PDF-4 2014;
(3) Scanning electron microscopy: Using the German Zeiss SUPRA™ 55 thermal field emission scanning electron microscope, the acceleration voltage is 0.1 ~ 30 kV, the resolution is 1.0 nm@15 kV; 1.7 nm@1 kV; 4.0 nm@0.1 kV, the magnification is 12 ~ 900,000 ×;
(4) Energy spectrum analysis: Using the X-ray energy spectrometer attached to the scanning electron microscope, the range of elements that can be analyzed is 4Be ~ 94Pu. The instruments used for mechanism analysis are shown in Fig. 4.19.

2. Test method

(1) X-ray fluorescence spectroscopy (XRF) analysis

The main purpose of the XRF test is to analyze the differences in the components of the two raw materials, red mud. After pretreatment, the sample is placed in a sample cup with a non-reusable film for measurement. When operating this window film, be careful not to touch its surface to avoid contamination, and try to ensure that the measured surface is as flat as possible to reduce light scattering. After turning on the machine, take points evenly around the center of the sample, and do weighted average processing of the data that has passed the screening test, to reflect the objective types and content of elements in the sample as much as possible.

(2) X-ray diffraction (XRD) analysis

Since the red mud sample is a powder sample, it needs further processing after pretreatment. First, the sample is ground into a 320-mesh particle size (about 40 degrees), then the powder sample is mixed with an adhesive and placed in the sample holder groove, so that the powder surface is scraped flat with the frame plane, each sample weighs about 1–2 g, the prepared sample is shown in Fig. 4.20. Finally, the sample is placed in the instrument sample chamber and tested according to the operation manual.

(1) Scanning electron microscopy and energy spectrum (SEM&EDX) analysis

After pretreatment, the samples are sorted according to the ratio and age, and the sample preparation is carried out 1 h before observation under the microscope. First, the sample is ultrasonically treated to clean the sample, then it is glued to the conductive glue for low-temperature vacuum drying. Since the red mud is powdery, when

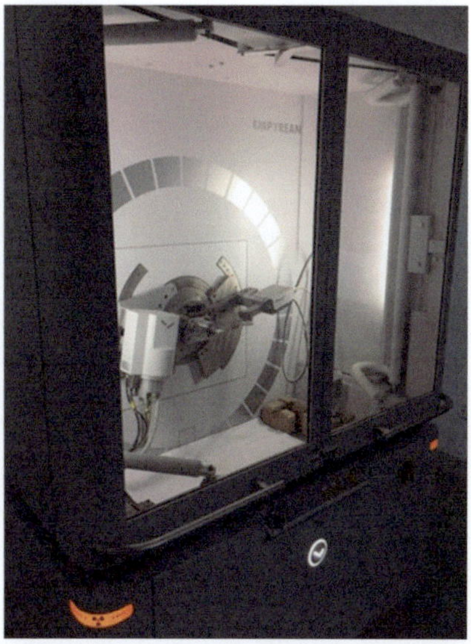

(a) XRD

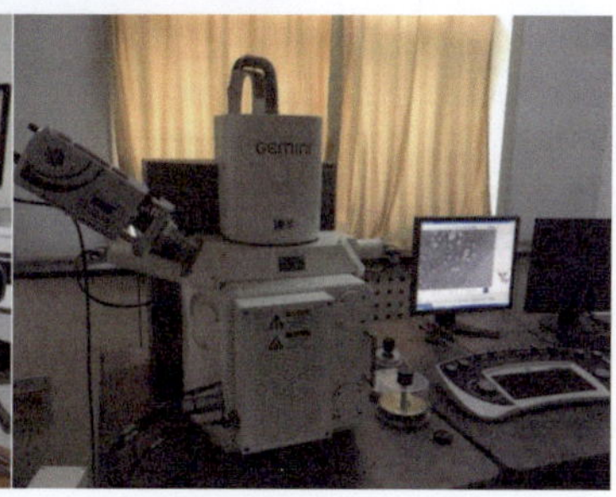

(b) XRF (c) SEM&EDX

Fig. 4.19 Mechanism analysis methods

Fig. 4.20 XRD sample

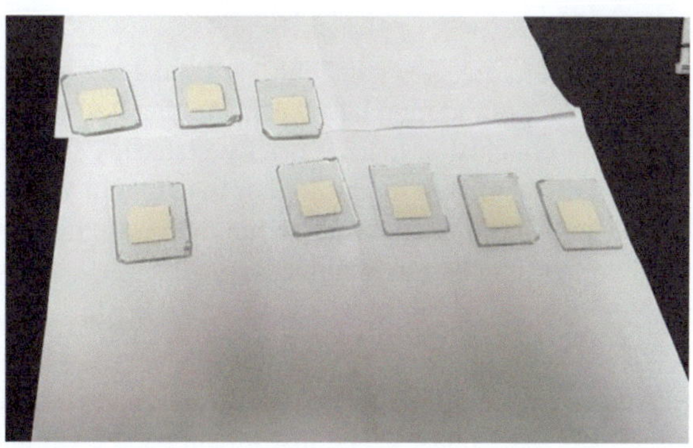

Fig. 4.21 SEM sample

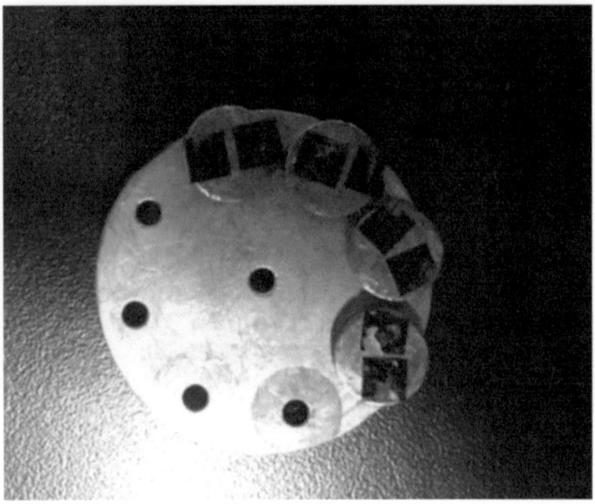

sprinkling on the conductive glue, tap the sample table lightly around to evenly distribute the sample, place the sample table sideways, and shake off the excess sample, as shown in Fig. 4.21.

Non-conductive samples will accumulate charge during observation, affecting the stability of observation. In order to enhance the imaging effect, the sample needs to be gold sprayed. Gold spraying can reduce the damage of the electron beam to the sample, and can also increase the secondary electron yield, making the high-magnification imaging photos clearer. Choose typical positions for each sample to take 1000 ×, 2000 ×, 5000 ×, 20,000 × magnified images.

Energy spectrum analysis is carried out with the X-ray energy spectrometer configured with the scanning electron microscope. During the analysis, first determine the position through SEM, then switch the lens, and then analyze the position.

3. Results and analysis

(1) X-ray Fluorescence Spectroscopy Analysis

The test results are shown in Tables 4.8 and 4.9. The differences in the composition of the two types of red mud have been summarized earlier, so they will not be repeated here. The conclusion can be briefly stated as follows: the content of silicon and calcium in Shanlu red mud is higher than that in Weiqiao red mud, while the content of iron and aluminum in Weiqiao red mud is higher, with the iron content nearly one third higher than that in Shanlu.

(2) X-ray Diffraction Analysis

Shanlu and Weiqiao red mud raw materials were selected for XRD analysis to determine their main phases. The instrument used a Cu target, and the diffraction peaks measured were mainly crystals. The results were processed using JADE, and the X-ray diffraction analysis diagrams are shown in Figs. 4.22 and 4.23.

Table 4.8 Elemental analysis of prismatic crystalline substance

Element	Mass percentage/%	Atomic percentage/%
C	0.41	0.67
O	54.15	66.98
Al	42.75	31.35
Ti	0.60	0.25
Fe	2.10	0.74
Total	100.00	100.00

Table 4.9 Elemental analysis of bulk crystalline material

Element	Mass percentage/%	Atomic percentage/%
C	0.64	1.08
O	51.06	64.61
Al	1.37	1.03
Si	45.42	32.74
Fe	1.51	0.55
Total	100.00	100.00

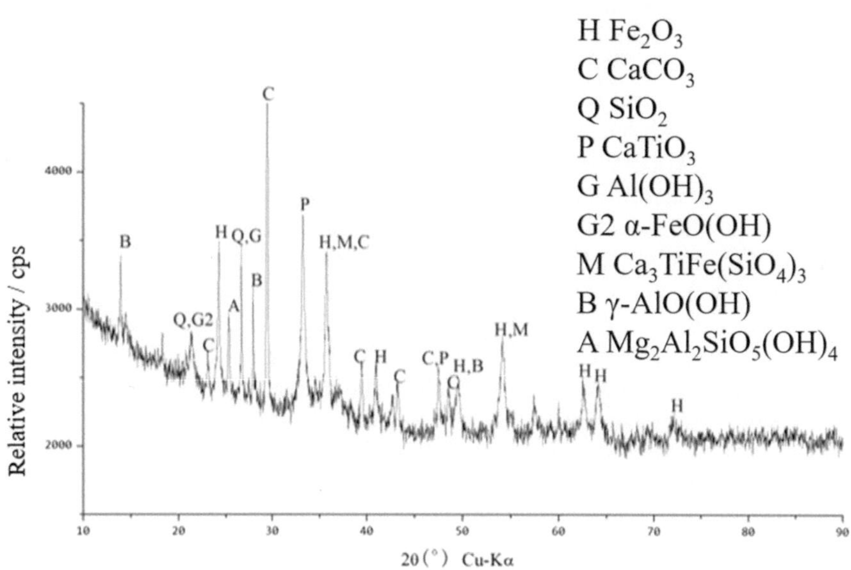

Fig. 4.22 X-ray diffraction analysis diagram of Shanlu red mud raw material

From Fig. 4.22, it can be seen that the Shanlu raw material mainly contains compounds of iron, aluminum, calcium, silicon, and titanium. The main phases are hematite (Fe_2O_3), calcite ($CaCO_3$), quartz (SiO_2), perovskite ($CaTiO_3$), hydrocalcite ($CaAl_2Si_2O_8 \cdot 4(H_2O)$), goethite (α-FeO(OH)), magnetite (γ-Fe_2O_3), boehmite (γ-AlO(OH)), and chamosite ($(Mg_2Al)[SiAlO_5](OH)_4$), etc. The iron compounds

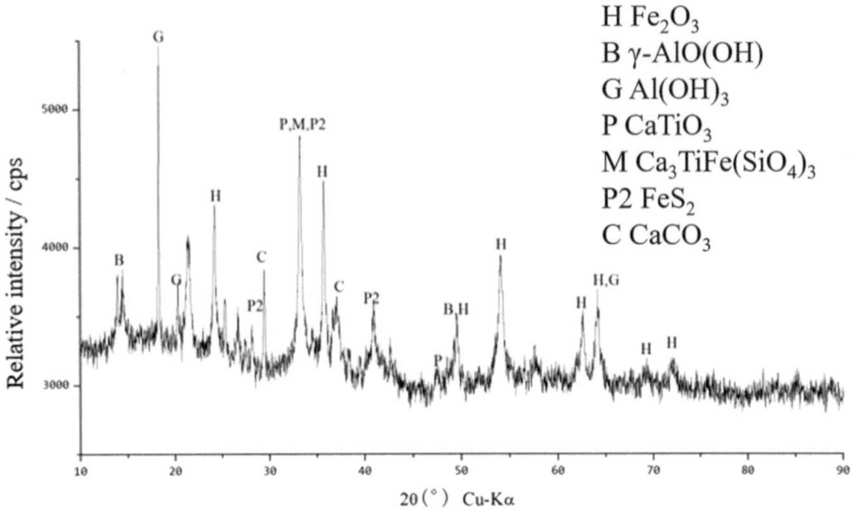

Fig. 4.23 X-ray diffraction analysis diagram of Weiqiao red mud raw material

include hematite, goethite, and magnetite, all of which are produced by red mud exposed to air oxidation or by dehydration of hydrated iron oxide minerals during the sedimentation process. The type of product is greatly related to the composition of the ore. Goethite and boehmite have a certain hydration ability, and Shanlu red mud contains some calcite to provide strength.

From Fig. 4.23, it can be seen that the Weiqiao raw material also mainly contains compounds of iron, aluminum, calcium, silicon, and titanium, but the phase composition is different from that of Shanlu red mud. The main phases are hematite (Fe_2O_3), pyrite (FeS_2), calcite ($CaCO_3$), perovskite ($CaTiO_3$), gibbsite ($Al(OH)_3$), garnet ($Ca_3TiFeSi_3O_{12}$), and boehmite (γ-AlO(OH)), etc. Comparing the phase compositions of the two types of red mud, it is found that Weiqiao red mud does not contain goethite and magnetite, and contains pyrite, which is not present in Shanlu red mud. This is a normal phenomenon due to the different mineral compositions in different bauxite ores, which can also explain the situation of minerals such as chamosite and garnet in the two types of red mud. Weiqiao red mud contains gibbsite but not boehmite, which may indicate that it has undergone a certain hydration reaction. In addition to the phase composition, the relative strengths of the same phases in the two types of red mud also vary greatly. From the peak height and half-peak width, it can be seen that the main oxide of iron—hematite, has a significantly stronger crystal strength in Weiqiao red mud than in Shanlu red mud, indicating that the crystallization degree of hematite in Weiqiao red mud is higher.

The ratio of $A_3B_2C_3$ was chosen for the red mud solidified body samples. Samples cured for 7 and 28d were prepared for testing, and the X-ray diffraction analysis diagrams are shown in Figs. 4.24 and 4.25.

From Figs. 4.24 and 4.25, it can be seen that after the red mud is mixed with lime and fly ash, the mineral composition undergoes significant changes, indicating that

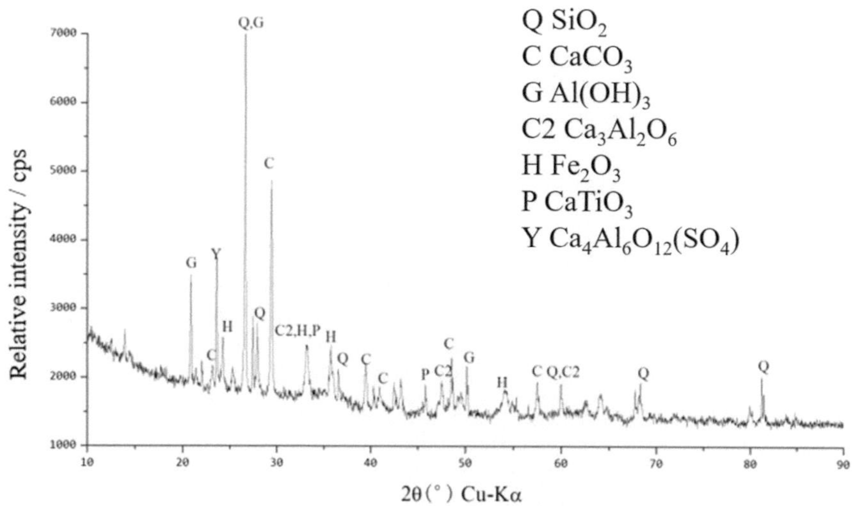

Fig. 4.24 X-ray diffraction analysis diagram of 7d cured $A_3B_2C_3$ specimen

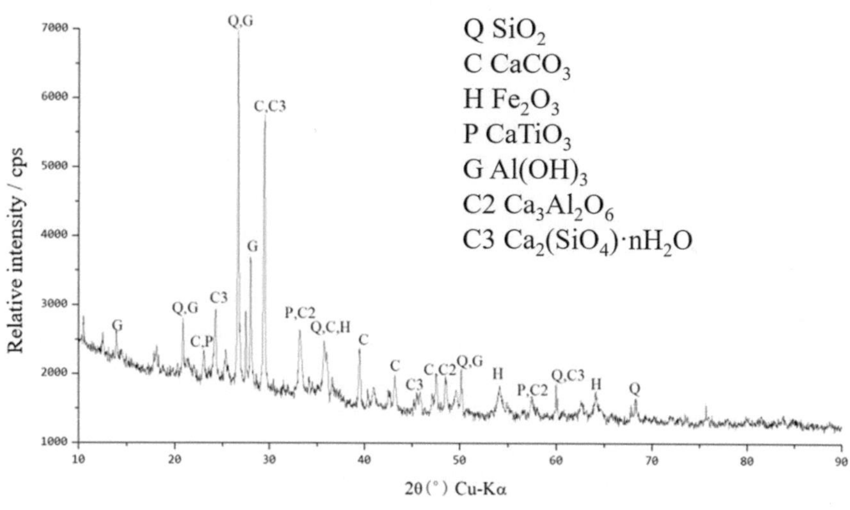

Fig. 4.25 X-ray diffraction analysis of 28d $A_3B_2C_3$ specimen

there are chemical reactions in the mixture. In the figure On April 24, the diffraction peaks of phases such as Goethite (α-FeO(OH)), Magnetite (γ-Fe_2O_3), Boehmite (γ-AlO(OH)), and Chamosite ((Mg_2Al)[$SiAlO_5$]($OH)_4$) have disappeared, while new substances such as Tricalcium Aluminate ($Ca_3Al_2O_6$) and Calcium Sulfoaluminate ($Ca_4Al_6O_{12}SO_4$) have appeared. This indicates that substances like Goethite participated in the hydration reaction of the solidifier and red mud raw materials, generating new crystalline minerals and inevitably amorphous minerals. Hematite

(Fe_2O_3), Calcite ($CaCO_3$), Quartz (SiO_2), Perovskite ($CaTiO_3$), and Hydrocalumite ($CaAl_2Si_2O_8 \cdot 4(H_2O)$) can be found in both red mud raw materials and solidified red mud, indicating that they may not have reacted or the reaction was incomplete. Another possibility is that after the formation of red mud, due to the presence of NaOH and water, some hydration reactions may have occurred during storage, generating several of the above substances.

Comparing the X-ray diffraction spectra of red mud before and after solidification in Figs. 4.24 and 4.25, the differences in the positions of the diffraction peaks are not significant, but the diffraction intensity is weakened, and scattered X-ray peaks appear, indicating that the hydration products are mainly some amorphous aluminosilicates. The phase compositions of samples at different curing ages are similar, but some new crystalline phases, such as Dicalcium Silicate ($Ca_2(SiO_4)$), were also detected in the 28d age samples. For the amorphous substances produced by hydration, the most likely substance they contain is Calcium Silicate Hydrate ($CaO \cdot SiO_2 \cdot H_2O$, C-S-H), which is common in cement hydration products. Since the reaction process and products of red mud have many similarities with it, it is very likely that C-S-H gel is present in the solidified red mud. Therefore, including the substances produced during the storage of red mud and the substances generated by hydration after adding the solidifier, various substances can provide a positive impact on the mechanical strength of the solidified body. These substances include Calcite ($CaCO_3$), Zeolite ($CaAl_2Si_2O_8 \cdot iCa_2O$), Tricalcium Aluminate ($Ca_3\ Al_2O_6$), and Dicalcium Silicate ($Ca_2\ (SiO_4)$). In terms of composition, it is similar to aluminous cement, so it is speculated that solidified Bayer process red mud belongs to hydraulic cementitious materials, which may have the advantages of fast setting speed, high early strength, certain resistance to sulfate corrosion, and heat resistance. The long-term strength growth trend is relatively slow. Since XRD analysis is more accurate for crystalline substances and difficult to distinguish for amorphous substances, SEM analysis is needed.

(3) Scanning Electron Microscopy and Energy Spectrum Analysis

From the analysis in the previous section, it is known that hydration products can be divided into crystalline and amorphous substances. XRD analysis has shown that there are crystalline substances in the solidified red mud that can provide mechanical strength, and they gradually increase with the extension of reaction time. Using SEM and EDX as observation methods, we directly look for the impact of the reaction at the nanoscale microscopic level. According to the aforementioned steps, samples were made for testing and divided into four groups: Shan Aluminum, Wei Bridge, 7d solidified red mud, and 28d solidified red mud. The magnification multiples are 2000 ×, 5000 ×, and 20,000 × respectively. Typical structures were selected for imaging, as shown in Figs. 4.26 and 4.27.

From Fig. 4.26, it can be seen that the particles of Shan Aluminum red mud are composed of complex components and contain various substances. Its surface particles are rough and loose with many voids, which leads to the low mechanical strength of the original red mud. Comparing Fig. 4.26 with Fig. 4.27, it can be found that there are certain differences in the microstructure of Shan Aluminum red mud

Fig. 4.26 SEM analysis of Shan Aluminum red mud **a** 2000 ×, **b** 5000 ×

and Wei Bridge red mud. First, the particle composition of the two types of red mud is different: Shan Aluminum red mud has many prismatic, flaky, and layered structures, which are rarely seen in Wei Bridge red mud. After magnifying 20,000 × , it can be found that the particles of Wei Bridge red mud are mostly clumpy and bud-shaped; secondly, the particle sizes of the two types of red mud are different: in Shan Aluminum red mud, these structures range in diameter from 200 nm to 1–2 μm, while the particle size of Wei Bridge red mud is larger than the former, reaching up to 4 μm. But they also have some common features, such as being loose with many voids, which should also be the reason for their low strength.

Comparing Figs. 4.27 and 4.28, it can be observed that the microstructure of red mud has undergone significant changes after the addition of fly ash for solidification. Firstly, as the particle size increases to about 5–10 μm, the surface voids decrease; Secondly, the surface of the structure is smoother, with more plate-like and sheet-like structures interwoven together, as shown in Fig. 4.28c; Finally, there is an increase in prismatic dark substances, which have smooth surfaces and regular shapes. They are present in small amounts in the original red mud, but their density significantly increases after solidification.

Comparing Figs. 4.28 with 4.29, it can be found that under longer curing time, the structure of the solidified red mud becomes denser, some large particles reach 20–30 μm in diameter, the surface is smooth, large voids are filled by newly generated crystalline and amorphous substances, flaky, plate-like, prismatic substances continue to grow, and are connected by filamentous, network-like substances, further enhancing the structural strength, as shown in Fig. 4.29b. This network-like substance widely present in the structure is likely to be the calcium silicate hydrate gel (C-S–H) generated by hydration, which, like glue, connects large structures together, greatly enhancing the mechanical strength of the solidified body. It can be seen that prismatic and network-like substances play an important role in the solidification process of Bayer red mud. It is speculated that this prismatic substance may be calcium hydroxide ($Ca(OH)_2$), tricalcium aluminate ($Ca_3Al_2O_6$), or calcium carbonate ($CaCO_3$), and its actual composition needs to be explored through energy spectrum analysis.

Fig. 4.27 SEM analysis of Wei Bridge red mud **a** 2000 ×, **b** 5000 ×, **c** 20,000 ×

Fig. 4.28 SEM analysis of 7d solidified red mud **a** 2000 ×, **b** 5000 ×, **c** 20,000 ×

Fig. 4.29 SEM analysis of 28d solidified red mud **a** 2000 ×, **b** 5000 ×

Through X-ray energy spectrum (EDX) analysis, the elemental composition and content of the typical structure were analyzed, as shown in Table 4.8 and Fig. 4.30. It was found that the atomic ratio of aluminum to oxygen is about 1:2, and it contains almost no other common elements (except hydrogen, which cannot be detected by energy spectrum analysis), so it can be guessed that this substance is likely to be boehmite (γ-AlO(OH)). As mentioned above, boehmite and goethite are consumed in the hydration reaction process, which can also be corroborated by microscopic analysis. The flaky structure is likely to be an accumulation of substances such as calcite and tricalcium aluminate. At the same time, there is a certain amount of quartz in Shan Aluminum red mud, which can also be proven by EDX, as shown in Table 4.9 and Fig. 4.31, with an oxygen-silicon atomic ratio close to 2:1.

In summary, after adding lime and fly ash, a series of hydration reactions occur, greatly improving the strength of the solidified body. Firstly, boehmite and goethite participate in a series of reactions to form crystalline substances such as tricalcium aluminate, which is the basis for the early strength of red mud. As the curing time

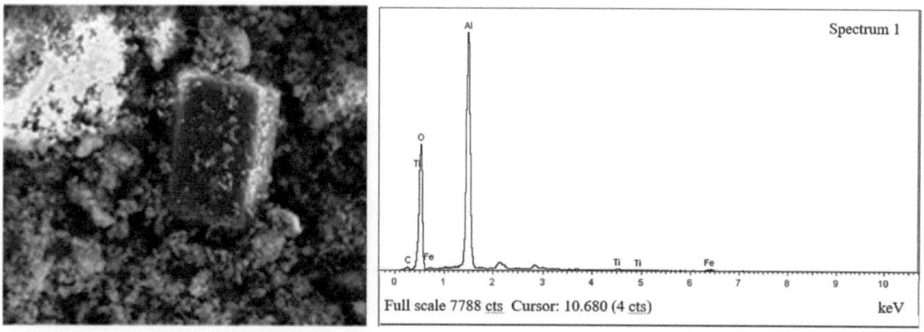

Fig. 4.30 EDX analysis of prismatic crystalline material

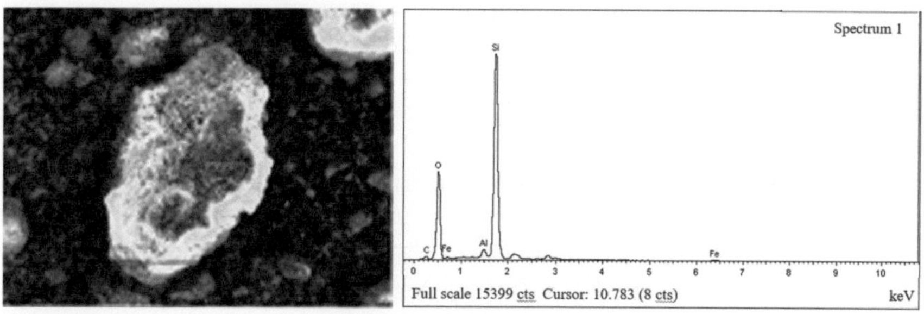

Fig. 4.31 EDX analysis of bulk crystalline material

extends, some amorphous substances are formed. These networked, filamentous, and fibrous substances wrap, fill, and bond structures with certain strengths such as calcium carbonate, tricalcium aluminate, and garnet, greatly enhancing the strength of the solidified body structure. The main possible substance in these amorphous substances is hydrated calcium silicate gel. From the reactants and products, the reaction mechanisms of the two types of red mud have many similarities. However, as can be known from the unconfined compressive strength test, there is a certain difference in the solidification strength of the two types of red mud. The main reason for this difference may be that the raw material sources and storage times of the two types of red mud are different, so the chemical compositions are different, as shown in the previous XRF test results. Due to the differences in the storage time and conditions of the two types of red mud, this leads to differences in the amount of reactants contained in the test. If a large amount of amorphous gel substance that can wrap and fill is not generated, it is impossible to have a strong mechanical strength based on the loose and porous structure of the red mud itself. These factors may be the main reasons for the difference in strength.

4.2.3 *Improved Solidified Red Mud Strength Formation Mechanism*

1. Strength formation mechanism of red mud

After adding lime and fly ash, a series of hydration reactions occur, greatly improving the strength of inorganic stabilized red mud. By observing the microstructure diagrams of stabilized red mud samples at different ages, it is found that the strength formation process of inorganic stabilized red mud can be roughly divided into the following three stages.

(1) Early strength formation

Firstly, boehmite and goethite participate in a series of reactions to form crystalline substances such as tricalcium aluminate, which is the basis for the early strength of red mud. By observing the XRD spectra of different age-stabilized red mud samples $A_3B_2C_3$ (as shown in Fig. 4.32), it is found that the goethite (α-FeO(OH)), magnetite (γ-Fe_2O_3), boehmite (γ-AlO(OH)), and magnesia green mudstone ($(Mg_2Al)[SiAlO_5](OH)_4$) and other phase diffraction peaks disappear, tricalcium aluminate ($Ca_3Al_2O_6$) and calcium sulfoaluminate ($Ca_4Al_6O_{12}SO_4$) and other new substances appear, indicating that substances such as goethite participate in the hydration reaction of red mud stabilization, generating new mineral components; the figure also shows hematite (Fe_2O_3), calcite ($CaCO_3$), quartz (SiO_2), perovskite ($CaTiO_3$), and hydrocalcite ($CaAl_2Si_2O_8 \cdot 4(H_2O)$), which are the same as the mineral components of the red mud sample, and may be the mineral composition of the red mud itself, or may be the NaOH contained in the red mud and water have already undergone partial hydration reaction during the storage process, generating several of the above substances.

(2) Strength enhancement stage

As the curing time extends, some amorphous substances are generated. These networked, filamentous, and fibrous substances wrap, fill, and bond structures with

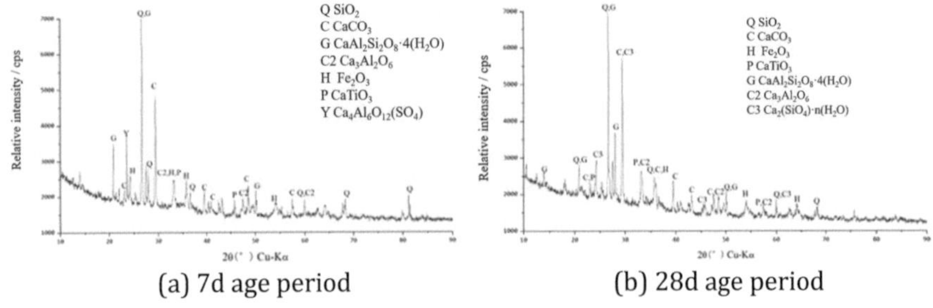

Fig. 4.32 $A_3B_2C_3$ test piece X-ray diffraction spectrum

certain strengths such as calcium carbonate, tricalcium aluminate, and garnet, greatly enhancing the strength of the solidified structure.

Comparing the XRD spectra of stabilized red mud at 7 and 28d age periods, it is found that the differences in the positions of the diffraction peaks are not significant, but the diffraction intensity weakens and diffuse X-ray peaks appear, indicating that the hydration products are mainly some amorphous aluminosilicates, hydrated calcium silicate gels, etc. The content of dicalcium silicate ($Ca_2(SiO_4)$) in the 28d age period sample increases. The substances produced by red mud during storage and the substances hydrated after the addition of the solidifier can all provide a positive impact on the mechanical strength of the solidified body, including calcite ($CaCO_3$), zeolite ($CaAl_2Si_2O_8 \cdot Ca_2O$), tricalcium aluminate ($Ca_3Al_2O_6$), dicalcium silicate ($Ca_2(SiO_4)$), etc. From the composition point of view, it is similar to aluminosilicate cement, so it is speculated that the solidified Bayer red mud belongs to hydraulic cementitious materials, which may have the advantages of fast setting speed, high early strength, certain resistance to sulfate corrosion and heat resistance, and the long-term strength growth trend is relatively slow.

(3) Late strength growth

As the age increases, the structure of the solidified red mud becomes more compact, some large particles reach 20–30 μm in diameter, the surface is smooth, large voids are blocked by newly generated crystalline and amorphous substances, and plate-like, slab-like, and prismatic substances further grow, and are connected between structures by filamentous, networked substances (hydrated calcium silicate gel), further enhancing the structural strength. Figure 4.33 shows the SEM images of the stabilized red mud samples at different ages. The figure shows that the microstructure of the red mud has changed significantly, manifested as: the particle size increases to about 5–10 μm, the surface voids decrease; the structure surface becomes smoother, and the plate-like and slab-like structures increase, as shown in Fig. 4.33a; there are more prismatic dark substances, these structures have smooth surfaces, regular shapes, rarely appear in the original red mud samples, but the density significantly increases after solidification, as shown in Fig. 4.33b. It can be seen that in the solidification process of Bayer red mud, prismatic substances and networked substances play an important role.

To determine the types of prismatic and blocky substances in Fig. 4.33, X-ray energy spectrum (EDX) tests were conducted on typical structures to analyze their elemental composition, as shown in Table 4.10 and Fig. 4.34. The results show that the atomic ratio of aluminum to oxygen is about 1:2, and it contains almost no other common elements except hydrogen (which cannot be detected by spectral analysis), so it is judged that the blocky substance is boehmite (γ-AlO(OH)). As mentioned above, boehmite and goethite are consumed in the hydration reaction; the flaky structures are mostly accumulations of calcite, tricalcium aluminate, etc., and there is a certain amount of quartz in the bauxite red mud, which can also be proven by EDX, with an oxygen-silicon atomic ratio close to 2:1.

(a) 7d age period

(b) 28-day period

Fig. 4.33 SEM image of stabilized red mud

Table 4.10 EDX elemental analysis results

Element	Mass percentage/%		Atomic percentage/%	
	Prismatic substance	Blocky substance	Prismatic substance	Blocky Substance
C	0.41	0.64	0.67	1.08
O	54.15	51.06	66.98	64.61
Al	42.75	1.37	31.35	1.03
Ti	0.60	–	0.25	–
Fe	2.10	1.51	0.74	0.55
Si	–	45.42	–	32.74
Total	100.00	100.00	100.00	100.00

2. Mechanism of strength formation of composite stabilized red mud

The factors that constitute the strength of stabilized red mud mixed materials include two aspects: ① the internal friction and interlocking forces between red mud particles. ② The bonding and cohesion formed by the chemical reaction between the stabilizing material and the red mud particles. After the red mud and stabilizing material are mixed with water and compacted, it can be assumed that no chemical reaction has occurred in the early stage, and its strength mainly comes from the internal friction between the compacted overall particles and the "original cohesion" formed by the molecular attraction between the water film between the particles and the adjacent particles. As time goes on, a series of physicochemical actions occur between the solid and liquid phases in the mixture, generating a series of substances with a binding

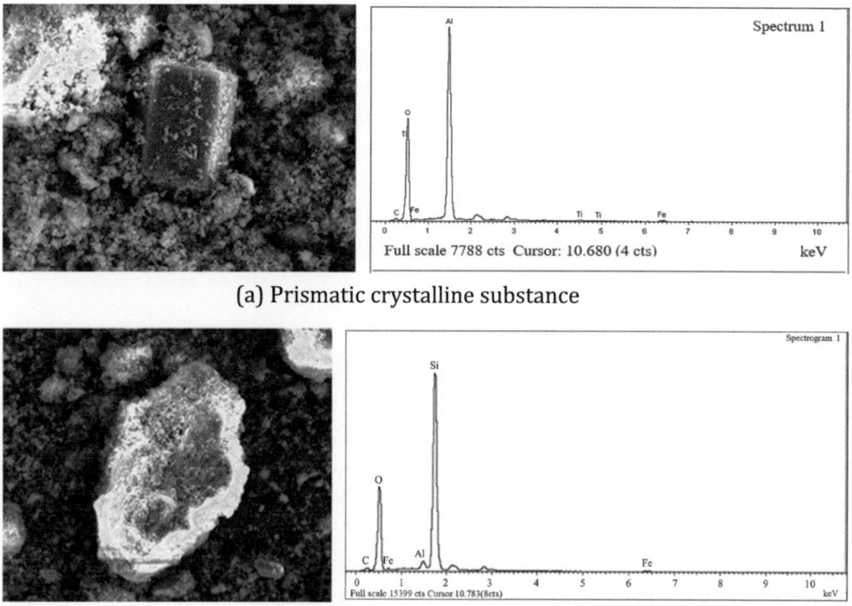

(a) Prismatic crystalline substance

(b) Blocky crystalline substance

Fig. 4.34 EDX spectrum

effect, which strengthens the connection between particles in the mixture, forming "solidified cohesion", which is the main source of strength formation in stabilized red mud.

(1) Formation of early strength

After the red mud and stabilizing material are mixed and compacted, no chemical reaction occurs in the early stage. The formation of early strength mainly relies on the bond force between red mud particles and the linking effect of mineral crystalline structures, as well as the ion exchange and crystallization effects produced by the addition of lime.

1) Mineral crystalline structure

Bayer red mud contains a large amount of elements such as aluminum, calcium, iron, silicon, etc., and the main mineral phase is gibbsite $Al(OH)_3$, diaspore γ-AlOOH, hematite Fe_2O_3, goethite FeO(OH), etc. In the mineral crystal structure, there are a large number of silicon-oxygen tetrahedra and gibbsite octahedra. In each silicon-oxygen tetrahedron, four oxygen atoms form an equilateral tetrahedron, the tetrahedron base is on a plane, and the oxygen atoms at the top of the tetrahedron are shared, forming a hexagonal network structure on the plane; each gibbsite octahedron is arranged by six oxygen or hydroxide ions at equal distances, with aluminum ions in the middle, forming a grid-like layered structure.

The silicon-oxygen tetrahedron sheets and gibbsite octahedron sheets are stacked in different combinations to form various layered structures. In the layered structure, the tetrahedron sheets and octahedron sheets share an atomic plane, so the bond between the tetrahedron sheets and the octahedron sheets is the main bond, providing a strong linking effect. When the electrolyte concentration of the external solution of the structure is high, the interparticle repulsion can be weakened, the particles gradually agglomerate, and the red mud particles form a gel-like pore frame structure, mainly composed of structure-agglomerates, structure-granules, and structure-agglomerates.

2) Lime ion exchange, crystallization action

The micro-particles of red mud have certain colloidal properties. They generally carry negative charges and adsorb a certain amount of low-valence cations such as sodium and potassium (Na^+, K^+). Lime is a strong electrolyte. After adding lime to the red mud, the calcium ions (Ca^{2+}) ionized from the lime in the solution have an ion exchange effect with the sodium and potassium ions in the soil. The sodium (potassium) soil of the raw material becomes calcium soil, and the ions adsorbed on the particle surface change from monovalent to divalent, reducing the thickness of the water film adsorbed on the surface of the red mud particles, making the particles closer to each other, and the molecular attraction increases. Many single particles agglomerate into small granules, forming a stable structure.

After mixing lime with red mud, only a part of the slaked lime $Ca(OH)_2$ undergoes ion exchange, and most of the saturated lime $Ca(OH)_2$ crystallizes itself. The slaked lime reacts with water to form a slaked lime crystal grid, and its chemical reaction formula is:

$$Ca(OH)_2 + nH_2O \rightarrow Ca(OH)_2 \cdot nH_2O$$

The slaked lime crystal grid has cementitious properties and can harden in a two-phase environment of solid and water. The hardening process forms a stable protective film on the periphery of the red mud micro-particle group, fills the particle voids, causes the particles to gel, reduces the voids and permeability between the particles, and at the same time improves the compactness.

(2) Strength growth process

1) Hydration reaction

After adding cement to the red mud, C_3S (tricalcium silicate), C_3A (tricalcium aluminate), C_4 AF (tetracalcium aluminoferrite) quickly hydrate, and at the same time, gypsum quickly dissolves, forming a saturated solution of $Ca(OH)_2$ and $CaSO_4$, the hydration products first appear as hexagonal plate-shaped $Ca(OH)_2$ and needle-shaped AFt phase ($3CaO \cdot Al_2O_3 \cdot 3CaSO_4 \cdot 32H_2O$) and amorphous C-S-H (calcium silicate hydrate). Afterwards, due to the continuous generation of AFt phase, $SO_4{}^{2-}$ continuously decreases, and then AFm phase and C-A-H crystals and $C_4(A \cdot F) \cdot H_{13}$ crystals are formed. Its chemical reaction formula is as follows:

$$3CaO \cdot SiO_2 + nH_2O = xCaO \cdot SiO_2 \cdot yH_2O + (3 - x)Ca(OH)_2$$

$$\begin{aligned} & 2(3CaO \cdot Al_2O_3) + 27H_2O \\ & = 4CaO \cdot Al_2O_3 \cdot 19H_2O + 2CaO \cdot Al_2O_3 \cdot 8H_2O \end{aligned}$$

$$\begin{aligned} & 4CaO \cdot Al_2O_3 \cdot Fe_2O_3 + 4Ca(OH)_2 + 22H_2O \\ & = 2(4CaO \cdot (Al_2O_3, Fe_2O_3)) \cdot 13H_2O \end{aligned}$$

During the hydration reaction of C_3A, when there is sufficient gypsum, C_3A will undergo the following reaction:

$$\begin{aligned} & 3CaO \cdot Al_2O_3 + 3(CaSO_4 \cdot 2H_2O) + 26H_2O \\ & = 3CaO \cdot Al_2O_3 \cdot 3CaSO_4 \cdot 32H_2O \end{aligned}$$

Among them, $3CaO \cdot Al_2O_3 \cdot 3CaSO_4 \cdot 32H_2O$ is called trisulfate hydrated sulphoaluminate (AFt), also known as ettringite.

The $4CaO \cdot Al_2O_3 \cdot 19H_2O$ produced by the hydration reaction of C_3A will lose 6 mol of crystal water to become $4CaO \cdot Al_2O_3 \cdot 13H_2O$ at a relative humidity below 85%. When C_3A is not fully hydrated and gypsum is exhausted, the $4CaO \cdot Al_2O_3 \cdot 13H_2O$ produced by the hydration of C_3A reacts with the previously formed AFt as follows:

$$\begin{aligned} & 3CaO \cdot Al_2O_3 \cdot 3CaSO_4 \cdot 32H_2O + 2(4CaO \cdot Al_2O_3 \cdot 19H_2O) \\ & = 3(3CaO \cdot Al_2O_3 \cdot CaSO_4 \cdot 12H_2O) + 2Ca(OH)_2 + 20H_2O \end{aligned}$$

Among them, $3CaO \cdot Al_2O_3 \cdot CaSO_4 \cdot 12H_2O$ is called monosulfate hydrated sulphoaluminate (AFm).

The hydration reaction of C_2S (dicalcium silicate) starts early, and the surface becomes uneven quickly after contact with water, but due to the tight structure, the release of $Ca(OH)_2$ is slow, and the supersaturation of Ca^{2+} during hydration is low, making the reaction extremely slow. C_3 The CaO in S is easily released, thus the hydration speed of tricalcium silicate is relatively fast. The lattice of C_3A has a large porosity, water molecules can easily enter the lattice, and each C_3A cell contains aluminum atoms with coordination numbers of 4 and 6. The unsaturated valence bond with a coordination number of 4 can easily accept two water molecules or hydroxyl ions to form a more stable coordination. Therefore, C_3A can quickly undergo a violent hydration reaction when it encounters water, and the hydration speed is the fastest.

The products of cement hydration reaction mainly include C-S-H gel, $Ca(OH)_2$, AFt, AFm. The C-S-H gel is amorphous and has a poor degree of crystallization; the presence of $Ca(OH)_2$ maintains the alkalinity of the cement stone system and can stabilize the C-S-H gel; AFt is needle-like, its formation and growth will cause a certain expansion, which is beneficial to the formation of the structural connection

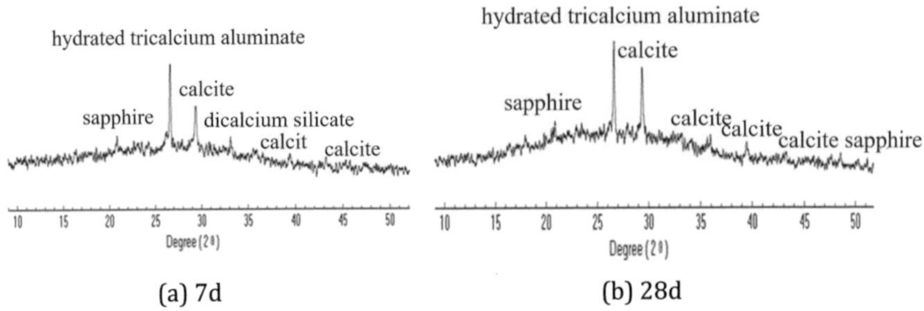

Fig. 4.35 XRD patterns of composite material stabilized red mud samples

of the internal particles of the material; AFm is hexagonal thin plate (flake) shape, irregular petal shape, with a good degree of crystallization. In the early stage of the hydration reaction, the hydration crystal is too small to connect into a network structure. As the hydration reaction accelerates over time, C-S-H also grows long fiber crystals from the mature material particles, and at the same time, the AFt crystals gradually grow larger, interweaving and connecting with each other in the cement slurry to form a network structure, and begin to generate strength. The network structure formed by the overlapping of hydration crystals, combined with the red mud particles to form a solid, dense whole in three-dimensional space. At this time, the porosity is significantly reduced, the network structure is continuously densified, and the strength continues to increase. The XRD patterns of 7 and 28d stabilized red mud are shown in Fig. 4.35.

2) Alkali-activation effect

The addition of phosphogypsum will generate free Ca^{2+}, SO_4^{2-} in the mixture, and the red mud attached liquid contains a large amount of OH^-, which can react with SiO_2, Al_2O_3 in the red mud to generate calcium silicate hydrate, calcium aluminate hydrate or calcium sulfoaluminate hydrate, that is, the pozzolanic reaction. The reaction formula is:

$$x\mathrm{Ca(OH)_2} + \mathrm{SiO_2} + n\mathrm{H_2O} \rightarrow x\mathrm{CaO}\cdot\mathrm{SiO_2}(n+1)\mathrm{H_2O}$$

$$x\mathrm{Ca(OH)_2} + \mathrm{Al_2O_3} + n\mathrm{H_2O} \rightarrow x\mathrm{CaO}\cdot\mathrm{Al_2O_3}(n+1)\mathrm{H_2O}$$

The mineral composition of red mud basically belongs to silicate and aluminate, containing a large amount of silicon tetrahedron and aluminum octahedron. Under normal circumstances, these minerals have relatively high stability. However, the Bayer process red mud attached liquid contains a large amount of OH^-. When the pH value around the red mud particles increases to a certain level, the activity of some SiO_2 and Al_2O_3 in the minerals will be stimulated, and they will react with Ca^{2+} in the solution to form new minerals. These minerals are mainly calcium silicate and calcium aluminate series, such as $4CaO\cdot 5SiO_2\cdot 5H_2O$, $4CaO\cdot Al_2O_3\cdot 19H_2O$,

$3CaO \cdot Al_2O_3 \cdot 16H_2O$, $CaO \cdot Al_2O_3 \cdot 10H_2O$, etc. The composition and structure of these minerals are similar to the hydration products of cement, and they also have cementing ability. These cementing substances generated wrap around the surface of red mud particles, together with the hydration products of cement, they agglomerate the red mud particles into a whole, further enhancing the strength and water stability of the stabilized red mud.

In addition to the above-mentioned water-hardening cementing minerals, there is also a silicon-aluminum-sodium-calcium substance, which is likely to be a silicon-aluminum polymer after research. Silicon-aluminum inorganic polymer materials contain a variety of amorphous to semi-crystalline three-dimensional aluminosilicate mineral polymers. Under the action of an activator, the silicon-aluminum-oxygen chain depolymerizes and re-polymerizes into a network silicon-aluminum oxide compound under alkaline conditions. Its chemical formula is:

$$(SiO_2, Al_2O_3) + nH_2O \xrightarrow{NaOH} n(OH)_3 - Si - O - Al^{(-)} - (OH)_3$$

$$n(OH)_3 - Si - O - Al^{(-)} - (OH)_3 \xrightarrow{NaOH} (Na, K)\left(-SiO - Al^{(-)} - O-\right)_n + 3nH_2O$$

$$(Si_2O_5, Al_2O_2) + nH_2O \xrightarrow{NaOH} n(OH)_3 - Si - O - Al^{(-)} - O - Si - (OH)_3$$

$$n(OH)_3 - Si - O - Al^{(-)} - O - Si - (OH)_3 \xrightarrow{NaOH} (Na, K)\left(-Si - O - Al^{(-)} - O - Si - O-\right)_n + 3nH_2O$$

The stabilized red mud liquid phase contains a large number of Na^+, AlO^{2-}, $SiO_3{}^{2-}$ ions, and the pH value is high. The increase in the number of Na^+ ions accelerates the formation of the diffusion double layer. A large amount of NaOH promotes negative ions to enter the fixed layer, compresses the double layer, reduces the potential, reduces the repulsion potential energy, and accelerates the coagulation of the silicate sol. The increase in the number of AlO^{2-}, $SiO_3{}^{2-}$ ions increases the concentration of nuclei, which is conducive to the formation of gel nuclei. The alkaline solution reacts with the particle surface to form a long chain of silicoaluminates, causing the mineral particles to stick together and form a material with a certain strength.

Based on the above chemical reactions, the stabilized red mud forms a network spatial structure mainly bonded by cementation and secondarily by crystallization after the action of cement hydration and alkali activation. Although red mud itself has the characteristics of an overhead structure and a relatively small natural density, after solidification and hardening, its structure changes from an unstable state to a stable state, and the structural strength is greatly improved.

(3) Late strength growth

1) Carbonation

The $Ca(OH)_2$ in the mixture can react with CO_2 in the gas phase or dissolved in water to produce carbonation:

$$Ca(OH)_2 + CO_2 + nH_2O \rightarrow CaCO_3 + (n + 1)H_2O$$

The $CaCO_3$ crystal has high strength and water stability. Its cementing effect on particles strengthens the red mud. When $CaCO_3$ crystals are deposited in the interstices of the stabilized red mud mixture particles, a certain cementing strength is produced. Since CO_2 enters from the pores of the mixture, the hard shell formed by the surface carbonation of the stabilized red mud mixture will hinder the further infiltration of CO_2, so the carbonation of $Ca(OH)_2$ is quite slow, which is manifested as the formation of late strength of stabilized red mud.

The stabilized red mud also contains a large amount of active oxides, which react with CO_2 in the air to form carbonate precipitates or colloidal substances. As the storage time increases, the mineral phase of red mud transforms from vaterite to calcite, which also makes the stabilized red mud water-hardening, but this process is relatively slow. The following chemical reaction formula is the possible carbonation reaction during the drying process of stabilized red mud, revealing the chemical reaction mechanism of the formation of late cementing strength of stabilized red mud.

$$NaAl(OH)_4 + CO_2 = NaAlCO_3(OH)_2 + H_2O$$

$$NaOH + CO_2 = NaHCO_3$$

$$Na_2CO_3 + CO_2 + H_2O = 2NaHCO_3$$

$$3Ca(OH)_2 \cdot 2Al(OH)_3 + CO_2 = 3CaCO_3 + Al_2O_3 \cdot 3H_2O + 3H_2O$$

$$Na_6[AlSiO_4]_6 \cdot 2NaOH + CO_2 = Na_6[AlSiO_4]_6 + 2NaHCO_3$$

2) Secondary hydration

The main hydration products of cement are C-S-H gel, $Ca(OH)_2$, Aft and AFm. C-S-H gel and $Ca(OH)_2$ are the main components of hydration products, constituting the main structure of cement stone. The C-S-H gel is amorphous and colloidal, while the crystalline $Ca(OH)_2$ often appears as thin hexagonal plates in the early stage of hydration, and becomes layered as it grows, with pore structures often appearing at the junction of crystals. Figure 4.36 is the SEM image of the stabilized red mud of the 28d composite material. From the figure, it can be clearly seen that the sample

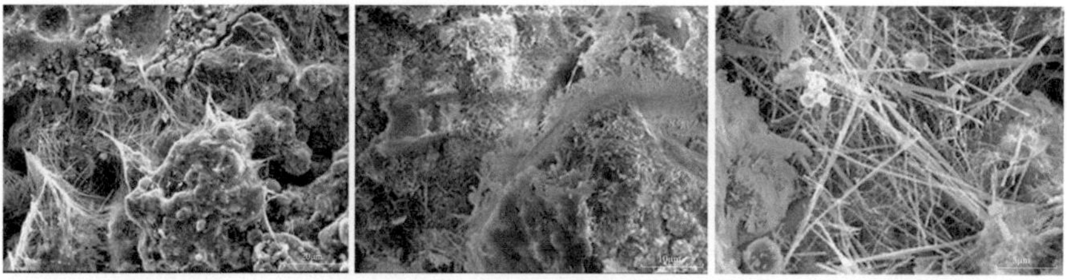

(a) ×1064 (b) ×2000 (c)×4000

Fig. 4.36 SEM image of stabilized red mud of 28d composite material

has produced a lot of hydrated cementitious substances and formed a stable spatial network structure of colloidal-crystal combination.

The $Ca(OH)_2$ produced by hydration diffuses to the surface of red mud particles through the liquid phase, and Ca^{2+}, OH^- can erode the surface of the particles, forming a permeation layer with Na^+, K^+ and other ions in the attached liquid. The permeation layer causes the wrapping layer of red mud particles to rupture due to expansion, and the alkaline solution reaches the surface of the particles, dissolving the silico-aluminate components, and forming hydrated calcium silicate and hydrated calcium aluminate, that is, secondary hydration reaction. The hydration products will fill the pores, increase the C-S-H gel, refine the pore structure, and further improve the compactness of the stabilized red mud.

This section uses microscopic test techniques to explore the solidification mechanism of lime and fly ash stabilized red mud, and the main conclusions are as follows:

(1) The difference in chemical composition between Shan Aluminum red mud and Wei Bridge red mud is the main reason for the difference in strength of the mixed material after stabilization. The main manifestations are: Shan Aluminum red mud contains more calcium and silicon substances, and contains more active substances such as boehmite and goethite.
(2) After the red mud reacts with lime and fly ash, the mineral composition has changed significantly. The diffraction peaks of goethite (α-FeO(OH)), magnetite (γ-Fe_2O_3), boehmite (γ-AlO(OH)), and chlorite ($(Mg_2Al)[SiAlO_5](OH)_4$) have disappeared. The hydration reaction of the solidifier and the raw material of red mud has produced new crystalline substances such as tricalcium aluminate ($Ca_3Al_2O_6$), calcium sulfoaluminate ($Ca_4Al_6O_{12}SO_4$), and dicalcium silicate ($Ca_2(SiO_4)$), and inevitably also produced hydrated calcium silicate ($CaO \cdot SiO_2 \cdot H_2O$) and other amorphous substances.
(3) The substances produced by red mud during storage and the substances produced by hydration after adding solidifying agent and other substances have promoted the mechanical strength of the solidified body. These substances include calcite ($CaCO_3$), zeolite ($CaA_{12}Si_2O_8 \cdot 4(H_2O)$), tricalcium aluminate ($Ca_3Al_2O_6$), dicalcium silicate ($Ca_2(SiO_4)$) and so on.

(4) After adding lime and fly ash to red mud, the strength of the solidified body has greatly improved through a series of hydration reactions. Firstly, boehmite and goethite participated in a series of reactions to produce tricalcium aluminate and other crystalline substances, which are the basis of the early strength of red mud. With the extension of curing time, some amorphous substances are generated. These network, filamentous, and fibrous substances wrap, fill, and bond structures with certain strength such as calcium carbonate, tricalcium aluminate, garnet, etc., greatly improving the strength of the solidified body structure.

4.3 Evaluation of the Effect of Modified Solidification of Red Mud

4.3.1 Inorganic Material Solidification of Red Mud

1. Compaction Characteristics

The process of soil compaction is to use mechanical energy to reduce the pore volume between soil particles and make the contact more compact. Compacted soil can improve its strength and resistance to deformation. Table 4.11 shows the results of the standard heavy compaction test of red mud treated with various ratios of inorganic solidification technology. It can be seen that the optimal moisture content of stable red mud is between 31 and 37%, and the maximum dry density is between 1.29 and 1.43 g/cm^3; Through the range and variance analysis of the performance indicators of the orthogonal test with the optimal moisture content and maximum dry density, the results show that: the type of red mud (factor C) has a significant effect on the optimal moisture content and maximum dry density, and the optimal moisture content of Shanalco red mud is larger than that of Weiqiao red mud, and the maximum dry density is smaller than that of Weiqiao red mud; the total amount of two ashes (factor B) has a significant effect on the maximum dry density, but the effect on the optimal moisture content is small; the ratio of two ashes (factor A) has a small effect on the optimal moisture content and maximum dry density; adding clay to red mud will affect its optimal moisture content and maximum dry density, which is manifested as after adding 30% clay, the optimal moisture content decreases, and the maximum dry density increases. When using red mud for roadbed filling, the compaction density and moisture content suitable for local working conditions can be obtained by adding clay.

2. Strength Characteristics

The unconfined compressive strength test is a strength test conducted under conditions where the sample is not subject to lateral constraints. Table 4.12 shows the results of the unconfined compressive strength test of red mud treated with various ratios of inorganic solidification technology at different ages (7, 28, and 90d).

Table 4.11 Results of standard heavy compaction test for each ratio of inorganic solidification technology

Ratio number	Proportion of each component				Optimal moisture content/%	Maximum dry density/g/cm^3
	Lime/%	Fly ash/%	Red mud/%	Clay/%		
1	2.3	22.7	Shan aluminum 75.0	0.0	37.00	1.377
2	4.6	45.4	Weiqiao 50.0	0.0	31.98	1.429
3	6.8	68.2	Shan aluminum 17.5	7.5	32.21	1.309
4	8.3	16.7	Wei bridge 75.0	0.0	31.35	1.544
5	16.7	33.3	Shan aluminum 35.0	15.0	33.00	1.359
6	25.0	50.0	Shan aluminum 25.0	0.0	33.53	1.285
7	15.0	10.0	Shan aluminum 52.5	22.5	34.28	1.388
8	30.0	20.0	Shan aluminum 50.0	0.0	36.00	1.260
9	45.0	30.0	Wei bridge 25.0	0.0	31.00	1.328

Through range and variance analysis, the results show: For the unconfined compressive strength of the 7d age, the lime-ash ratio (Factor A) is the most important influencing factor. For the strength of the 28d age, the type of red mud (Factor C) is the most important influencing factor. For the strength of the 90d age, the lime-ash ratio (Factor A) is also the most important influencing factor. Based on the strength of the 90d age, the best ratio is $A_3B_2C_1$, that is, lime: fly ash: Shan Aluminum red mud = 30:20:50; this is slightly different from the best ratio of $A_3B_2C_3$ at the 7d age.

For the strength of specimens at different ages, they are influenced by similar factors, but the degree of influence is different. As the proportion of lime in the lime-ash ratio increases (i.e., the amount of lime increases), the strength of each age increases; as the total amount of lime-ash increases, the strength of each age shows a trend of first increasing and then decreasing. The 7d age is least affected by this factor, and the 28 and 90d ages are more affected by it, indicating that this factor mainly affects long-term strength; under the stabilization of lime-ash, the strength of Wei Bridge red mud is smaller than that of Shan Aluminum red mud and red mud soil. For the 7 and 28d ages, the strength of red mud soil is the highest, but at the 90d age, the strength of pure Shan Aluminum red mud is significantly higher than that of red mud soil.

Table 4.12 Results of unconfined compressive strength test at different ages

Ratio sequence number	7d Unconfined compressive strength/MPa	28d Unconfined compressive strength/MPa	90d Unconfined compressive strength/MPa
1	0.265	0.701	0.721
2	0.336	0.035	0.034
3	0.619	0.515	0.457
4	0.448	0.520	0.687
5	1.175	2.244	2.277
6	0.913	1.222	1.284
7	1.544	2.288	2.312
8	1.710	2.056	2.107
9	1.182	0.497	0.681

3. Thermal Shrinkage Properties

When inorganic binder stabilized materials are used in the road base layer, temperature stress is generated in winter due to temperature differences. The surface temperature is lower than the internal temperature, and there is tensile stress at the top of the base layer. Combined with the tensile stress generated by the vehicle load at the bottom of the base layer, it can lead to structural layer cracking and other diseases. Thermal shrinkage refers to the performance of inorganic binder stabilized materials shrinking in size after cooling when the ambient temperature drops. Thermal shrinkage is characterized by the thermal shrinkage coefficient, which is the shrinkage coefficient of inorganic binder stabilized materials when the water content remains unchanged as the temperature drops. Figure 4.37 shows the relationship curve between the thermal shrinkage coefficient and temperature for each mix ratio.

Overall, in the low temperature zone (less than 10 °C), the thermal shrinkage coefficient of raw materials of bauxite red mud and stabilized red mud is larger, and the rate of change decreases with the decrease of temperature; in the high temperature zone (greater than 10 °C), the thermal shrinkage coefficient is smaller, and the rate of increase of the thermal shrinkage coefficient is less than that in the low temperature zone. For mix ratio one, the thermal shrinkage coefficient in the high temperature zone hardly changes.

The thermal shrinkage coefficient of lime-stabilized red mud is higher than that of red mud raw materials, and mix ratio three is higher than mix ratio four, indicating that the thermal shrinkage change of stabilized red mud is larger than that of raw materials, and adding clay to red mud will increase its thermal shrinkage coefficient. Replacing 30% of the red mud in the mix ratio with clay increases the thermal shrinkage coefficient by about 30.9%. Therefore, depending on the specific conditions during construction, it can be flexibly decided whether to add clay. In terms of thermal shrinkage properties, it is suitable to use pure red mud as filler in areas with low winter temperatures.

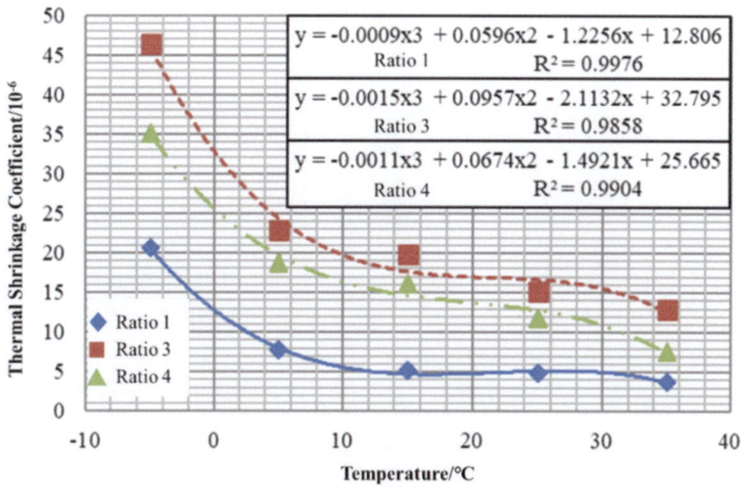

Fig. 4.37 Relationship curve between thermal shrinkage coefficient and temperature

4.3.2 Composite Material Stabilization of Red Mud

1. Compaction Characteristics

Figure 4.38 shows the results of compaction tests performed on remolded red mud samples treated with cement, lime, phosphogypsum, and new high polymer composite materials.

From the graph, it can be seen that: as the dosage of lime increases, the optimal moisture content of lime-stabilized red mud continues to rise, while the maximum dry density gradually decreases, especially when the dosage exceeds 8%, the decrease in maximum dry density is significant; as the dosage of cement increases, the optimal moisture content of cement-stabilized red mud continues to decrease, and the maximum dry density gradually increases. When the cement dosage reaches more than 10%, the maximum dry density is significantly increased; as the dosage of phosphogypsum increases, the optimal moisture content of phosphogypsum-stabilized red mud first increases and then decreases, and the maximum dry density first decreases and then increases, with the turning point of the trend occurring at around 10% dosage; as the dosage of composite stabilizing material increases, the optimal moisture content of red mud slightly increases, while the maximum dry density also gradually increases and the increase is quite significant. Overall, when the dosage of stabilizing material is within the range of 4–10%, the compaction characteristics of stabilized red mud show a strong regularity. When the stabilizing material exceeds 10%, the changes in optimal moisture content and maximum dry density significantly increase, indicating that the physical properties of the stabilizing material itself have a significant impact on the test results under high dosage conditions.

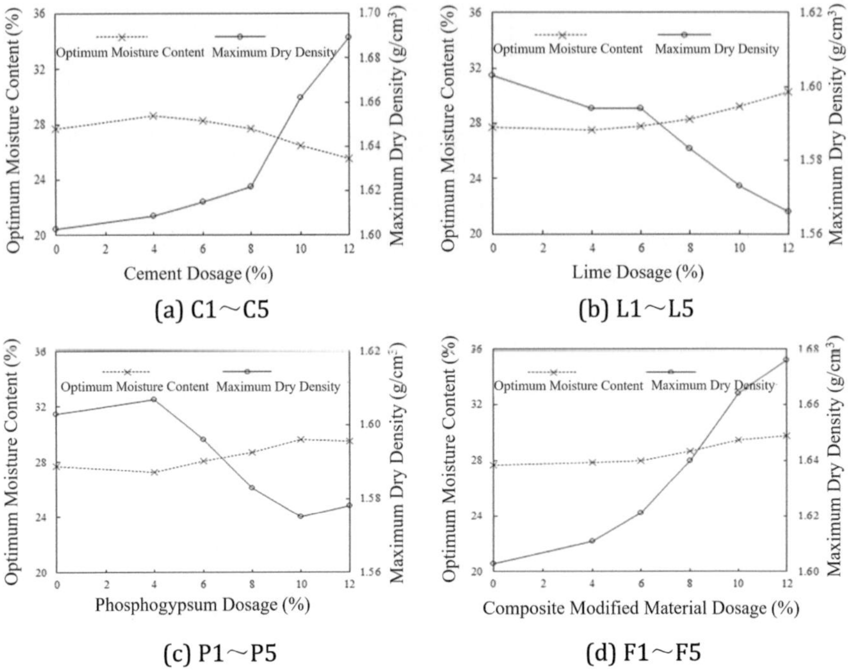

Fig. 4.38 Compaction test results of red mud stabilized by composite curing technology

2. Strength Characteristics

According to the basic physical and mechanical characteristics of red mud determined by the indoor geotechnical test results of the "Highway Geotechnical Test Procedures" (JGT-E40), the test red mud can be defined as high liquid limit clay.

According to the test results, red mud is difficult to meet the construction and performance requirements of roadbed filling. In order to improve the road performance of red mud, the study used a cement-based composite modified curing material CPS-12, which is a composite material formed by using cement as the basic material and adding phosphogypsum and other activators.

The test results of the seven-day unconfined compressive strength of different dosages of cement-based composite materials show that the strength loss after soaking for 24 h is very small (Fig. 4.39).

Indoor evaluations are mainly conducted at the Shandong Province Transportation Science Research Institute, considering the applicability, safety, economy, reliability, and other aspects of modified cured red mud as roadbed filling material, and conducting indoor test testing and analysis evaluation.

The unconfined compressive strength test is a strength test conducted under conditions where the sample is not subject to lateral restrictions. Figures 4.40, 4.41, 4.42

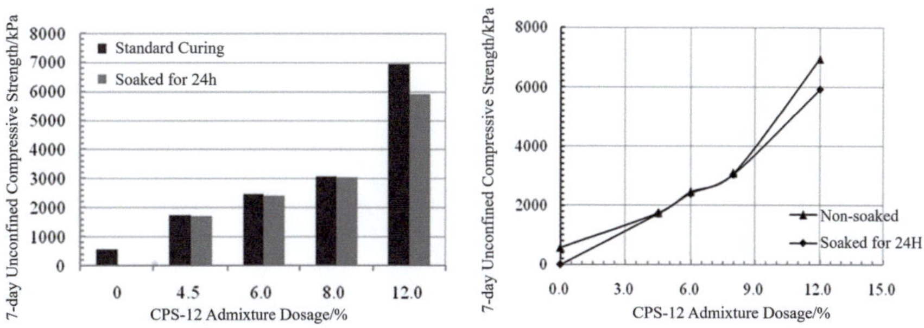

Fig. 4.39 Unconfined compressive strength of red mud

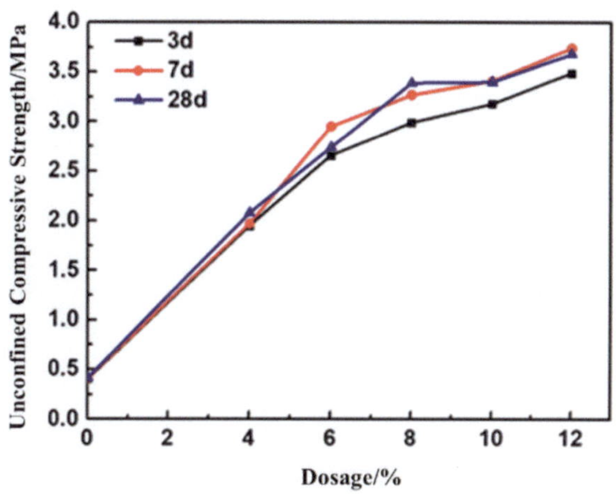

Fig. 4.40 Comparison of cement-stabilized red mud strength

and 4.43 respectively show the changes in the unconfined compressive strength of Bayer red mud stabilized by cement, lime, phosphogypsum, and polymer composite materials with the dosage and curing age.

From Fig. 4.40, it can be seen that cement as a stabilizer has a good stabilizing effect. Under different ages and cement dosages, Bayer red mud shows different strength characteristics. Under 7d standard curing, the unconfined compressive strength of 4% ~ 12% cement dosage is 1.97 MPa, 2.95 MPa, 3.27 MPa, 3.41 MPa, 3.74 MPa respectively; under 28d standard curing, the unconfined compressive strength of 4% ~ 12% cement dosage is 2.08 MPa, 2.74 MPa, 3.39 MPa, 3.40 MPa, 3.68 MPa respectively. The unconfined compressive strength of red mud generally shows a trend of increasing with the increase of dosage and age.

From Fig. 4.41, it can be seen that under the same dosage conditions, the unconfined compressive strength of cement-stabilized red mud increases with the increase of curing age; while under the same age and different lime dosages, Bayer red mud shows different strength characteristics. Under 7d standard curing, the unconfined compressive strength of 4% ~ 12% lime dosage stabilized red mud is 1.54 MPa,

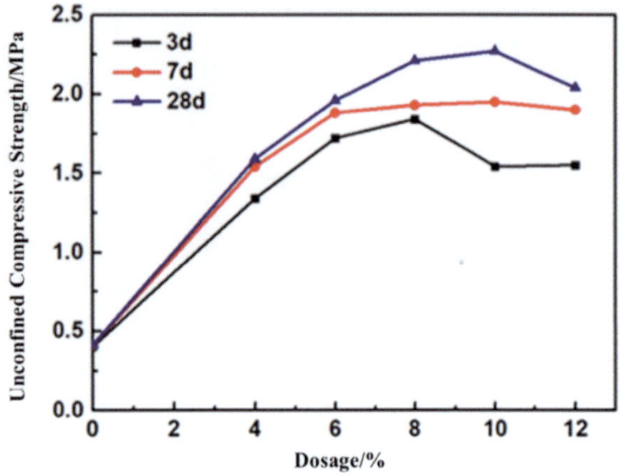

Fig. 4.41 Comparison of lime-stabilized red mud strength

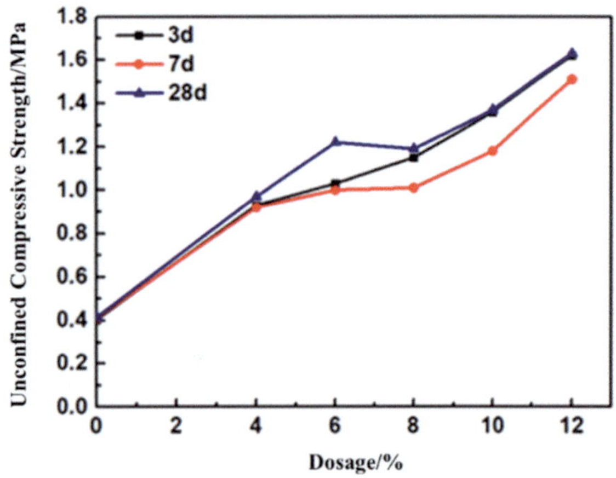

Fig. 4.42 Comparison of phosphogypsum-stabilized red mud strength

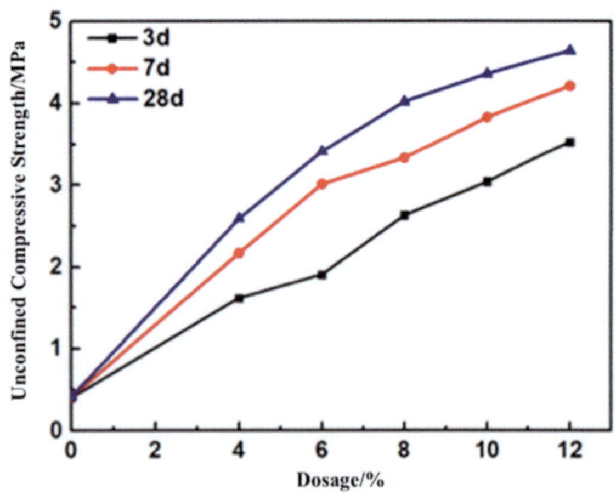

Fig. 4.43 Comparison of composite material-stabilized red mud strength

1.88 MPa, 1.93 MPa, 1.95 MPa, 1.90 MPa respectively, and under 28d standard curing, the unconfined compressive strength of 4% ~ 12% lime dosage stabilized red mud is 1.59 MPa, 1.96 MPa, 2.21 MPa, 2.27 MPa, 2.04 MPa respectively. The strength of stabilized red mud first increases and then decreases with the increase of dosage, but compared with unstabilized Bayer red mud, lime as a red mud stabilizer has a better stabilizing effect. Under the same curing age and dosage conditions, the unconfined strength of cement-stabilized red mud is greater than the result of lime-stabilized red mud under the same conditions; overall, the stabilizing effect of lime is slightly worse than that of cement.

As can be seen from Fig. 4.42, under the same aging conditions, the strength of stable red mud increases with the increase of phosphogypsum content, specifically under 7d standard curing, the unconfined compressive strength of stable red mud with 4%–12% phosphogypsum content is 0.92 MPa, 1.00 MPa, 1.01 MPa, 1.18 MPa, 1.51 MPa respectively, under 28d standard curing, the unconfined compressive strength of stable red mud with 4% to 12% lime content is 0.97 MPa, 1.22 MPa, 1.19 MPa, 1.37 MPa, 1.63 MPa respectively, and the unconfined compressive strength of phosphogypsum stabilized red mud at 28d curing age is greater than that at 7d age, indicating that compared with unstabilized Bayer red mud, lime as a red mud stabilizer has a better stabilizing effect, and in the later stage of curing, the stabilizing effect of phosphogypsum is better. Under the same curing age and content conditions, the unconfined strength of cement and lime stabilized red mud is greater than that of phosphogypsum stabilized red mud under the same conditions, overall, the effect of stabilizing red mud can be expressed as: phosphogypsum < lime < cement.

From the results in Fig. 4.43, compared with unstabilized Bayer red mud, stabilizing materials as stabilizers have a better stabilizing effect, under the same content conditions, the unconfined compressive strength of stabilized red mud gradually increases with the increase of age, under different composite stabilizing material content conditions, Bayer red mud shows different strength characteristics, under 7d standard curing, the unconfined compressive strength of 4% to 12% stabilizing material content is 2.17 MPa, 3.01 MPa, 3.33 MPa, 3.83 MPa, 4.21 MPa respectively; under 28d standard curing, the unconfined compressive strength of 4% to 12% stabilizing material content is 2.59 MPa, 3.41 MPa, 4.02 MPa, 4.36 MPa, 4.64 MPa respectively, with the increase of stabilizing material content, the unconfined compressive strength of stabilized red mud shows a gradually increasing trend; and compared with the stabilizing effect of cement, lime and phosphogypsum, under the same age and content conditions, the unconfined compressive strength of stabilized red mud is: composite stabilizing material > cement > lime > phosphogypsum, the stabilizing disposal effect of composite material is the best.

3. Water stability performance

Figure 4.44 shows the change of the unconfined compressive strength of stabilized red mud under soaked and unsoaked conditions with the change of cement content. As can be seen from the figure, soaking damage conditions have a significant impact on cement stabilized Bayer red mud, mainly manifested as a decrease in compressive strength, the strength of different age cement stabilized Bayer red mud increases with the increase of content, after soaking for one night, the strength of the same cement content stabilized Bayer red mud decreases, this indicates that soaking conditions have a certain impact on its internal structure and the chemical reactions it produces, thereby affecting its mechanical properties.

Figure 4.45 shows the soaking strength retention rate of stabilized Bayer red mud under the same age and equal cement content. Compared with unsoaked conditions, the unconfined compressive strength of Bayer red mud still has a certain guarantee rate after soaking; the soaking strength guarantee rate varies with different curing ages, the 7d strength guarantee rate is between 46 and 55%, the 28d strength guarantee rate is between 51 and 60%, indicating that with the increase of age, the cement stimulates the stabilizing effect of red mud gradually strengthens, the stabilizing effect of cement gradually comes into play, making the stabilized Bayer red mud more resistant to water damage.

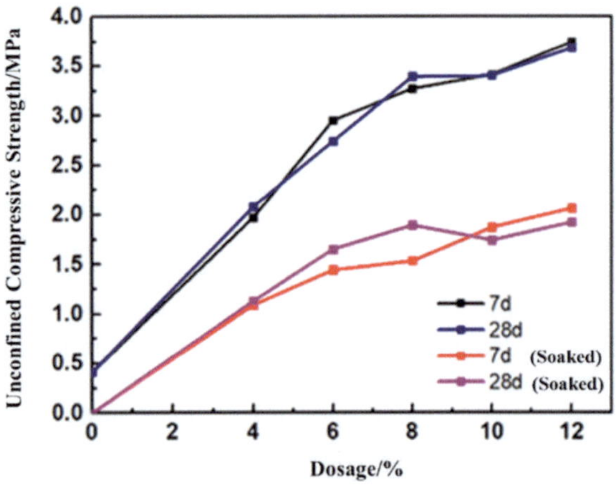

Fig. 4.44 Unconfined compressive strength of cement stabilized red mud under soaked and unsoaked conditions

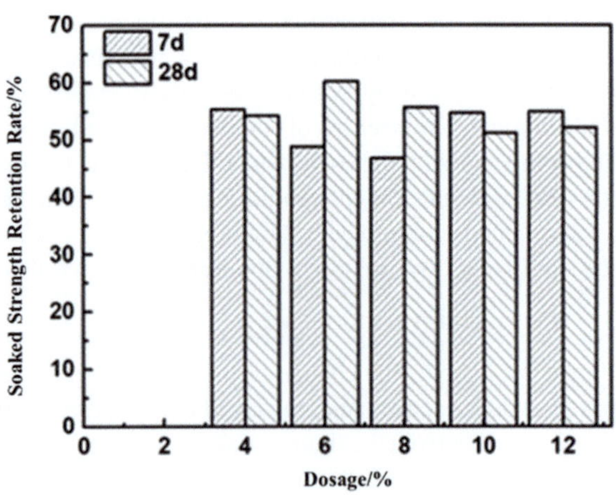

Fig. 4.45 Soaking strength retention rate of cement stabilized red mud

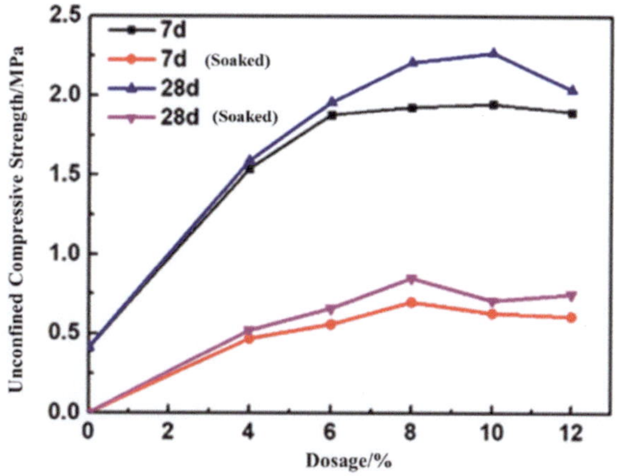

Fig. 4.46 Unconfined compressive strength of lime-stabilized red mud with and without water immersion

Figure 4.44 shows the change of the unconfined compressive strength of lime stabilized red mud under soaked and unsoaked conditions with the change of lime content. As can be seen from the figure, soaking damage conditions have a significant impact on cement stabilized Bayer red mud, mainly manifested as a decrease in compressive strength; after soaking for one night, the unconfined compressive strength of lime stabilized red mud increases first and then decreases with the increase of content under different curing ages, but the strength increase after soaking is less than that under unsoaked conditions, this indicates that soaking conditions have a certain impact on its internal structure and the chemical reactions it produces, thereby affecting its mechanical properties (Fig. 4.46).

Figure 4.47 represents the water immersion strength retention rate of stable Bayer method red mud under the same age and equal lime content. Compared with the non-immersed condition, the unconfined compressive strength of the Bayer method red mud still has a certain guarantee rate after immersion. The water immersion strength guarantee rate varies with different curing ages, with a 7d strength guarantee rate between 30 and 36%, and a 28d strength guarantee rate between 11 and 39%. This indicates that as the age increases, the lime's effect on stimulating the stability of the red mud gradually strengthens, and the stabilizing effect of the lime gradually comes into play, increasing the water damage resistance of the stabilized red mud. However, compared with cement, the water immersion strength guarantee rate of lime-stabilized red mud is slightly lower than that of cement, indicating that the stabilizing effect of lime is slightly worse than that of cement.

Figure 4.48 represents the change in the unconfined compressive strength of phosphogypsum-stabilized red mud under water immersion and non-immersion conditions with the change in lime content. As can be seen from the figure, the water damage condition has a significant impact on the Bayer method red mud stabilized by red mud, and the compressive strength of the red mud significantly decreases after immersion, especially when the content exceeds 10%. After being treated with water for one night, the unconfined compressive strength of phosphogypsum-stabilized red

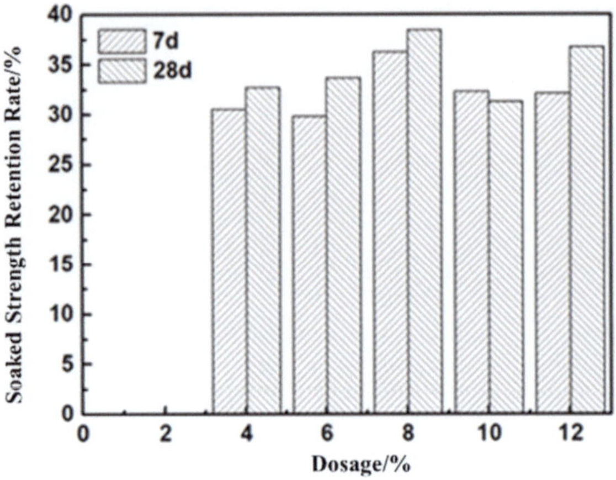

Fig. 4.47 Water immersion strength guarantee rate of lime-stabilized red mud

mud increases first and then decreases with the increase in the amount of lime under different curing ages, and the strength growth after immersion is less than that without immersion. This indicates that the water immersion condition has a certain impact on its internal structure and the chemical reactions it produces, thereby affecting its mechanical properties.

Figure 4.49 represents the water immersion strength retention rate of stable Bayer method red mud under the same age and equal phosphogypsum content. Compared with the non-immersed condition, the unconfined compressive strength of the Bayer method red mud still has a certain guarantee rate after immersion. The water immersion strength guarantee rate varies with different curing ages, with a 7d strength guarantee rate between 38 and 59%, and a 28d strength guarantee rate between 17 and 37%. This indicates that water immersion has a strong weakening effect on the stabilizing effect of phosphogypsum, and as the age increases, the weakening effect

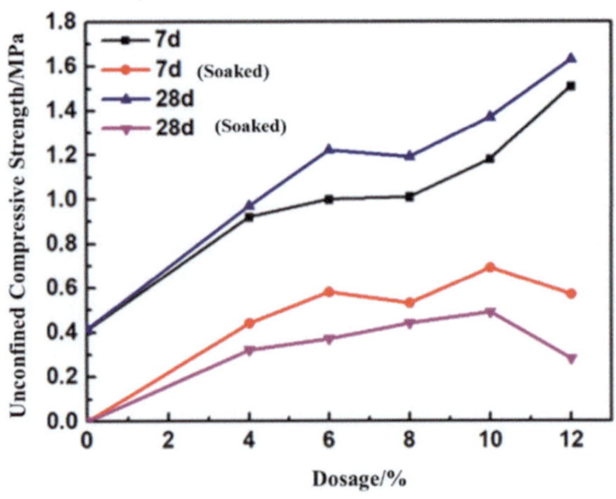

Fig. 4.48 Unconfined compressive strength of phosphogypsum-stabilized red mud with and without water immersion

of water on the stabilizing effect of phosphogypsum gradually strengthens, and the stabilizing effect of lime gradually weakens, reducing the water damage resistance of the stabilized Bayer method red mud.

Figure 4.50 represents the unconfined compressive strength of composite stabilized material under water immersion and non-immersion conditions at the same content of 7 and 28 d curing ages. From the figure, it can be seen that the water damage condition also has an impact on the Bayer method red mud stabilized by the stabilized material. After being treated with water for one night, the strength of the composite stabilized material stabilized Bayer method red mud decreases, indicating that the water immersion condition has a certain impact on its internal structure and the chemical reactions it produces, thereby affecting its mechanical properties, but the degree of strength reduction is significantly reduced.

Fig. 4.49 Water immersion strength guarantee rate of phosphogypsum-stabilized red mud

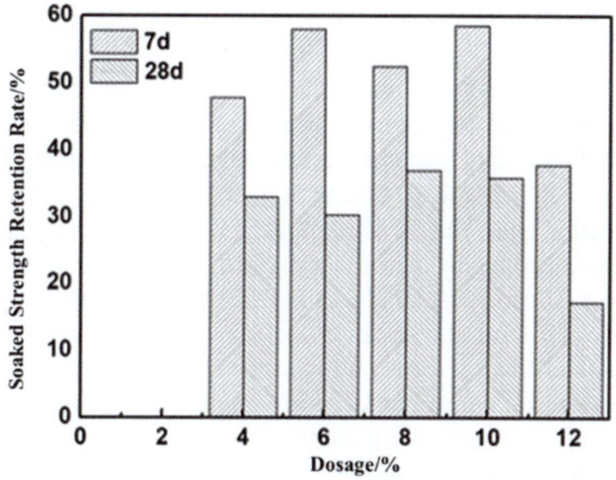

Fig. 4.50 Unconfined compressive strength of composite material-stabilized red mud with and without water immersion

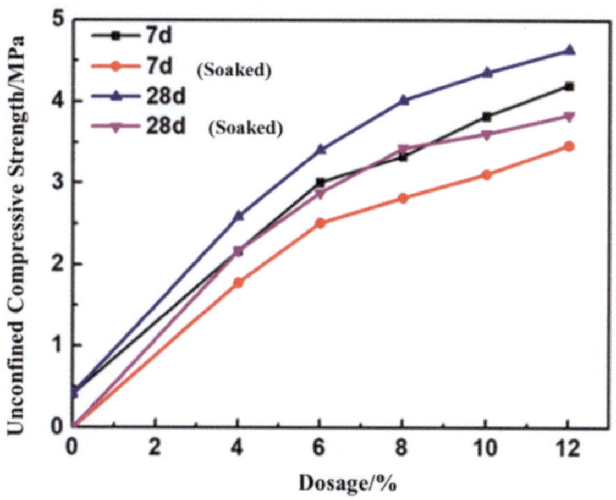

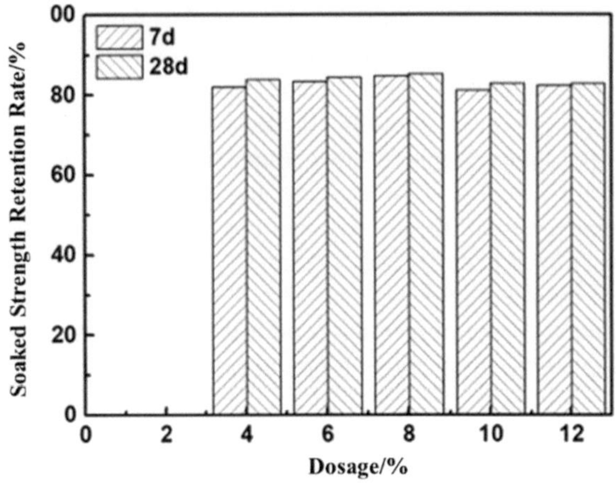

Fig. 4.51 Water immersion strength guarantee rate of composite material-stabilized red mud

Figure 4.51 represents the water immersion strength retention rate of stable Bayer red mud under the same age and equal phosphogypsum content. Compared with red mud under non-immersion conditions, its strength has a certain guarantee rate after immersion. The water immersion strength guarantee rate varies with different curing ages, the 7d strength guarantee rate is between 81% and 85%, and the 28d strength guarantee rate is between 82% and 86%, indicating that as the age increases, the composite stabilizing material gradually enhances the excitation stability effect on red mud, the stabilizing effect of the composite stabilizing material is gradually exerted, and the ability of the stabilized Bayer red mud to resist water damage increases; compared with cement, lime and phosphogypsum, the water immersion strength guarantee rate of the composite material stabilized red mud is the highest, indicating that the ability of the composite stabilizing material to stabilize red mud to resist water damage is significantly stronger than cement, lime and phosphogypsum. For red mud, the composite stabilizing material is the best stabilizing and solidifying material.

Considering the applicability, safety, economy, reliability and other aspects of modified solidified red mud road materials as subgrade filling materials, relevant physical and mechanical index indoor tests were conducted on the original red mud and modified solidified red mud road materials respectively, and the analysis test comparison results of the original red mud and modified red mud are shown in Table 4.13.

Based on the current highway industry subgrade design, construction, acceptance standards and environmental protection department related standards, the indoor test detection analysis results of the sample system show:

(1) The original red mud has characteristics such as high alkalinity, high liquid limit, low strength and modulus (no strength when encountering water). From the perspective of current highway design, acceptance and other standard requirements, the unconfined compressive strength, rebound modulus, California bearing ratio (CBR value) and other indicators of the original red mud

Table 4.13 Test of road performance of modified solidified red mud road materials

	Original red mud	Modified solidified red mud road material
Optimum moisture content (%)	28.5	26.8
Maximum dry density (g/cm^3)	1.74	1.71
Internal friction angle φ (°)	24.2	36.8
Cohesion c (kPa)	31.0	143.2
7-day non-immersion compressive strength (kPa)	506.0	3080.6
7-day immersion compressive strength (kPa)	0 (Collapse)	3047.1
Liquid limit W_L (%)	45.4	36.4
Plastic limit W_P (%)	29.6	21.5
Plasticity Index	15.8	14.9
Compression coefficient a_{1-2} (MPa^{-1})	0.19	0.07
Free swelling rate (%)	17.0	5.0
No-load swelling rate (%)	0.05	0.0
Resilient modulus (MPa)	26.5	526.5
CBR (%)	2.8	94.0
pH value	12.2	9.4

cannot meet the relevant specification requirements. Therefore, red mud cannot be used as subgrade filling material for highways and municipal roads.

(2) The test results of modified solidified red mud road materials have significantly improved compared with the original red mud. Among them, the PH value has significantly decreased, the unconfined compressive strength (immersed and non-immersed) and rebound modulus values have increased significantly, and are higher than the general lime (or cement) improved soil. The liquid limit and plastic limit decrease at the same time, the plasticity index remains basically unchanged, which is conducive to the controllability of on-site construction. The 7d immersion strength loss is small, and the CBR value is high, indicating that the material has high waterproof and water resistance.

4.4 Summary of This Chapter

This chapter aims to meet the requirements of highway subgrade and base layer use, improve the comprehensive road performance of red mud, propose inorganic solidification and composite solidification of red mud comprehensive stabilization treatment technology, study the stabilization treatment effect of red mud, and explore the stabilization solidification mechanism of red mud. The main conclusions are as follows:

(1) Using orthogonal experimental design to study the comprehensive treatment of red mud with different inorganic solidifiers, a large number of indoor test results show.

① The type of red mud (factor C) has a significant impact on the optimum moisture content and maximum dry density; the total amount of two ashes (factor B) has a significant impact on the maximum dry density, but the impact on the optimum moisture content is small; the ratio of two ashes (factor A) has a small impact on the above two indicators.

② For the unconfined compressive strength of the 7d age, the ratio of two ashes (factor A) is the most important influencing factor, and the total amount of two ashes (factor B) and the type of red mud (factor C) have a weak influence; for the strength of the 28d age, the type of red mud (factor C) is the most important influencing factor, the total amount of two ashes (factor B) is the least influencing factor, and the ratio of two ashes (factor A) is between the two; for the strength of the 90d age, similar to the 7d age, the ratio of two ashes (factor A) is the most important influencing factor, the total amount of two ashes (factor B) is the second, and the total amount of two ashes (factor B) is the least factor. According to the strength of the 90d age, the best ratio is lime: fly ash: mountain aluminum red mud = 30:20:50.

③ In terms of thermal shrinkage performance, the thermal shrinkage performance of two ash stabilized red mud is worse than that of red mud raw materials. When the parts are made according to the ratio of lime: fly ash: mountain aluminum red mud = 30:20:50, its thermal shrinkage coefficient is about 112% higher than that of the original red mud, and the maximum thermal shrinkage coefficient appears in the range of 0 to − 10 °C, which is 17.90×10^{-6}; Adding clay to red mud will increase the thermal shrinkage coefficient. If 30% of the red mud in the ratio is replaced with clay, the thermal shrinkage coefficient will increase by about 30.9%. Therefore, in terms of thermal shrinkage characteristics, it is suitable to use pure red mud as filler in areas with low winter temperatures.

(2) Select lime, cement, phosphogypsum and new high polymer composite materials for the stable treatment of Bayer method red mud, a large number of indoor test results show.

① The addition of cement can reduce the optimum moisture content of red mud and increase its maximum dry density. When the cement content exceeds 8%, the maximum dry density significantly increases; the strength of red mud also significantly improves after the addition of cement; the strength increase of red mud after the addition of cement is large, but the strength growth slows down after 7d or even the strength decreases in the later stage; after soaking for one night, the strength guarantee rate of cement stabilized red mud at 7d (28d) age is between 46 and 55% (51% and 60%), showing certain water stability performance.

② The addition of lime causes the optimum moisture content of red mud to continuously increase, while the maximum dry density gradually decreases, and when the cement content exceeds 8%, the decrease in maximum dry density is significant; the addition of lime can improve the strength of red mud, but as the content increases, the strength value shows a trend of first increasing and then decreasing; after soaking for one night, the strength guarantee rate of lime stabilized red mud at 7d (28d) age is between 30 and 36% (11% and 39%), compared with cement, the water strength guarantee rate of lime stabilized red mud is slightly lower than that of cement, indicating that the stabilization and solidification effect of lime is slightly worse than that of cement.

③ The addition of phosphogypsum causes the optimal moisture content of red mud to first increase and then decrease, while the maximum dry density first decreases and then increases, with the turning point occurring around 10% addition. After adding phosphogypsum, the increase in the strength of red mud is relatively small, and it shows a trend of first decreasing and then slowly increasing in 7d strength, but the strength is less than the results of cement and lime stabilized red mud under the same conditions. After immersion, the strength guarantee rate of phosphogypsum stabilized red mud at 7d (28d) age is between 38 and 59% (17 and 37%), and as the curing age increases, the weakening effect of water on the stabilization effect of phosphogypsum gradually increases, the stabilizing effect of phosphogypsum gradually weakens, and the ability of red mud to resist water damage decreases.

④ After the addition of polymer composite materials, the optimal moisture content of red mud slightly increases, and the maximum dry density also gradually increases and the increase is relatively obvious. When using polymer composite materials for treatment, the strength of red mud continuously increases with the increase of material addition and age, and is higher than the results of cement, lime and phosphogypsum stabilized red mud under the same conditions. Compared with cement, lime and phosphogypsum, the immersion strength guarantee rate of composite material stabilized red mud is the highest, reaching over 80% at 7d age, indicating that the ability of polymer composite materials to stabilize red mud to resist water damage is significantly stronger than cement, lime and phosphogypsum.

Overall, the effect of polymer composite materials in stabilizing red mud is the best, making it the best material for stabilizing red mud.

(3) The stabilization and solidification mechanism of red mud is obtained through microscopic analysis.

① When stabilizing red mud with inorganic solidification technology, after adding lime and fly ash, boehmite and goethite participate in a series of reactions to form crystalline substances such as tricalcium aluminate, which is the basis for the early strength of inorganic stabilized red mud. As the curing time extends, some amorphous substances are formed. These networked, filamentous, and fibrous substances wrap, fill, and bond structures with certain strengths such as calcium carbonate, tricalcium aluminate, and garnet, greatly enhancing the strength of the solidified structure.

② The early strength of red mud stabilized by composite solidification technology mainly comes from the internal friction between compacted whole particles and the "original cohesion" formed by the molecular attraction between the water film between particles and adjacent particles. The hydration reaction of cement and the alkali-activation effect are the main sources of strength formation for red mud. The strength of stabilized red mud continues to slowly increase with carbonation and secondary hydration reactions.

(4) The environmental control mechanism of red mud mainly includes the alkali reduction treatment of red mud and the suppression of heavy metal ions:

① The Bayer process red mud contains a large amount of alkali, which will cause alkali leaching phenomenon in actual use, so reducing the alkali content of red mud is very important for the utilization of red mud. There are two forms of alkali in red mud, one is the soluble alkali solution brought in during the production of alumina; the other is sodium aluminosilicate hydrate in the form of cancrinite, which is an insoluble alkali. Only the soluble alkali in red mud is harmful alkali. Using the principle of alkali activation of fly ash to form sodium aluminosilicate hydrate, mixing red mud with fly ash can transform harmful soluble alkali into insoluble non-soluble alkali, which is a new idea.

② The suppression of heavy metal ions in red mud is mainly achieved through solidification and stabilization: Solidification is the process where amorphous aluminosilicate dissolves quickly under the action of alkali activator during the reaction process of stabilizing red mud, resulting in gelation. A low-polymer gel is formed and gradually forms a network structure with the aggregation effect. Heavy metal ions are sealed in the gel body by the encapsulation of the low-polymer gel. Stabilization is the process where some harmful ions enter the lattice of the hydration product during the hydration process of stabilized red mud through chemical reactions. The hydration product binds to harmful ions through chemical bonds, and harmful metal ions can play a role similar to alkali metal ions in balancing the valence inside the polymer network.

Chapter 5
Application of Improved Solidified Red Mud in Subgrade Engineering

5.1 Overview

Based on the unique physical and chemical properties and mineralization of red mud, the comprehensive utilization of red mud mainly includes the following three aspects: first, the extraction and recovery of valuable metals; second, the preparation of building materials; and third, the use of red mud adsorbents for wastewater purification. Although there are many studies and applications related to red mud, the above research on red mud can only consume a very small part of red mud, which is a drop in the bucket compared to the annual output of hundreds of millions of tons of red mud. The physical properties and composition of red mud are similar to soil, which is a special soil with a high alkalinity. Applying red mud to road engineering, after modifying and solidifying red mud using additives, it can be used for road subgrade filling materials and base layers, which opens up an effective way for the comprehensive utilization of red mud and is also the focus of large-scale treatment of bulk industrial solid waste such as red mud.

The physical and mechanical properties, particle composition, chemical composition and other characteristics of red mud cannot meet the applicability requirements of road materials. In order to study the on-site engineering application characteristics of red mud, based on the preliminary understanding of the modification and solidification of a large amount of red mud and the modification rules of inorganic binders, different modified red mud, different filling thicknesses, and different environmental filling forms are determined through comprehensive analysis and selection from technical, economic, construction, environmental and other aspects. According to the detection, testing and monitoring of the typical filling structure of modified red mud, the engineering characteristics of different filling structure forms of modified and solidified red mud are comprehensively evaluated, and a detection and evaluation system for different age subgrade structures is established, providing a basis and guarantee for the filling of modified red mud and other modified red mud subgrade engineering in the process of large-scale resource utilization of red mud.

Y. Cheng and X. Zhu, *Utilization and Environmental Impact Assessment of Red Mud Resources*, https://doi.org/10.1007/978-981-96-7411-4_5

5.2 Modified and Solidified Red Mud Road Subgrade

5.2.1 Experimental Section Implementation Plan

The verification of the on-site road reliability and pollution controllability of the modified and solidified red mud road materials was implemented through the red mud road demonstration project of the first phase of the Binzhou Beihai Vein Industrial Park.

As a professional park built by Binzhou Beihai Economic Development Zone, Binzhou Beihai Vein Industrial Park mainly carries out projects such as comprehensive utilization of industrial waste and red mud. After the park is completed, it will form an environmentally friendly, resource-saving, and technology-intensive modern green environmental protection park, which is conducive to accelerating the development and construction of Beihai New District and the implementation of ecological environmental protection projects, and is very important for Beihai New District and Binzhou City's emphasis on efficient ecology and green development. The first phase of the Vein Industrial Park includes 5 roads with a total length of 5.5 km. Among them, the main road of the park is 3.3 km long and 24 m wide. The secondary roads of the park are 2.2 km long and 18 m wide. The modified and solidified red mud subgrade project, as a demonstration road for the utilization of solid waste new materials in the park, uses modified and solidified red mud as a new type of subgrade filling material in its construction. It is a new exploration of red mud road technology, focusing on solving regional red mud large-scale storage, land occupation, waste of resources, environmental pollution, safety hazards and other issues.

However, there are no relevant norms, standards and mature experiences for red mud roads at home and abroad. In order to ensure the safety and reliability of the later project, according to the requirements of relevant technical specifications and the opinions of the owner, the work of the modified and solidified red mud test section was carried out in advance. The length of the test section subgrade is 150 m (100 m modified and solidified red mud section + 50 m conventional subgrade filling material section), the width is 24 m, and the thickness of the modified and solidified red mud subgrade is 40 cm (as shown in Fig. 5.1). The implementation of this project is the first in the world, and there are no relevant norms, standards and mature experiences in China. According to the requirements of Shandong Provincial Transportation Planning and Design Institute, the construction and detection evaluation standards meet the following requirements.

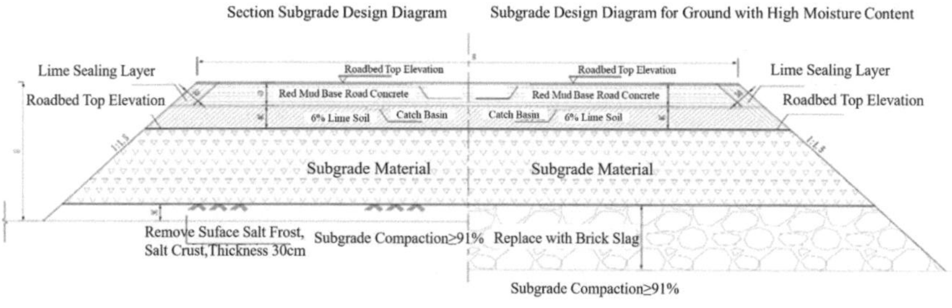

Fig. 5.1 Schematic diagram of the cross section of the modified and solidified red mud test road

5.2.2 *Test Section Construction Process and Standards*

(1) Preliminary experiment requirements

In order to determine the road reliability of the preliminary test section, the modified and solidified red mud road materials must meet the relevant requirements of road performance in the subgrade design specifications, and the following test evaluation work should be carried out:

① Carry out chemical composition and mineral composition analysis tests to determine its chemical composition, mineral composition, harmful substance content in leachate, pH value, loss on ignition, etc.
② Complete the compaction test to determine the maximum dry density and optimal moisture content; through the test, determine the internal friction angle, cohesion, compression coefficient, expansion coefficient, rebound modulus and California Bearing Ratio (CBR) value.
③ Submit a highway construction technical guide that matches the red mud modification materials, technology, and methods related to Beihai New District. Including construction methods, construction requirements, testing standards that can match the on-site construction capabilities, and the mixing ratio and technical parameters of the modified materials.

(2) Material parameters

Considering the applicability, safety, economy, reliability, and other aspects of modified solidified red mud as a subgrade filling material, indoor experimental tests and analysis evaluations are conducted. Based on the current municipal and highway industry subgrade design, construction, acceptance standards, and environmental protection department related standards, the indoor experimental test analysis results of the sample system show:

① The original red mud has characteristics such as high alkalinity, high liquid limit, low strength and modulus (no strength when encountering water). From the perspective of current highway design, acceptance and other standard requirements, the unconfined compressive strength, rebound modulus, California bearing ratio (CBR value) and other indicators of the original red mud cannot meet the relevant specification requirements. Therefore, red mud cannot be used as a subgrade filling material for highways and municipal roads.

② The test results of modified solidified red mud have significantly improved compared to the original red mud. Among them, the pH value has significantly decreased, the unconfined compressive strength and rebound modulus values have increased significantly, and are higher than general lime (or cement) improved soil. The liquid limit and plastic limit decrease at the same time, the plasticity index remains basically unchanged, which is conducive to the controllability of on-site construction. The 7d immersion strength loss is small, and the CBR value is high, indicating that the material has high water resistance and water resistance. After the red mud is modified and solidified, the 7-day unconfined compressive strength of the indoor test is greater than 1.5 MPa, and the indoor rebound modulus is higher than 60 MPa.

③ Through the comparative sampling test of the original red mud and the modified solidified red mud, the leaching toxicity test results of the pollutants in the original red mud are lower than the hazardous waste limit, and it is not a hazardous waste, but it is higher than the groundwater quality standard value. The leaching toxicity test results of the pollutants in the modified solidified red mud material can be lower than the groundwater Class III standard value. Compared with the original red mud, the leaching of hexavalent chromium, total chromium, nickel, arsenic, vanadium and other ions is significantly reduced.

(3) Test section construction process

① Pre-construction preparation. As shown in Fig. 5.2, the test site is set out, the original surface of the fill section is cleared and compacted.

② Original surface treatment. As shown in Fig. 5.3, after the test section surface is cleared, the 40 cm depth range is treated with 6% lime soil modification and

Fig. 5.2 Pre-construction preparation diagram

Fig. 5.3 Original surface treatment construction diagram

compacted to a compaction degree of 93%. After compaction, the sample should meet the compaction standard specified in this procedure.

③ Push the fill material. The prepared red mud is pushed to the fill section by a bulldozer from both sides of the subgrade or within a short distance.

④ Spreading the fill material. As shown in Fig. 5.4, the red mud fill is spread and leveled by a bulldozer, or leveled by a bulldozer in conjunction with a grader. After the bulldozer spreads, each layer thickness should not exceed 25 cm. The loose spreading thickness can be controlled by measuring the elevation before and after the fill.

Fig. 5.4 Spreading fill construction diagram

Fig. 5.5 Initial pressure construction of adding solidification materials

⑤ Adding solidification materials. As shown in Fig. 5.5, the moisture content of the filled soil body is tracked and detected at any time. The optimal moisture content of the modified red mud used in this test road is 27.3%. The road mixing method is used to add solidification materials, and the amount of additives is strictly controlled.

⑥ Mixing. First, the piled solidifier is evenly spread with a grader, and then mixed with a BMW road mixer. Ensure no interlayer during the mixing process, overlap 20 cm for each mixing width, check the mixing depth and overlap width at any time during the mixing process, check whether the mixing is uniform at any time, and whether the soil particle size meets the design and specification requirements (the maximum particle size is not more than 50 mm). If it does not meet the requirements, mix it twice. If it still does not meet the requirements after two mixes, mix it three times.

⑦ Stabilization. As shown in Fig. 5.6, during stabilization, follow the compaction process of general earthwork subgrade, roll from the edge of the subgrade to the inside, the overlap width of the wheel track during rolling should not be less than ½ of the single wheel width, a single side wheel track covers one working surface as one pass, stabilize 2 times, and you can also use a staggered rolling method longitudinally and horizontally during stabilization.

⑧ Vibratory roller compaction. As shown in Fig. 5.7, use a vibratory roller with a drive of 18 T or more for compaction. Roll slowly first and then quickly, using strong vibration for vibratory compaction. The front and rear adjacent sections should overlap longitudinally by more than 1.0 m, to achieve no missed pressure, no dead corners, and ensure uniform compaction.

⑨ Final pressure. The final pressure compaction process is the same as the stabilization, the overlap width of the wheel track should not be less than ½ of the single wheel width, and the final pressure is 2 times.

Fig. 5.6 Subgrade compaction construction diagram

Fig. 5.7 Watering and maintenance of red mud subgrade

⑩ Detection. Detect the compaction degree, rebound modulus, top surface deflection, etc., the detection results should meet the standards specified in this procedure, if it does not meet the requirements, re-pressure should be carried out until it is qualified.

5.2.3 Quality Control and Evaluation of Subgrade Engineering in the Test Section

The evaluation of the road mechanical properties of the modified solidified red mud subgrade is directly related to the quality of the pavement structure layer, the quality of road operation and maintenance workload after construction. In road construction, the assessment of this issue is of great significance. For the performance evaluation of the subgrade and the evaluation of the road mechanical properties of the modified solidified red mud material used in highway engineering, systematic field tests and analyses were carried out using on-site load plate tests, deflection detection and comprehensive probing DCP technology.

(1) Resilient modulus of the top surface of the modified solidified red mud subgrade

The resilient modulus of the top surface of the subgrade is the most direct reflection of the overall bearing capacity of the subgrade, and it is also the direct description of the stiffness of the subgrade under the vertical load of the corresponding test layer. It is related to the performance of the overlying structure layer under the action of vehicle load. It can even be said that the compaction test of the subgrade filling is, to a certain extent, to ensure that the top surface of the subgrade has sufficient stiffness, so as to provide a guarantee for the durability of the pavement structure under the repeated load of the vehicle after the pavement structure layer is laid. Therefore, under the premise of ensuring the water stability of the structure of the modified solidified red mud subgrade, the use of the resilient modulus to evaluate the quality of the filling of the modified solidified red mud subgrade is more direct and effective.

The resilient modulus of the modified solidified red mud subgrade is determined by on-site load plate, as shown in Fig. 5.8, which is suitable for on-site measurement of the resilient modulus of any layer surface of the modified solidified red mud subgrade. The test is mainly to read the corresponding subgrade rebound deformation value under each level of load by gradually loading and unloading the bearing plate, and to calculate the resilient modulus of the subgrade. The resilient modulus of the subgrade measured by this method can be used as a pavement design parameter.

The resilient modulus E_i under each level of load pressure is calculated according to formula (5.1)

$$E_i = \frac{\pi D}{4}\frac{p_i}{L_i}(1 - \mu_0^2) \tag{5.1}$$

The calculation of the resilient modulus of the on-site load plate is based on the rebound deformation values before the end of the test, and the subgrade resilient modulus E_0 is calculated by the linear regression method according to formula (5.2):

$$E_0=\frac{\pi \mathrm{D}}{4} \cdot \frac{\sum P_i}{\sum L_i}(1-\mu_0^2) \tag{5.2}$$

where

Fig. 5.8 On-site load plate resilient modulus test diagram

E_0 Subgrade resilient modulus (MPa);
μ_0 Poisson's ratio of soil, selected according to the pavement design specifications issued by the department;
L_i The measured rebound deformation values at each level before the end of the test;
P_i Pressure values at each level relative to L_i.

On the surface of the lime soil layer and the surface of the modified solidified red mud subgrade, two parallel test areas were arranged in the north and south, numbered NK and SK respectively. After the completion of the subgrade filling with the modified solidified red mud material, the backfill modulus tests were carried out on the 3rd and 7th days respectively, and the test results are shown in Table 5.1.

From the table, it can be seen that the rebound modulus of the second layer of lime soil surface is less than 60 Mpa, and the rebound modulus E of the surface of the modified solidified red mud road material (modified solidified red mud road material) base measured for 3 and 7 days $E_0 > 130$ Mpa, which is much higher than the requirements of the design document. The growth of the age can effectively promote the improvement of the rebound modulus of the modified solidified red mud road material base.

(2) Dynamic Cone Penetration Test

Dynamic Cone Penetrometer (DCP), also known as penetrometer. DCP is a small and portable in-situ soil testing dynamic probe, with a hammer weight of 8 kg, a drop distance of 575 mm, a penetration rod length of 1000 mm, a rod diameter of Φ16 mm,

Table 5.1 Measured results of resilient modulus of modified red mud test section

Test location	Layer position	Resilient modulus/MPa	Test location	Layer position	Resilient modulus/Mpa
NK0 + 010	Top of lime soil 3 days	41.3	SK0 + 010	Top of lime soil 3 days	32.6
	Top of red mud 3 days	181.1		Top of red mud 3 days	93.1
	Top of red mud 7 days	196.2		Red mud top 7 days	160.3
NK0 + 030	Lime soil top 3 days	45.5	SK0 + 030	Lime soil top 3 days	44.5
	Red mud top 3 days	200.9		Red mud top 3 days	161.7
	Red mud top 7 days	201.9		Red mud top 7 days	174.5
NK0 + 050	Lime soil top 3 days	42.1	SK0 + 050	Lime soil top 3 days	50.8
	Red mud top 3 days	179.1		Red mud top 3 days	164.2
	Red mud top 7 days	131.8		Red mud top 7 days	298.0
NK0 + 075	Lime soil top 3 days	–	SK0 + 075	Lime soil top 3 days	32.8
	Red mud top 3 days	243.3		Red mud top 3 days	167.8
	Red mud top 7 days	149.8		Red mud top 7 days	175.0
NK0 + 090	Lime soil top 3 days	44.0	SK0 + 090	Lime soil top 3 days	60.1
	Red mud top 3 days	176.9		Red mud top 3 days	89.4
	Red mud top 7 days	148.9		Red mud top 7 days	62.4

a cone head diameter of Φ20 mm, a cone angle of 60°, and a 1000 mm reading scale connected to the side of the penetration rod, directly reading the penetration value for each hit. The in-situ testing DCP penetrometer has been widely used in engineering abroad, and a lot of practical experience has been accumulated in relation to the corresponding soil property indicators. Generally, the penetration value for each hammer hit is set as DN (mm/blow), then the relationship between the penetration value DN and the California Bearing Ratio (CBR) is:

$$CBR = \frac{292}{DN^{1.12}} \quad \text{or} \quad CBR = \frac{405.3}{DN^{1.259}} \tag{5.3}$$

Considering the Dynamic Cone Penetration Test (DCP) is simple to operate on-site, does not require soil sampling, and is easy to conduct at a high sampling rate. At the same time, the penetration depth and continuity are more advantageous than traditional compaction tests. Therefore, this study uses DCP in-situ testing technology, applied to the quality inspection and evaluation of modified solidified red mud road materials, and the test results are shown in Tables 5.2, 5.3, 5.4 and 5.5.

Table 5.2 DCP penetration values of each layer of the red mud test section subgrade (top of the lime soil layer)

Test location	Penetration depth/mm	0 ~ 20/cm	20 ~ 40/cm	40 ~ 60/cm	60 ~ 80/cm	Weighted average
SK0 + 090	450	10.0	8.9			9.4
SK0 + 075	580	15.6	7.7	10.3		10.3
SK0 + 050	660	13.9	10.7	11.8		12.0
SK0 + 030	750	17.1	9.6	8.2		10.5
SK0 + 010	820	20.0	11.1	13.5		14.0
NK0 + 010	650	14.7	8.4	8.9		10.0
NK0 + 030	650	13.3	8.7	13.1		11.3
NK0 + 050	630	16.5	8.7	9.4		10.7
NK0 + 075	410	14.7	8.4			10.7
NK0 + 090	490	18.8	8.2			11.4

Table 5.3 Average CBR values of each layer in the DCP test of the subgrade (top of the lime soil layer)

Test location	Penetration depth/mm	0 ~ 20/cm	20 ~ 40/cm	40 ~ 60/cm	60 ~ 80/cm	Weighted average
SK0 + 090	450	22.2	25.3			23.7
SK0 + 075	580	13.4	29.7	21.4		21.4
SK0 + 050	660	15.3	20.5	18.5		18.1
SK0 + 030	750	12.1	23.2	27.8		20.9
SK0 + 010	820	10.2	19.7	15.8		15.2
NK0 + 010	650	14.3	27.1	25.3		22.2
NK0 + 030	650	16.0	25.9	16.4		19.4
NK0 + 050	630	12.7	25.7	23.7		20.6
NK0 + 075	410	14.3	27.1			20.6
NK0 + 090	490	10.9	27.6			19.1

Table 5.4 DCP penetration values of each layer of the modified red mud test section subgrade (top of the red mud layer)

Test location	Penetration depth/mm	0 ~ 20/cm	20 ~ 40/cm	40 ~ 60/cm	60 ~ 80/cm	Weighted average
SK0 + 090	636	8.6	9.6	9.0		9.0
SK0 + 075	637	5.7	10.1	6.6		7.0
SK0 + 050	610	6.1	12.2	5.6		7.1
SK0 + 030	617	4.3	8.9	6.8		6.1
SK0 + 010	616	5.9	9.9	8.2		7.6
NK0 + 010	499	2.9	7.9			4.3
NK0 + 030	616	6.4	9.5	7.4		7.6
NK0 + 050	647	7.4	8.8	7.4		7.8
NK0 + 075	597	4.9	11.5			6.9
NK0 + 090	613	6.9	12.0	8.4		8.6

Table 5.5 Average CBR value of each layer of the modified red mud test section roadbed (top of the red mud layer)

Test location	Penetration depth/mm	0 ~ 20/cm	20 ~ 40/cm	40 ~ 60/cm	60 ~ 80/cm	Weighted average
SK0 + 090	636	26.2	23.3	25.0		24.8
SK0 + 075	637	41.3	22.0	35.5		32.8
SK0 + 050	610	38.2	17.7	42.4		32.5
SK0 + 030	617	57.2	25.3	34.0		38.6
SK0 + 010	616	40.3	22.5	27.7		30.1
NK0 + 010	499	87.5	29.0			57.5
NK0 + 030	616	36.3	23.5	31.2		30.3
NK0 + 050	647	31.1	25.5	31.1		29.2
NK0 + 075	597	49.5	18.9			33.8
NK0 + 090	613	33.5	18.1	26.9		26.1

It can be seen that the DCP comprehensive probing test, applied to the quality and performance evaluation of the roadbed fill, can more directly reflect the mechanical properties of the detected layer of the roadbed, and has good feasibility.

Using the above regression empirical formula, the CBR value of the layered modified solidified red mud road material can be calculated, see Table 5.5. The test results show that the CBR value of the modified solidified red mud road material roadbed is distributed between 25 and 60, and the measured rebound modulus of the top surface of the modified solidified red mud road material roadbed is distributed between 100 and 250; the CBR value of the lower lime soil layer is distributed between 10 and 15, and the measured rebound modulus of the top surface of the lime soil layer is distributed between 20 and 50 Mpa. The CBR obtained by the on-site DCP is consistent with the results of the on-site load plate test at the corresponding layer position, which fully meets the roadbed design control requirements.

(2) FWD Deflection

The direct test method of subgrade rebound modulus generally uses the subgrade load plate test. As a direct test technology of subgrade test layer rebound modulus, the obtained subgrade rebound modulus is relatively accurate. However, this test technology is relatively complex compared to FWD deflection testing, with a long cycle, large workload, high cost, and the stronger the plate structure of the layered structure, the greater the impact of the plate structure settlement basin on the load plate deformation test accuracy. At the same time, although deflection is an indirect test indicator for determining the rebound modulus of the pavement structure layer, a large amount of existing practical data provides many empirical methods for this indicator to evaluate road performance. The load plate test not only has the number of test samples that cannot reach the magnitude of the deflection test, but also has the influence of the layered structure plate settlement basin, so it is less used in the rebound modulus test on the surface of the highway fill layer, and more is the use of FWD deflection testing. It can be seen that deflection testing occupies a dominant position in road performance evaluation. The purpose of FWD deflection test results is to indirectly evaluate the rebound modulus of the road structure under vehicle load through the empirical relationship representing deflection.

The test was conducted using the Falling Weight Deflectometer (FWD) from Dynatest of Denmark. The FWD applies a pulse load to the road base by lifting and dropping a heavy hammer controlled by a computer-controlled hydraulic system. The size of the load can be adjusted within a considerable range by changing the weight of the hammer and the lifting height, and it is applied to the road base through a rigid disc. The deformation of the road base is measured by 5 ~ 9 sensors. The FWD is fast (about 40 s per test point), highly accurate (resolution is 0.1 μm), and simulates the dynamic effect of vehicle loads well. It is currently considered an ideal non-destructive road testing equipment. In particular, the FWD can accurately measure multi-point deflection, providing a basis for the back analysis of road structures (Fig. 5.9).

According to the test results, as shown in Tables 5.6, 5.7, 5.8, 5.9, 5.10 and 5.11, the deflection values of the top surface of the two layers of lime soil are all above 300, most of which exceed the detection limit of 327.7. The deflection of the top surface of the modified solidified red mud road material layer significantly decreases with the completion time of construction, from the initial 300 to about 50 after 7 days, which is far greater than the design standard of far deflection.

A systematic field test analysis was conducted through the load plate test of the modified solidified red mud subgrade, the DCP comprehensive dynamic probing, and the FWD deflection detection. Through the field test of the different ages of the red mud base concrete subgrade after construction, it can be seen that:

Fig. 5.9 On-site inspection diagram of the falling weight deflectometer

① The rebound modulus of the surface of the second layer of lime soil is less than 60 MPa. After the modified solidified red mud (CPS 6.5%) subgrade surface is compacted for 3 and 7 days, the measured rebound modulus of the surface is $E_0 >$ 130 MPa, which is much higher than the design requirements. With the increase of age, it can effectively promote the gradual increase of the rebound modulus of the modified solidified red mud subgrade.

② The CBR value of the modified solidified red mud subgrade is distributed between 25 and 60, and the measured rebound modulus of the top surface of the subgrade is distributed between 100 and 250; the CBR value of the lower lime soil layer is distributed between 10 and 15, and the measured rebound modulus of the top surface of the lime soil layer is distributed between 20 and 50 MPa. The CBR obtained from the on-site DCP is consistent with the results of the on-site load plate test at the corresponding layer position, which fully meets the subgrade design control requirements.

③ The on-site load plate test, DCP lightweight comprehensive probing, and FWD deflection test show that the rebound modulus of the 20 cm modified solidified red mud top surface is distributed between 120 and 200 MPa, the CBR value is distributed between 25 and 60, and the subgrade deflection is distributed between 50 and 100.

On-site load plate tests, DCP lightweight comprehensive probing, and FWD deflection tests show that the top rebound modulus of 20 cm modified solidified red mud road materials is distributed between 120 and 200 MPa, the CBR value is distributed between 25 and 60, and the subgrade deflection is distributed between 50 and 100.

Table 5.6 Summary of standard load deflection test results (unit 0.01 mm)

Test location	Test layer	Test time/days	NK0 + 6 m	NK0 + 2 m	SK0 + 2 m	SK0 + 6 m
K0 + 000	Top of lime soil	3	327.7	327.7	327.7	327.7
		3	324.2	321.3	310.1	310.1
	Top of red mud	1	320.6	307.7	308.9	327.7
		3	151.8	81.2	231.7	190.5
		7	100.7	202.3	317.1	310.8
K0 + 010	Top of lime soil	3	327.7	327.7	308.1	327.7
		3	327.7	327.7	314.8	279.4
	Top of red mud	1	293.3	320.1	322.0	311.3
		3	91.8	57.3	151.0	119.0
		7	55.4	52.4	149.7	144.3
K0 + 020	Grey soil top	3	327.7	327.7	327.7	327.7
		3	323.7	327.7	321.9	309.2
	Red mud top	1	306.4	315.1	317.6	306.3
		3	83.2	68.6	155.5	128.9
		7	53.8	57.0	327.7	133.6
K0 + 030	Grey soil top	3	327.7	327.7	327.7	327.7
		3	327.7	327.7	319.6	327.7
	Red mud top	1	298.8	313.3	316.1	298.5
		3	75.0	74.0	123.6	85.1
		7	54.9	66.2	305.7	50.1
K0 + 040	Grey soil top	3	327.7	327.7	327.7	320.4
		3	319.4	321.4	327.7	319.2
	Red mud top	1	291.8	314.1	327.7	308.8
		3	73.6	84.8	71.3	84.4
		7	57.6	55.2	86.7	50.1
K0 + 050	Grey soil top	3	312.2	327.7	327.7	318.8
		3	301.8	323.0	327.7	309.8
	Red mud top	1	273.0	302.1	308.4	295.8
		3	86.2	68.5	124.7	76.0
		7	63.3	58.1	155.6	50.2
K0 + 060	Grey soil top	3	313.8	318.7	311.2	327.7
		3	301.7	311.7	315.4	322.7
	Red mud top	1	305.0	281.4	312.8	303.7
		3	96.0	73.9	90.1	82.7
		7	–	76.2	78.7	54.2

(continued)

Table 5.6 (continued)

Test location	Test layer	Test time/ days	NK0 + 6 m	NK0 + 2 m	SK0 + 2 m	SK0 + 6 m
K0 + 070	Grey soil top	3	327.7	309.6	315.1	327.7
		3	308.7	303.1	323.6	312.0
	Red mud top	1	284.9	300.5	327.7	276.6
		3	99.3	83.8	99.5	88.2
		7	71.5	64.5	79.8	54.4
K0 + 080	Grey soil top	3	310.3	321.9	327.7	319.5
		3	304.6	327.7	317.0	327.7
	Red mud top	1	271.4	272.9	311.7	289.5
		3	97.8	73.3	144.0	82.7
		7	71.3	55.2	73.2	57.6
K0 + 090	Gray soil top	3	313.6	327.7	313.1	327.7
		3	324.2	316.3	327.7	327.7
	Red mud top	1	256.7	321.4	327.7	312.6
		3	107.2	81.6	119.5	223.8
		7	73.7	65.5	66.8	171.6

Note The gray deflection value exceeds the limit of the deflection device, greater than 327.7 (0.01 mm)

Moreover, as the age of the modified solidified red mud road material increases after filling, the index continues to rise. Therefore, the modified solidified red mud road material is used to fill the subgrade, the road performance is good, and it can meet the requirements of road stiffness.

5.3 Modified Solidified Red Mud Highway Subgrade

The verification of the reliability of the performance of the modified solidified red mud road material for highways and the feasibility of the construction process was implemented through the demonstration project of the modified red mud subgrade in the test section of the Jinan to Qingdao highway expansion project.

5.3.1 Implementation Background and Plan

The Jinan to Qingdao highway expansion project starts from the new mainline toll station in Zhujia Guanzhuang, Jimo City, Qingdao, and passes through Weifang, Zibo, and Binzhou from east to west, connecting with the G35 (Ji'nan-Guangzhou

Table 5.7 Results of the first layer of lime soil top deflection (3 days after construction)

Test location	Pressure/kPa	Load/kN	NK0 + 6 m	NK0 + 2 m	SK0 + 2 m	SK0 + 6 m
K0 + 000	462.0	32.7	319.4	319.8	313.3	327.7
	563.3	39.8	313.7	309.8	327.7	327.7
	659.0	46.6	327.7	327.7	327.7	327.7
K0 + 010	462.7	32.7	267.8	327.7	273.0	323.4
	578.3	40.9	282.1	327.7	316.1	327.7
	725.7	51.3	327.7	327.7	308.1	327.7
K0 + 020	457.0	32.3	320.6	315.8	319.0	327.7
	579.7	41.0	313.8	322.1	327.7	327.7
	729.0	51.5	327.7	327.7	327.7	327.7
K0 + 030	472.3	33.4	318.3	309.8	316.6	315.1
	584.0	41.3	308.8	327.7	312.3	320.2
	742.7	52.5	327.7	327.7	327.7	327.7
K0 + 040	476.7	33.7	316.5	327.7	317.4	252.1
	588.7	41.6	323.5	327.7	314.1	261.0
	750.3	53.0	327.7	327.7	327.7	320.4
K0 + 050	471.3	33.3	229.9	327.7	317.4	316.8
	584.7	41.3	253.8	327.7	327.7	312.2
	740.7	52.4	312.2	327.7	327.7	318.8
K0 + 060	459.0	32.4	276.1	305.4	273.6	327.7
	578.7	40.9	296.7	307.2	309.7	327.7
	728.0	51.5	313.8	318.7	311.2	327.7
K0 + 070	344.3	24.3	323.0	308.5	310.4	317.8
	600.3	42.4	320.6	306.0	313.7	325.0
	736.3	52.1	327.7	309.6	315.1	327.7
K0 + 080	482.0	34.1	272.3	314.6	327.7	317.8
	591.0	41.7	280.2	306.3	327.7	311.8
	765.7	54.1	310.3	321.9	327.7	319.5
K0 + 090	460.0	32.5	314.3	321.0	287.3	321.5
	579.7	41.0	299.8	320.6	316.1	321.3
	730.0	51.6	313.6	327.7	313.1	327.7

Expressway) at the Tangwang Interchange, and ends at the Jinan Zero Point Interchange. The total length of the route is 309.2 km, of which the Qingdao-Yin Expressway section is 286.947 km, and the Jinan-Guangzhou Expressway section is 22.210 km. The existing highway is a two-way four-lane road, which is expanded to an eight-lane highway standard, with a design speed of 120 km per hour, using a widening method of "mainly widening on both sides, supplemented by widening on one side or both sides separately", a standard subgrade width of 42 m, and a

Table 5.8 Results of the second layer of lime soil top surface bending sink test (3 days after construction)

Test location	Pressure/kPa	Load/kN	NK0 + 6 m	NK0 + 2 m	SK0 + 2 m	SK0 + 6 m
K0 + 000	462.0	32.7	287.9	279.5	249.1	245.9
	563.3	39.8	291.0	317.8	278.5	253.5
	659.0	46.6	324.2	321.3	310.1	310.1
K0 + 010	462.7	32.7	309.6	274.2	273.7	193.6
	578.3	40.9	327.7	311.9	314.4	209.4
	725.7	51.3	327.7	327.7	314.8	279.4
K0 + 020	457.0	32.3	296.3	259.4	283.1	224.0
	579.7	41.0	317.7	280.8	319.7	229.7
	729.0	51.5	327.7	327.7	321.9	309.2
K0 + 030	472.3	33.4	301.1	321.5	240.1	257.8
	584.0	41.3	317.5	327.7	266.5	272.7
	742.7	52.5	323.7	327.7	319.6	327.7
K0 + 040	476.7	33.7	257.2	246.2	263.3	244.1
	588.7	41.6	279.6	281.4	300.7	246.0
	750.3	53.0	327.7	321.4	327.7	319.2
K0 + 050	471.3	33.3	259.1	289.7	301.2	257.1
	584.7	41.3	281.3	327.7	327.7	268.5
	740.7	52.4	319.4	323.0	327.7	309.8
K0 + 060	459.0	32.4	272.9	290.8	268.6	252.4
	578.7	40.9	296.7	315.9	288.6	313.2
	728.0	51.5	301.8	311.7	315.4	322.7
K0 + 070	344.3	24.3	284.9	264.4	266.3	276.0
	600.3	42.4	302.5	297.1	296.7	308.1
	736.3	52.1	301.7	303.1	323.6	312.0
K0 + 080	482.0	34.1	189.2	297.2	295.2	287.7
	591.0	41.7	299.3	299.6	327.7	315.5
	765.7	54.1	308.7	327.7	317.0	327.7
K0 + 090	460.0	32.5	240.2	290.3	294.3	269.4
	579.7	41.0	254.3	305.0	319.3	305.3
	730.0	51.6	304.6	316.3	327.7	327.7

new and widened bridge culvert design. The vehicle load level adopts the highway I level. Due to the promotion of the highway's functional status, the uniqueness of the traffic flow, the diversity of technical standards and technical indicators, and the long construction period, the design and implementation stages involve complex and changing specialties, mainly including: traffic and construction coordination

Table 5.9 Results of the modified red mud test section top surface bending settlement (1 day after construction completion)

Test location	Pressure/kPa	Load/kN	NK0 + 6 m	NK0 + 2 m	SK0 + 2 m	SK0 + 6 m
K0 + 000	469.67	33.18	248.0	192.5	285.3	327.7
	579.00	40.94	241.8	245.0	314.7	327.7
	762.00	53.86	320.6	307.7	308.9	327.7
K0 + 010	489.33	34.60	176.0	226.4	263.3	248.2
	588.00	41.56	212.7	270.7	280.9	252.7
	755.33	53.39	293.3	320.1	322.0	311.3
K0 + 020	476.33	33.66	181.1	260.1	312.5	248.7
	588.67	41.60	222.7	264.7	318.4	254.8
	764.33	54.01	306.4	315.1	317.6	306.3
K0 + 030	466.67	32.97	189.5	271.6	252.9	246.7
	589.33	41.65	242.2	292.0	272.1	248.8
	763.67	53.98	298.8	313.3	316.1	298.5
K0 + 040	465.67	32.91	167.5	228.5	312.7	251.7
	579.00	40.93	206.9	276.0	310.8	255.8
	750.33	53.04	291.8	314.1	327.7	308.8
K0 + 050	454.67	32.13	151.7	250.2	260.5	225.1
	581.00	41.04	191.5	250.4	265.6	231.5
	765.67	54.09	273.0	302.1	308.4	295.8
K0 + 060	465.33	32.88	200.7	179.5	263.1	238.4
	579.00	40.94	232.2	195.2	273.6	231.3
	762.33	53.91	305.0	281.4	312.8	303.7
K0 + 070	457.00	32.31	168.3	212.4	243.7	215.5
	576.67	40.74	203.8	230.4	243.6	219.9
	759.00	53.64	284.9	300.5	327.7	276.6
K0 + 080	461.67	32.63	164.6	172.7	259.1	254.7
	579.67	40.96	199.2	215.1	275.9	241.5
	749.67	53.00	271.4	272.9	311.7	289.5
K0 + 090	453.00	32.03	159.6	220.5	251.2	296.9
	579.00	40.92	187.8	248.6	261.3	312.0
	744.00	52.57	256.7	321.4	327.7	312.6

organization without interrupting traffic construction, old subgrade pavement detection and evaluation, bridge and culvert structure detection evaluation and load test, comprehensive geological survey, engineering survey, safety and compliance evaluation, civil engineering design and construction (subgrade pavement, bridge splicing), traffic engineering and along-line facilities, greening and environmental protection, operation management system, engineering demolition and utilization, and project

Table 5.10 Results of the modified red mud test section top surface bending (3 days after construction)

Test location	Pressure/kPa	Load/kN	NK0 + 6 m	NK0 + 2 m	SK0 + 2 m	SK0 + 6 m
K0 + 000	491.00	34.71	108.8	59.8	178.5	127.2
	579.00	40.95	123.6	63.7	181.0	143.3
	697.00	49.23	151.8	81.2	231.7	190.5
K0 + 010	482.00	34.06	58.3	42.4	106.1	78.1
	566.00	39.99	70.6	45.3	114.6	89.4
	703.00	49.69	91.8	57.3	151.0	119.0
K0 + 020	482.00	34.07	52.7	47.5	113.5	88.8
	562.00	39.75	62.6	52.3	119.2	98.5
	701.00	49.54	83.2	68.6	155.5	128.9
K0 + 030	482.00	34.04	47.3	51.8	93.1	57.0
	563.00	39.82	58.6	57.2	99.9	65.4
	709.00	50.14	75.0	74.0	123.6	85.1
K0 + 040	491.00	34.69	46.3	56.6	49.8	57.4
	564.00	39.83	56.5	61.6	55.9	65.1
	700.00	49.44	73.6	84.8	71.3	84.4
K0 + 050	487.00	34.39	52.1	46.0	91.9	52.2
	570.00	40.26	65.8	52.0	102.3	58.5
	711.00	50.28	86.2	68.5	124.7	76.0
K0 + 060	476.00	33.67	61.2	64.4	58.7	55.2
	564.00	39.83	74.5	71.4	68.1	61.7
	700.00	49.48	96.0	73.9	90.1	82.7
K0 + 070	485.00	34.25	60.1	59.8	64.4	58.6
	568.00	40.11	74.3	64.7	74.9	67.4
	706.00	49.89	99.3	83.8	99.5	88.2
K0 + 080	479.00	33.88	57.2	50.8	97.3	57.1
	564.00	39.83	73.3	55.9	108.4	63.4
	699.00	49.41	97.8	73.3	144.0	82.7
K0 + 090	467.00	33.03	60.9	54.9	84.4	150.4
	564.00	39.83	80.0	62.6	92.5	176.1
	701.00	49.54	107.2	81.6	119.5	223.8

comprehensive management, etc. Therefore, the highway expansion is different from new projects, with its own technical requirements and characteristics, it needs to "make full use of existing traffic facilities", "reduce the impact on existing traffic flow", and also make the highway expansion adapt to new technical standards and the development trend of the traffic industry. This project is located in the economically developed plain area. Due to the busy traffic volume, special geographical location,

Table 5.11 Results of modified red mud test section top surface bending sink (7 days after construction)

Test location	Pressure/kPa	Load/kN	NK0 + 6 m	NK0 + 2 m	SK0 + 2 m	SK0 + 6 m
K0 + 000	501.7	35.5	73.7	84.2	207.9	261.4
	579.3	40.9	81.6	92.0	250.0	267.0
	753.7	53.3	100.7	202.3	317.1	310.8
K0 + 010	518.7	36.7	35.3	37.0	56.2	61.5
	569.7	40.3	41.3	44.0	66.9	71.8
	740.7	52.3	55.4	52.4	149.7	144.3
K0 + 020	499.7	35.3	36.1	42.1	290.3	65.4
	570.7	40.3	48.0	45.7	271.0	67.5
	745.7	52.7	53.8	57.0	327.7	133.6
K0 + 030	510.0	36.0	31.6	47.9	252.1	–
	570.7	40.4	41.1	58.6	256.1	–
	738.0	52.2	54.9	66.2	305.7	–
K0 + 040	487.7	34.5	35.1	32.7	54.9	38.5
	573.3	40.5	47.8	41.8	60.6	44.6
	757.0	53.5	57.6	55.2	86.7	50.1
K0 + 050	487.7	34.5	48.0	36.3	62.8	38.2
	569.7	40.3	55.7	47.3	69.1	44.5
	745.7	52.7	63.3	58.1	155.6	50.2
K0 + 060	519.0	36.7	–	47.8	52.8	44.2
	569.3	40.2	–	56.3	55.4	48.7
	730.7	51.6	–	76.2	78.7	54.2
K0 + 070	507.3	35.9	51.8	44.8	52.6	44.2
	572.7	40.5	63.6	52.4	60.3	46.3
	735.3	52.0	71.5	64.5	79.8	54.4
K0 + 080	484.3	34.2	46.5	34.0	45.7	45.6
	574.3	40.6	59.9	46.8	55.7	45.6
	759.0	53.6	71.3	55.2	73.2	57.6
K0 + 090	484.3	34.2	50.3	40.7	43.6	140.3
	576.0	40.7	62.0	47.3	51.6	149.9
	752.7	53.2	73.7	65.5	66.8	171.6

engineering conditions, and important social and economic impacts, this project has a special status and demonstration significance in the country.

Using solid industrial waste red mud as raw material to fill the roadbed, fully utilizing the large amount of industrial waste red mud along the line, the project cost is relatively low, which will reduce land occupation, avoid environmental pollution, and save construction funds. The specific advantages include the following aspects:

① The application of industrial solid waste red mud for highway subgrade filling and expansion projects has realized the effective utilization of industrial solid waste, achieved environmental protection and resource conservation, and developed a circular economy.
② Lightweight industrial waste materials (red mud) are used for highway bridge abutment backfill and subgrade filling at the splicing part, which can not only realize the utilization of waste materials, but also ensure the engineering quality and effectively reduce the post-construction settlement.
③ The recycling and reuse of industrial waste red mud can not only reduce the cost of highway construction, but also the light density of red mud can effectively reduce post-construction settlement, while avoiding the destruction of non-renewable land resources by highway construction, alleviating the ecological environment pressure brought by the centralized storage of red mud in the area, and benefiting the people along the line.
④ Promoting the use of red mud in highway filling can not only save a large amount of road filling soil and reduce engineering costs, but also provide a new attempt for the resource utilization of red mud, thereby saving a large amount of land resources occupied by stacking, and achieving a high degree of unity of "economy", "functionality" and "environmental friendliness" of subgrade filling materials.

The modified solidified red mud subgrade implemented in Binzhou City, Shandong Province, and Zibo City, Shandong Province, ensures the construction quality and progress, reduces the project cost, shortens the construction period, and on this basis, summarizes the construction process, detection methods and evaluation standards of red mud, and breaks through the key technologies of red mud subgrade strength water stability and pollution control. The technology is mature.

In order to promote the application of red mud and other solid industrial wastes in high-grade highways, cooperate with the steady progress of the Ministry of Transport's science and technology demonstration projects, ensure that the project quality of the plan meets the requirements of standard specifications, the construction plan of red mud and other new materials is feasible, and ensure the quality and environmental safety of road projects of red mud and other new materials. According to the "Highway Subgrade Construction Technical Specifications" (JTG F10.2006) requires, "Special material embankments, new processes, and new material subgrades must undergo test section construction, carry out related tests and inspections, collect relevant data indicators, and provide suggestions for modifying the construction design drawings, optimizing the construction organization plan and process." Combining the reconstruction project of the Jinan-Qingdao Expressway, the current situation, project location, regional geology, topographic conditions, and related specifications and technical standards, the test section construction plan is determined as follows:

Two sections of Bayer process red mud were used in the roadbed filling project in the fourth and fifth contract sections of the Jinan-Qingdao Expressway expansion project.

The red mud test Sect. 5.1 project pile number K242 + 200 ~ K242 + 420 left side, located in Zhangdian District, Zibo City, 20 km away from the second stockyard of Shandong Aluminum Industry Company, is within the economic transportation range. The project uses road mixing construction, the road mixer is used on the construction site to mix the red mud with the modified material evenly, and then it is leveled, compacted, and maintained to complete the filling construction. The length of this application project is 240 m, the width is 8 m, the roadbed filling thickness is 0.8 m, replacing the original roadbed 0.8 m of lime soil. A total of 3100 t of red mud was consumed and used, saving 2074 m^3 of soil extraction.

The red mud test Sect. 5.2 project pile number K289 + 090.5 ~ K294 + 385 right side, located in Qingyang Town, Zouping City, Binzhou City, the construction site is 17 km away from the red mud stockyard of Weiqiao Aluminum and Electricity, and is within the economic transportation range. The project uses factory mixing construction, the red mud is mixed evenly with the modified material in the mixing station through factory mixing, and then the modified red mud is transported to the roadbed filling site, spread, compacted, and maintained to complete the filling construction. The length of this application project is 5000 m, the roadbed width is 11 m, the filling thickness is 0.2 m, replacing the upper layer 0.2 m of lime soil. The mass ratio of red mud to composite modified material is 92:8, a total of 22,100 t of red mud was consumed and used, saving 14,850 m^3 of soil extraction.

5.3.2 Roadbed Construction Process

1. Construction Preparation

(1) Basic requirements for red mud

① The red mud material used has been modified, and the strength and other indicators meet the basic requirements for highway roadbed filling.

② The modified solidified red mud material meets the requirements of relevant national laws and regulations after testing, where the leaching concentration of heavy metals is better than the Class III groundwater quality classification standard of the "Groundwater Quality Standard" (GB/T14848.9), and the content of the main pollutants in the leachate is better than the first-level discharge standard in the "Integrated Wastewater Discharge Standard" (GB8978.1996).

(2) Construction preparation work

① Carry out construction surveying work according to current relevant specifications before construction, including guide wire, centerline, re-measurement of level points, spot checks and supplementary measurements of cross-sections, additional level points, etc.;

② Before the official construction of the red mud roadbed, a representative section is paved for testing. The length of the test section should not be less than 200 m. During and after the construction of the test section, strengthen the detection of relevant indicators. After completion, write a test section summary report in time;

③ When the red mud for road mixing construction needs to be temporarily stored, plan the storage location reasonably, take windproof and dust suppression measures, and set up surface isolation to prevent rainwater surface runoff erosion; before factory mixing construction, the red mud storage in the factory (station) should be equipped with isolation walls, rain shelters and other closed isolation facilities;

④ Strengthen safety and environmental protection training for workers before construction, focusing on effective protective measures and emergency plans for red mud dust on the faces, eyes, and respiratory organs of workers during construction;

⑤ Treat the site before construction, and the treated ground meets the requirements for flatness and compaction.

(3) Test

Carry out the following test items for red mud and modified red mud used for roadbed filling, and the methods are implemented according to the current "Highway Geotechnical Test Procedures", "Highway Engineering Inorganic Binder Stabilized Material Test Procedures", "Highway Pavement Base Construction Technical Specifications", etc.

Test items for red mud or modified red mud filling sections.

a. Particle analysis test
b. Liquid limit, plastic limit, plasticity index, natural consistency or liquidity index
c. Natural moisture content test
d. Heavy compaction test (determine maximum dry density, optimum moisture content)
e. Strength test (CBR value)
f. Unconfined compressive strength

If changes in the test characteristics of red mud are found during construction, all red mud test items should be supplemented in time.

(4) Red mud storage and transportation

① The adjustment of the moisture content of red mud should be carried out in the red mud stockyard. Overly wet red mud should be spread out to dry, while overly dry red mud should be watered and left to sit in the stockyard for 2–3 days before spreading. Depending on the transportation distance and climate conditions, the moisture content should be adjusted to slightly higher than the optimal moisture content range of 2–3%.

② The mode of transportation should be adapted to local conditions, using self-discharging trucks and mechanized loading to reduce the number of transfers and lower transportation costs.

③ Measures should be taken to prevent dust from red mud during transportation or spillage pollution, and cover measures should be taken if necessary.

(5) Test Section

Before large-scale construction of the red mud subgrade, according to the provisions of the "Technical Specifications for Highway Subgrade Construction", different construction machinery is used for the construction of the test road to familiarize and master the basic methods and procedures of red mud subgrade construction, so as to guide large-scale construction.

The location of the test road section should be representative in terms of geological conditions and cross-section types, and the length of the road section should not be less than 100 m. The red mud materials and construction machinery used in the test are the same as those used in the construction of the entire line in the future. The construction of the red mud subgrade test section includes the following contents:

① Red mud filling test, inspection report, etc.;
② Main parameters of red mud compaction process: mechanical combination; compaction machinery specifications, loose paving thickness, number of rolling passes, rolling speed; optimal moisture content and allowable deviation of moisture content during rolling, etc.;
③ Process quality control methods, indicators;
④ Quality evaluation indicators, standards;
⑤ Optimized construction organization plan and process;
⑥ Original records, process records;
⑦ Suggestions for modifications to the construction design drawings, etc.

During and after the construction of the test section, strengthen the detection of relevant indicators; after completion, write the test report in time. If defects in the subgrade design or imperfections in the relevant provisions of the specifications are found, suggestions should be submitted to the higher-level department for approval.

2. Red Mud Subgrade Construction

(1) General Requirements

The selection of red mud follows the principle of economic rationality, considering the nearby red mud stockyard along the project line.

The stock of the red mud stockyard meets the borrow volume of the design fill, minimizing the construction variability caused by changes in red mud.

The minimum bearing ratio and particle size of red mud fill meet the basic requirements of the specifications.

For the construction of red mud subgrade filling, start from the lowest point and fill in layers, compacting each layer; when the ground slope is gentler than 1:5 and the base is treated to meet the requirements, it can be directly filled in layers on the treated foundation. When the original ground longitudinal slope is greater than 12% or the cross slope is steeper than 1:5, steps should be dug according to the design requirements, or steps with a slope inward and greater than 4%, and a width greater than 2 m should be set.

If the fill subgrade is constructed in several sections, at the junction of the two sections, if they are not filled at the same time, the first filled section should be dug into steps with a width of not less than 2 m; if they are filled at the same time, the layers should overlap each other, and the overlap length should not be less than 2 m.

(2) Requirements and Selection of Construction Machinery

The construction of red mud subgrade adopts mechanized construction, and the construction machinery selected can freely move in the red mud, such as crawler bulldozers, crawler shovels, and vibratory rollers with front and rear wheel drive, road mixers, etc.

In addition to meeting the technical requirements for compaction, the selection of construction machinery also needs to be determined comprehensively based on factors such as project scale, site size, compaction machinery efficiency, and project schedule requirements. Develop detailed and feasible mechanical construction work plans to maximize the requirements of mechanical capacity and fully utilize mechanical efficiency.

The construction machinery of the red mud subgrade is managed according to the following regulations:

① Formulate machinery use and management systems and fuel supply systems;
② Formulate machinery allocation measures;
③ Compile mechanical construction organization, technical plans, and comprehensive mechanical flow operation procedures, and guide mechanical construction according to different project contents;
④ Formulate daily maintenance, regular maintenance, and machinery warranty systems for machinery to ensure the normal operation of machinery and fully play the role of machinery;
⑤ Set up temporary machine repair workshops and machinery repair sites, install safety protection facilities, and appropriately allocate maintenance personnel according to the number and condition of machinery.

(3) Requirements for Red Mud Subgrade Compaction

The construction of red mud subgrade filling uses vibratory compaction machinery to fill in horizontal layers and sections, and compact in layers. It is filled layer by layer upwards according to the full width of the cross-section.

The compaction thickness of red mud needs to be determined by testing. The compaction thickness should be determined according to the type of compaction machinery and the size of the compaction function, and a rolling test should be carried out in advance. Generally, for a medium-sized vibrating roller of 20 ~ 30 t, the compaction thickness of each layer is controlled within 20 cm. For a medium-sized vibrating sheep foot roller or a heavy-duty vibrating roller of 40 ~ 50 t, the compaction thickness of each layer should not exceed 30 cm.

The red mud must be rolled in time after mixing and spreading to achieve immediate compaction after spreading. This is to prevent changes in moisture from affecting the compaction effect. When rolling, the red mud should be within the optimal moisture content range.

The width of each compacted fill layer should not be less than the design width.

The construction temperature of the red mud subgrade should be controlled above 0 °C. This is to prevent a large amount of ice crystals in the red mud from affecting the compaction effect and project quality.

(4) Road mixing construction

The process flow of road mixing construction: ① Preparation before construction → ② Surface treatment → ③ Inspection and approval → ④ Transportation of red mud → ⑤ Leveling → ⑥ Spreading of red mud modified materials → ⑦ Mixing with road mixer → ⑧ Stabilization → ⑨ Vibration compaction → ⑩ Final pressure → ⑪ Inspection and approval → ⑫ Curing → ⑬ Construction of the next layer (return to step ④ and repeat until the specified fill elevation is reached).

The original surface should be compacted with a vibrating roller. The compacted foundation can only proceed with the construction of the red mud subgrade after it has been tested and meets the foundation compaction standards specified in the specifications. If the test fails, additional pressure should be applied until it passes.

Leveling. The red mud transported to the fill section should be spread and leveled with a bulldozer or leveled with a bulldozer in conjunction with a grader. After spreading and leveling, the elevation should be measured at fixed points before and after filling to control the loose thickness. The loose thickness of each layer should be determined based on the preliminary test section.

Adding red mud modified materials. The moisture content of the filled soil body should be tracked and detected at all times. When it is generally 2 ~ 3 percentage points higher than the optimal moisture content, the addition of modified materials can be arranged. The grid method or other suitable spreading methods should be used to spread the red mud modified materials, and the amount of modifier added should be strictly controlled.

Mixing with a road mixer. After the modified materials are spread, a large-scale mixer should be used for mixing. The mixing should ensure no interlayer, the mixing width overlaps 30 cm each time, and the mixing depth and overlap width should be checked at any time during the mixing process. Check whether the curing material and red mud are uniform, and whether the soil particles meet the design and specification requirements (the maximum particle size is not more than 50 mm). If the requirements are not met, the second mixing should be carried out. If the second mixing still cannot meet the requirements, the third mixing should be carried out.

Stabilization. The compaction of red mud should follow the principle of light first and heavy later. First, use tracked equipment or light compaction machinery for stabilization. During stabilization, follow the compaction process of general earthwork subgrade, roll from the edge of the subgrade to the inside, the overlap width of the rolling track should not be less than 1/2 of the single wheel width, and one pass is considered when a single side track covers a working surface, stabilize twice.

The following requirements should be followed when compacting the red mud subgrade with vibration.

① Vibration compaction is carried out with a medium or heavy vibrating roller of 20 ~ 50 t, the rolling speed is slow first and then fast, and strong vibration is used for vibrating rolling;
② The number of compaction passes should be controlled according to the number of compaction passes determined by the red mud subgrade test section. If the controlled number of compaction passes exceeds 10 times, consider reducing the thickness of the fill layer. Only after passing the compaction test can it be transferred to the next process. If it fails, additional pressure should be applied and the test should be done again until it passes;
③ In the absence of test section compaction pass data, it is recommended that the number of passes of the vibrating roller should not be less than 6 times;
④ The rolling tracks should overlap each other, and the overlap width should not be less than 1/3. One pass is considered when a track covers a working surface;
⑤ The driving speed of the roller during rolling should not exceed 2.5 km/h; during rolling, the straight section is carried out from both sides to the middle, the small radius curve section is from the inside to the outside, and it is carried out in a longitudinal advance and retreat manner. The front and rear adjacent sections (the leveling pre-pressure section before the rolling section and the inspection section after it) should overlap longitudinally by more than 2.0 m, to achieve no missed pressure, no dead corners, and ensure uniform rolling;
⑥ For parts of the subgrade where the roller cannot be used for rolling, such as around inspection wells, behind bridge piers or retaining walls, and backfilling of trenches, use small handheld or vibrating rollers, frog compactors for rolling and compaction to achieve the specified compaction degree.

Final pressure. The final pressure compaction process is the same as stabilization, the overlap width of the track should not be less than 1/2 of the single wheel width, and the final pressure is applied twice.

Check. After the compaction of the red mud subgrade is completed, the compaction degree should be checked in time, and the sand filling method should be preferred for the compaction degree check. The compaction degree should meet the standard specified in the specification. If it does not meet the requirements, it should be re-compacted in time until it is qualified before the next process can be carried out. The sampling location and depth for the check should also comply with the specifications.

After all the red mud subgrade construction is completed and the inspection meets the requirements, continue to fill the upper layer.

Before the construction of the next layer, control the direction and speed of the dump truck, and do not turn around, drive at high speed, brake suddenly, etc. on the top surface of the lower layer of red mud to avoid loosening the compacted layer of red mud.

(5) Centralized factory mixing method construction

Modified red mud materials can be centrally mixed at the central plant station using factory mixing equipment. The centralized factory mixing method is used to produce modified red mud mixture (red mud base road concrete) for highways and first-class roads.

Before the official factory mixing construction of red mud mixed materials, choose the appropriate centralized factory mixing equipment and must first debug the used factory mixing equipment to meet the requirements of red mud mixing.

The centralized factory mixed red mud mixture meets the following requirements:

① The moisture content of the factory mixed red mud should be controlled slightly higher than the optimum moisture content + 3%, and the moisture content of the mixture should not be less than the optimum moisture content when it is transported to the site for spreading and rolling;
② The red mud clumps are uniformly crushed to a maximum size not exceeding 50 mm;
③ The dosage of modified curing material is accurately controlled, and the error is not more than ± 0.5%;
④ The mixing is uniform, without interlayers.

In humid and rainy areas or during the rainy season in other areas, take measures to protect the red mud and prevent rain or immersion.

The factory-mixed qualified red mud mixture should be transported to the paving site as soon as possible. Cover during transportation to reduce moisture loss and spillage.

For highways and first-class roads, use asphalt concrete pavers or dedicated stabilized soil pavers to spread the red mud mixture.

The production capacity of the centralized factory mixing equipment and the paver match each other and achieve continuous paving. When the production capacity of the centralized factory mixing equipment is small, use low-speed paving when using the paver to spread the red mud mixture, to reduce the situation where the paver is waiting for materials.

For secondary and lower grade roads, bulldozers or graders can be used to spread the mixture. The transportation of the red mud mixture starts from one end of the subgrade and gradually advances forward, spreading the material with a bulldozer, leveling with a grader, and carrying out the full width of the half-width subgrade at one time. An automatic grader can also be used to spread the mixture in the following steps:

① Calculate the spreading area of each truck of mixture according to the thickness of the paving layer and the required compaction dry density;
② Unload the mixture evenly in the middle of the road width, and when the road width is wide, the mixture can also be unloaded into two rows;
③ Use a grader to spread the mixture evenly according to the loose spreading thickness.

After the red mud mixture is spread, compact it in time according to the principle of light first and heavy later. First use a light roller to follow the paver for timely rolling, and then use a 20 ~ 50 t medium and heavy vibratory roller or tire roller to continue rolling and compacting.

The rolling and inspection acceptance after the completion of the centralized factory mixed red mud spreading are the same as the road mixing method construction.

The transverse joints during the construction of the centralized factory mixing method must meet the following requirements:

① When spreading the mixture with a paver, it should not be interrupted. If the interruption time exceeds 4 h, set a transverse joint, and the paver should drive away from the end of the mixture;
② Manually tidy up the end mixture with appropriate moisture content, put two square woods close to the mixture, and the height of the square woods should be the same as the compacted thickness of the mixture; level the mixture close to the square woods;
③ The paver returns to the end of the already compacted layer and starts spreading the mixture again;
④ If the spreading is interrupted and the transverse joint is not handled according to the above method, and the interruption time has exceeded 4 h, then shovel the mixture near the paver and the uncompacted mixture, and dig the end that has been rolled and compacted and the elevation and flatness meet the requirements. The end is a section perpendicular to the centerline of the road and perpendicular to the bottom, and then spread the new mixture.

Try to avoid longitudinal joints. The base layer of highways and first-class roads is spread in two widths, using two pavers, one before and one after, about 10 ~ 20 m apart, synchronously spreading the mixture forward, and rolling together.

In cases where vertical seams cannot be avoided, the seams must be vertically joined, diagonal joining is strictly prohibited, and the following regulations must be complied with:

① When paving the previous piece, use square wood or steel templates for support on the side close to the center. The height of the square wood or steel template should be the same as the compacted thickness of the red mud mixture layer;
② After curing, remove the supporting square wood or steel template before paving another piece.

(6) Maintenance of red mud subgrade

The red mud subgrade that meets the quality requirements after inspection should be maintained by covering or sprinkling water. The number of times of watering each day should be determined according to the climate, and the surface of the red mud layer should be kept moist during the entire curing period.

After the construction and rolling of the red mud mixture is completed, the upper layer of red mud mixture is paved 3 days after curing.

During the curing period, traffic is closed or sections where the upper layer of red mud cannot be immediately paved are prohibited or restricted from heavy vehicle traffic. If it is necessary to open traffic, the speed limit should be below 30 km/h.

When the red mud subgrade cannot continue construction for a long time due to a fault, cover the surface with soil or felt, make good road arch cross slope for surface drainage. Maintain the integrity of the subgrade drainage facilities and strengthen daily maintenance work. Before resuming work, if the embankment is damaged, it needs to be repaired according to design requirements.

5.3.3 Implementation and Construction Quality Control

The Bayer method red mud subgrade test section of the Jiqing Expressway expansion project is constructed in two sections, as shown in Fig. 5.10. The red mud test Sect. 5.1 is located at K242 + 200 ~ K242 + 420 left side, the project is 240 m long, 8 m wide, the subgrade fill thickness is 0.8 m, replacing the original subgrade 0.8 m lime soil. A total of 3100 t of red mud was consumed and utilized, saving 2074 m 3 of soil extraction. The road mixing method was used for construction. The red mud and modified materials were mixed evenly with a road mixer at the construction site, and then leveled, compacted, and cured to complete the filling construction.

Red mud test Sect. 5.2 is located at K289 + 090.5 ~ K294 + 385 right side, the project is 5000 m long, the subgrade width is 11 m, the fill thickness is 0.2 m, replacing the upper 0.2 m lime soil, the mass ratio of red mud to modified material is 92:8, a total of 22,100 t of Bayer method red mud was consumed and utilized, saving 14,850 m 3 of soil extraction. The factory mixing method was used for construction. The Bayer method red mud was mixed evenly with the modified materials in the mixing station through the factory mixing method, and then the modified Bayer method red mud was transported to the subgrade filling site, spread, compacted, and cured to complete the filling construction.

Key control of the process flow during construction.

(1) Raw material preparation

Before construction, samples of red mud are taken for natural moisture content, liquid plastic limit, CBR value, standard compaction, unconfined compressive strength test, to fully grasp the index values of its main physical and mechanical properties and determine stable parameters. When the source of ore or production process changes, retesting is required.

(2) Optimal moisture content control

The optimal moisture content and maximum dry density of stable red mud are determined through heavy standard compaction tests. After the red mud is paved, the moisture content test must be carried out before entering the next process. The on-site measured moisture content should be controlled to be 2 ~ 3 percentage points higher than the optimal moisture content. If the moisture content is insufficient, water should be added in time, and if the moisture content is too high, appropriate turning and drying should be carried out.

(a) Paving (b) Leveling

(c) Compaction (d) Curing

(e) Density detection (f) On-site core extraction unconfined compressive strength test

(h) On-site FWD deflection detection (i) On-site bearing plate rebound modulus test

Fig. 5.10 Construction process of red mud subgrade test section

(3) Layer thickness

The control of the fill layer thickness directly affects the uniformity of the stable material mixing and the compaction quality. The compacted fill layer thickness is generally 25 ~ 30 cm, and the loose paving thickness is 30 ~ 35 cm. The layer thickness should be uniform, the surface should be flat, measured by the pit method or level, the loose paving thickness should be checked, and the allowable error should be controlled within ± 3 cm.

(4) Mixing uniformity

According to the optimal dosage of the stable material design, the mixing station should accurately control the dosage of the composite stable material, and the error should not be more than ± 0.5%. The surface of the stable red mud after mixing should be uniform in color, without gray lumps and bars, no flower surface and "raw soil" blocks, and the red mud particles should be controlled to be less than 5 mm. The EDTA titration method is used to detect the dosage of stable materials, and the deviation between the measured content and the standard value should not exceed the control range of 0.5 ~ 1%.

(5) Compaction Time Control

Since the surface of red mud easily dries and loses water, to ensure the uniformity within the range of the spreading layer thickness, the compaction process should be completed within 4 h after the mixing is finished. After the mixing is uniform, use a road roller to quickly statically press to prevent water loss. Then use a 15 ~ 20 t vibrating roller to weakly vibrate once, strongly vibrate 3 ~ 4 times, and finally statically press once more. After the compaction is completed and the inspection is qualified, curing should be carried out immediately. If a multi-layer continuous filling construction process is adopted, it is necessary to ensure that the construction layer can be compacted within the 4-h compaction time of the previous filling layer at the corresponding position.

(6) Curing Conditions and Age

Moist curing is a key process for the formation of stable Bayer method subgrade strength and the stable reaction of metal ion solidification. After each layer of layered filling construction is compacted or the last layer of multi-layer continuous filling construction is compacted, felt should be used to cover and sprinkle water for curing, and the curing period should not be less than 2 days. During the curing period, the surface should be kept moist. In summer, water should be sprinkled at least 4 times a day, and in winter, when the minimum temperature is below 0 °C, 20 cm thick soil should be covered. Traffic should be restricted during the curing period, the speed of the water truck should not exceed 15 km/h, and heavy vehicles are prohibited.

5.3.4 Subgrade Performance Evaluation

1. Basic Requirements

(1) General Provisions

According to relevant specifications, establish and improve the quality management system and construction quality management. The management system implements a quality management system that combines self-inspection by the construction unit and random inspection by the quality supervision department. The inspection should be methodical, with complete original data and reliable data. Remedial measures must be taken for parts that fail the inspection to meet the specified requirements, and only after passing the inspection can the next process be carried out. After the subgrade construction is completed, a comprehensive inspection and acceptance should be carried out.

Construction quality management includes standard tests of materials used, intermediate inspections, and final inspections. The engineering quality inspection and evaluation standards are implemented in accordance with the current "Highway Engineering Quality Inspection and Evaluation Standards" for earthwork subgrades.

(2) Material Test

Before organizing on-site construction and when the characteristics of the red mud material source change significantly, basic performance tests should be carried out on the materials to be used to evaluate whether the quality of the materials meets the requirements.

The performance test of the soil used for the soil slope protection and lime soil sealing layer should be carried out in accordance with the "Highway Geotechnical Test Procedures" issued by the department; when sand and gravel materials, slag materials, etc. are selected for the isolation layer and drainage blind ditch, the maximum particle size should not exceed 8 cm, and the fine material content below 5 mm should not exceed 50%, and the mud content should be less than 5%.

(3) Compaction Test

① The compaction degree of the red mud subgrade should preferably be tested by the sand filling method, and the sampling location is at the bottom of each compaction layer. When testing with the ring knife method, the middle of the ring knife is at 1/2 depth of the compaction layer thickness.
② During the construction process, the compaction degree of each compaction layer should be tested, and the detection frequency is at least 2 points for every 1000 m^2, and 2 points for less than 1000 m^2, and the inspection points can be increased as needed.

2. On-site Detection Scheme and Results

In order to evaluate the construction quality and long-term performance of the stabilized red mud test road, core strength tests were conducted on the red mud subgrade test section and the conventional lime soil subgrade test section, and the top of the subgrade was subjected to bearing plate rebound modulus, FWD, PFWD, DCP and compaction tests. The test was carried out in a "point-to-point" corresponding manner, with the centerline of the test road as the boundary, ten test points were arranged at intervals of 10 m at the midpoint of the left and right roads, a total of 20 points were tested, and 3 and 7d were used as control ages. The data of different ages of the same test were compared and analyzed to evaluate the road performance of the improved Bayer method red mud subgrade.

(1) On-site Core Unconfined Compressive Strength Test

"Point-to-point" corresponding on-site coring, the diameter of the drill bit used for coring is 150 mm, the on-site coring time is selected to be 27 days after on-site rolling, sprinkling water and laying felt for curing, to ensure that the unconfined compressive strength test is carried out on the 28th day. The specific test process is shown in Figs. 5.11 and 5.12.

As shown in Fig. 5.11, the extracted core sample is complete in appearance, well connected between layers, and solid on the surface. The results of the unconfined compressive strength test of the test specimens after cutting and forming are shown

Fig. 5.11 On-site coring diagram

Fig. 5.12 Indoor cutting and testing

in Table 5.12; the SPSS software is used to analyze the test results, and the specific results are shown in Table 5.13 (Fig. 5.13).

The average unconfined compressive strength of the on-site core is 2.4 MPa, which meets the specification requirements. The unconfined compressive strength of the on-site core sample is within the 95% confidence interval of 2.06 ~ 2.77 MPa, with a variance of 0.246 and a standard deviation of 0.49621.

(2) On-site Bearing Plate Rebound Modulus Test

As an evaluation index of subgrade bearing capacity and a design index of asphalt pavement in China, the on-site bearing plate method rebound modulus has great significance for the evaluation of subgrade quality. According to the test method of T0943.2008 in JTG E60.2008, the rebound modulus test is carried out on the corresponding points.

The rebound modulus results of the same point under different ages are shown in Table 5.14 and Fig. 5.14. The results show that with the increase of age, the on-site bearing plate method rebound modulus shows a growing trend at the same point, the 3d on-site bearing plate rebound modulus is between 63 and 154 MPa, and the 7d rebound modulus is between 110 and 230 MPa. This indicates that the on-site rebound modulus of the modified Bayer method red mud subgrade is not only much higher than the standard design standard, but also has a certain amount of increase with the increase of age, ensuring the stiffness and bearing capacity of the subgrade top.

(3) PFWD dynamic deformation modulus

PFWD, also known as the portable falling weight deflectometer, has the advantages of easy operation, high accuracy, and little influence from human operation compared to traditional testing equipment. The dynamic deformation modulus at different points

Table 5.12 On-site core unconfined compressive strength table

Left frame position	28d strength (MPa)	Right frame position	28d strength (MPa)
C1	1.76	D1	3.20
C2	2.37	D2	2.14
C3	2.81	D3	2.73
C4	1.98	D4	1.98
C5	2.13	D5	3.03

Table 5.13 Descriptive statistics table of on-site sampling test data

Mean	Median	Variance	Standard deviation	Maximum value	Minimum value	95% Confidence Interval	
						Lower limit	Upper limit
2.4130	2.2550	0.246	0.49621	3.2	1.76	2.0580	2.7680

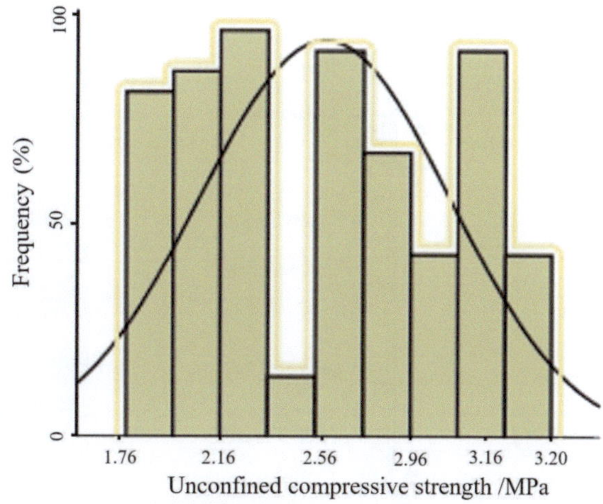

Fig. 5.13 Normal distribution histogram of on-site core unconfined compressive strength

Table 5.14 Summary table of on-site rebound modulus data of test section

Test point	3d rebound modulus (MPa)	7d rebound modulus (MPa)	Test point	3d rebound modulus (MPa)	7d rebound modulus (MPa)
L1	79.1	109.1	R1	91.1	134.1
L2	103.1	133.1	R2	83.1	123.1
L3	80.9	120.9	R3	90.9	117.9
L4	101.7	131.7	R4	103.7	129.7
L5	99.1	119.1	R5	89.1	132.1
L6	104.2	134.2	R6	84.2	115.2
L7	93.3	117.3	R7	103.3	139.3
L8	87.8	106.8	R8	97.8	134.8
L9	96.9	106.9	R9	96.9	128.9
L10	89.4	108.4	R10	109.4	144.4

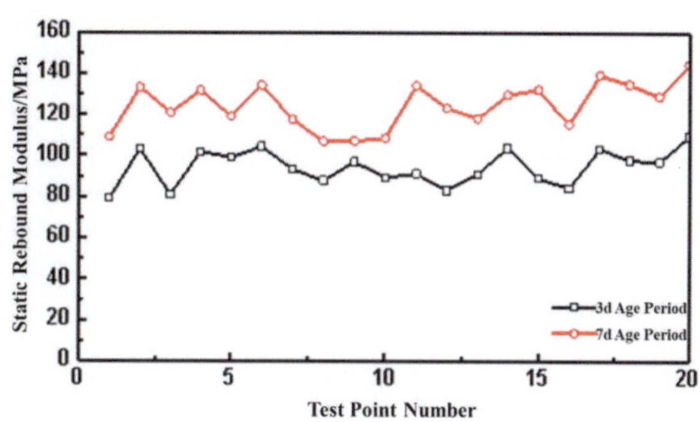

Fig. 5.14 Comparison of rebound modulus at the same point at different ages

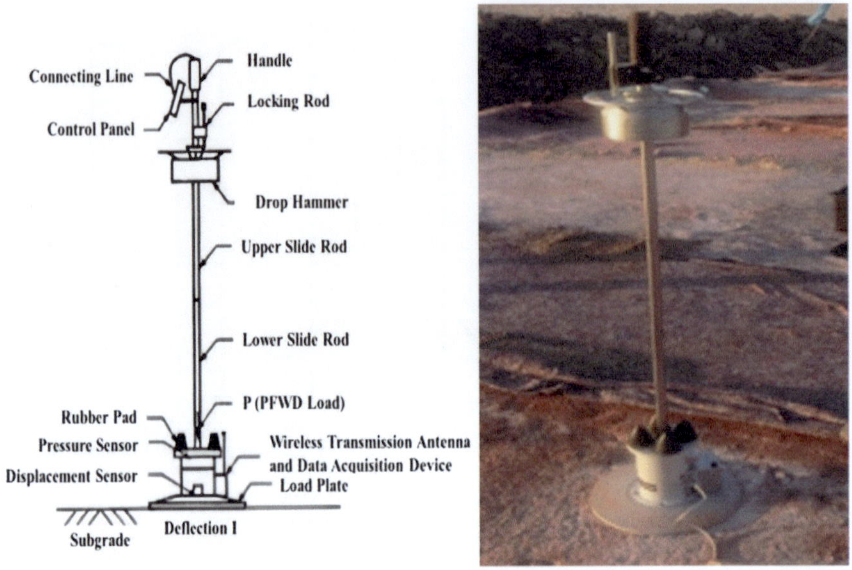

Fig. 5.15 Schematic diagram of PFWD

is calculated from the dynamic impact load and dynamic deflection value recorded during the PFWD test. The calculation formula is as follows:

$$E_p = \frac{\pi}{4} \cdot \frac{2p\delta(1 - \mu^2)}{l} \tag{5.4}$$

where E_p—Dynamic deformation modulus; δ—Radius of the bearing plate; p—Dynamic load; μ—Poisson's ratio; l—Dynamic deflection value.

The working schematic and field experiment of PFWD are shown in Fig. 5.15.

The dynamic deformation modulus values at different ages are shown in Table 5.15 and Fig. 5.16. The results show that the dynamic deformation modulus at the same point shows a growing trend with the increase of age. Although the calculation formula principle is the same, the dynamic deformation modulus at the same point is larger than the rebound modulus. This is because the PFWD impact load is small and the impact time is short, and no plastic deformation is produced in the deformation process of the subgrade. However, the rebound modulus has a large load and a long action time, which produces plastic deformation during the loading process, resulting in the difference in modulus.

(4) FWD Dynamic Deflection

The FWD test used in this study is a drop hammer deflection meter from Dynatest of Denmark. Its working principle is to lift and drop a heavy hammer under the control of a computer-controlled hydraulic system to apply a pulse load to the road base. FWD has the characteristics of fast measurement speed and high accuracy, and it

Table 5.15 Dynamic deformation modulus calculation table

Test point	3d dynamic deformation modulus (MPa)	7d Dynamic modulus (MPa)	Test point	3d Dynamic modulus (MPA)	7d Dynamic modulus (MPa)
L1	66	86	R1	98.2	138.2
L2	123	166	R2	67.1	87.1
L3	98.1	128.1	R3	109.5	149.5
L4	132.3	172.3	R4	138	178
L5	116.9	156.9	R5	120.6	150.6
L6	179.5	239.5	R6	99.6	149.6
L7	103.5	143.5	R7	161.1	201.1
L8	127.6	127.6	R8	109.8	149.8
L9	153.1	185	R9	117.2	147.2
L10	124.4	164.4	R10	147.9	197.9

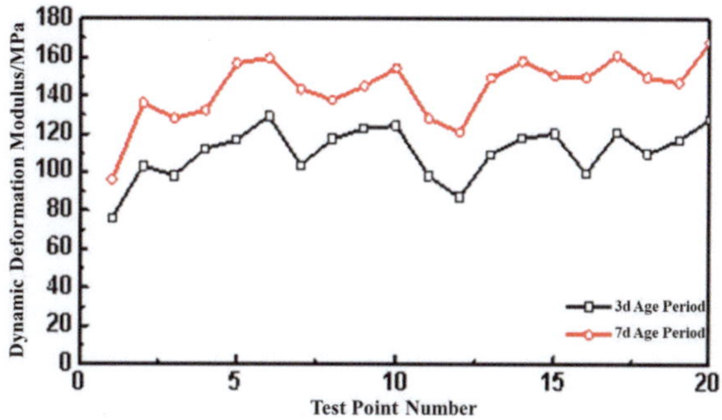

Fig. 5.16 Comparison of dynamic modulus at different ages

simulates the dynamic effect of vehicle load well, suitable for long-term tracking observation over a large range (Fig. 5.17).

The results of the standard load deflection test at different ages are shown in Table 5.16 and Fig. 5.18. The results show that with the increase of age, the deflection at the same test point shows a decreasing trend. This is because the improved red mud has a certain increase in strength and stiffness with the increase of curing age, which makes the vertical deformation at the same point under the same impact load decrease, thus leading to the decrease of deflection value.

Tables 5.17 and 5.18 show the bend sink results of the top of the modified red mud roadbed and the top of each layer in the demonstration section. The static rebound modulus of the red mud subgrade test section is greater than 120 MPa, and the bend sink values are mostly within the range of 50 ~ 100 (0.01 mm). The strength and modulus of the subgrade are higher than those of traditional lime or cement improved soil subgrade, providing application references for the design and quality control of modified red mud subgrade in highways.

Fig. 5.17 On-site FWD deflection test

Table 5.16 Standard load deflection test results

Test point	3d deflection value (0.01 mm)	7d deflection value (0.01 mm)	Test point	3d bend sink value (0.01 mm)	7d bend sink value (0.01 mm)
L1	111.8	100.7	R1	100.2	70.1
L2	91.8	60.8	R2	100.5	53.3
L3	83.2	65.4	R3	119	103.2
L4	85	54.3	R4	98.9	68.21
L5	93.6	59.8	R5	85.1	54.2
L6	80.2	43.6	R6	94.4	71.5
L7	96	74.9	R7	96	64.4
L8	99.3	80.1	R8	92.7	53.3
L9	92.8	67.6	R9	98.2	67.6
L10	111.8	100.7	R10	92.7	63.7

(5) On-site DCP test

The DCP in-situ testing technology is applied to the quality inspection of improved red mud subgrade. The test results are shown in Table 5.19. The SPSS software is used for data analysis, as shown in Tables 5.20, 5.21 and 5.22. With the increase of age, the mean and median of the rebound modulus, CBR value, and UCC of the improved red mud subgrade increase, while the variance and standard deviation decrease. Moreover, the upper and lower limits of the three indicators increase within

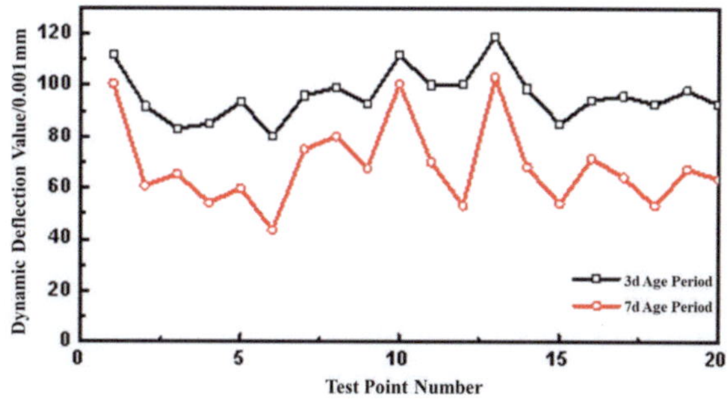

Fig. 5.18 Comparison of bend sink values at different ages

Table 5.17 Rebound modulus of modified red mud roadbed surface

Demonstration section pile number	K242 + 010	K242 + 060	K242 + 080	K242 + 110	K242 + 180	K242 + 020	K242 + 150
Rebound modulus (MPa)	223.9	121.9	155.4	117.5	133.0	249	216.3

Table 5.18 Results of the top surface bending of each layer of modified red mud

Position	Layer	No. 1	No. 2	No. 3	No. 4	No. 5	No. 6	No. 7	No. 8	No. 9
Inside (east to west)	First layer (1d)	303.1	251.9	256.1	96.0	327.6	327.6	327.6	327.6	327.6
	Second layer (1d)	327.6	153.0	200.8	327.6	181.5	327.6	207.1	101.7	327.6
	Third layer (1d)	212.2	327.6	327.6	80.0	164.4	196.7	162.3	–	–
	Fourth layer (1d)	302.7	242.7	248.7	311.7	227.3	327.6	227.6	196.2	173.2
	Fourth layer (2d)	65.0	57.1	101.1	68.2	97.4	133.0	266.1	195.5	139.5
	Fourth layer (3d)	125.7	78.9	60.0	39.6	88.1	58.8	116.6	95.6	114.2
Outside (east to west)	First layer (1d)	163.1	169.6	219.4	145.0	202.4	157.2	137.0	161.6	327.6
	Second layer (1d)	317.4	139.4	327.6	158.0	184.4	307.5	122.9	237.0	226.5
	Third layer (1d)	303.2	327.6	131.9	327.6	262.0	327.6	89.2	94.1	–
	Fourth layer (1d)	69.4	58.0	57.5	233.6	86.5	86.9	92.4	214.9	100.3
	Fourth layer (2d)	56.8	42.6	46.8	188.6	41.7	61.7	107.7	47.2	68.1
	Fourth layer (3d)	53.4	42.9	98.8	94.3	45.9	110.5	94.0	83.4	55.1

the 95% confidence interval. This indicates that as the age increases, the quality control indicators and design indicators of the improved red mud subgrade increase, the dispersion decreases, the overall uniformity of the subgrade is improved, and the performance of the subgrade is significantly improved (Figs. 5.19, 5.20 and 5.21).

(6) On-site compaction test

The compaction degree of the subgrade, as a quality control index in the construction process of the subgrade, has certain guiding significance for the road performance

Table 5.19 Summary of DCP results

Test point	3d E (MPa)	7d E (MPa)	3d CBR (%)	7d CBR (%)	3d UCC (KPa)	7d UCC (KPa)
1	103.37	108.03	23.71	24.83	233.21	244.69
2	93.69	140.69	21.38	32.79	209.56	326.17
3	79.84	139.68	18.06	32.55	176.07	323.64
4	91.77	164.37	20.91	38.63	204.87	386.34
5	67.62	129.54	15.17	30.06	146.94	298.13
6	96.91	239.67	22.15	57.45	217.40	582.44
7	85.41	130.43	19.39	30.28	189.46	300.37
8	90.48	126.18	20.61	29.24	201.75	289.74
9	90.48	144.83	20.61	33.81	201.75	336.63
10	84.08	113.13	19.07	26.07	186.26	257.27
11	108.03	127.18	24.83	29.49	244.69	292.23
12	140.69	121.44	32.79	28.09	326.17	277.90
13	139.68	127.42	32.55	29.55	323.64	292.84
14	164.37	146.67	38.63	34.26	386.34	341.29
15	129.54	115.59	30.06	26.67	298.13	263.38
16	239.67	150.99	57.45	35.33	582.44	352.25
17	130.43	176.35	30.28	41.60	300.37	417.10
18	126.18	157.81	29.24	37.01	289.74	369.60
19	144.83	115.97	33.81	26.76	336.63	264.30
20	113.13	148.11	26.07	34.62	257.27	344.95

Table 5.20 Analysis of rebound modulus E at different ages

Mean	Median	Variance	Standard deviation	Maximum	Minimum	95% confidence interval	
						Lower limit	Upper limit
116.01	105.7	1511.395	38.88	239.67	67.62	97.82	134.2
141.20	135.06	861.52	29.35	249.21	108.03	127.47	154.94

Table 5.21 Analysis of different period CBR (%)

Mean	Median	Variance	Standard deviation	Maximum	Minimum	95% confidence interval	
						Lower limit	Upper limit
26.84	24.27	90.99	9.54	57.45	15.17	22.37	31.30
32.96	31.426	52.75	7.26	61.23	24.83	29.56	36.35

Table 5.22 Analysis table of different age UCC (KPa)

Mean	Median	Variance	Standard deviation	Maximum value	Minimum value	95% confidence interval	
						Lower limit	Upper limit
265.64	238.95	9628.81	98.13	534.12	146.94	219.71	311.56
328.06	312.01	5646.49	75.14	582.44	244.69	292.905	363.23

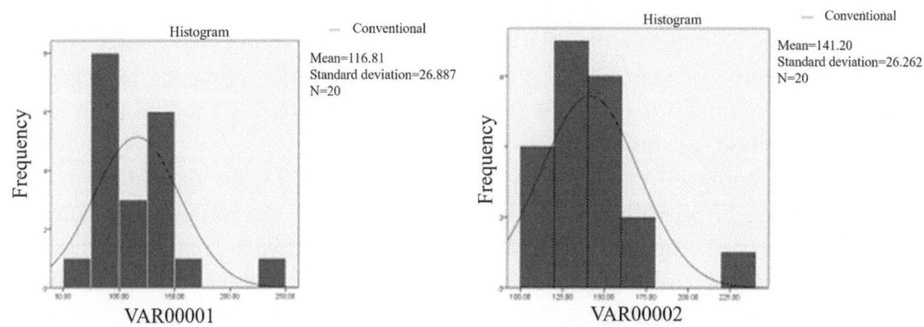

(a) 3d age period (b) 7d age period

Fig. 5.19 Normal distribution diagram of rebound modulus

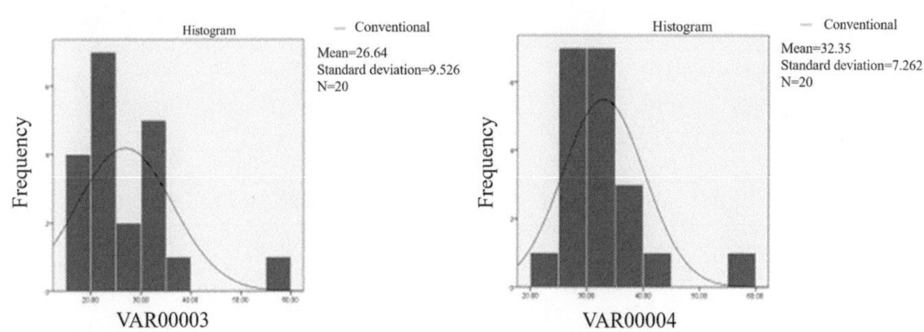

(a) 3d age period (b) 7d age period

Fig. 5.20 CBR normal distribution diagram

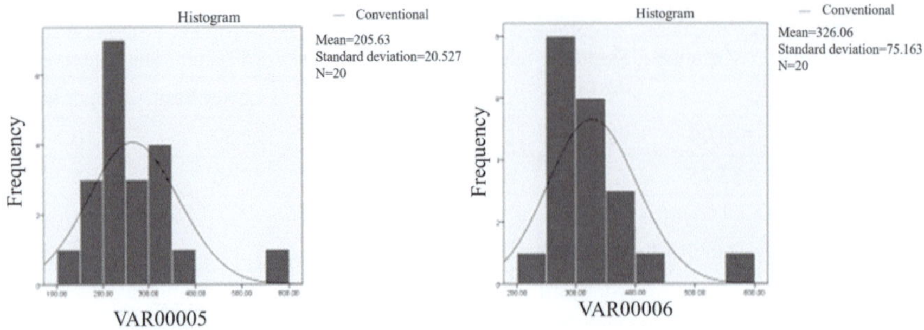

(a) 3d age period (b) 7d age period

Fig. 5.21 UCC normal distribution diagram

of the subgrade. To evaluate the relationship between the road performance and the compaction degree of the improved red mud subgrade, the compaction degree is tested after the completion of the water curing for 7d. The test method used is the sand filling method, as shown in Fig. 5.22.

The on-site compaction results are shown in Table 5.23; the data analysis was performed using SPSS software, as shown in Table 5.24. As can be seen from the table, the subgrade compaction degree is lower within the 95% confidence interval, with a lower limit of 91.3% and an upper limit of 93.2%, and the overall compaction degree is low, which does not meet the specification requirements.

Fig. 5.22 On-site sand filling compaction test

Table 5.23 On-site compaction test results

Left position	Compaction degree/%	Right position	Compaction degree/%
L1	90	R1	91
L2	93	R2	88
L3	91	R3	90
L4	95	R4	93
L5	94	R5	93
L6	97	R6	92
L7	91	R7	93
L8	91	R8	92
L9	94	R9	94
L10	91	R10	92

Table 5.24 Subgrade compaction analysis table

Mean	Median	Variance	Standard deviation	Maximum	Minimum	95% confidence interval	
						Lower limit	Upper limit
92.3%	92.01	0	0.021	88.3%	97.1%	91.3	93.2

Based on the above results, except for the compaction degree, all mechanical indicators, quality control indicators, and design control indicators of the modified red mud highway subgrade can meet the specification requirements. It can be considered that the strength, stiffness, and modulus of the improved red mud subgrade have been greatly improved. Therefore, the modified red mud has certain rationality and feasibility as a highway subgrade filling material.

5.4 Modified Solidified Red Mud Base Layer

The verification of the reliability of the road performance and the feasibility of the construction process of the modified solidified red mud road material used as the base layer of the highway is implemented through the G309 Qinglan Line Zhoucun Section Test Road Modified Red Mud Base Layer Paving Project.

5.4.1 Implementation Plan and Background

This project relies on the G309 Qinglan Line Zhoucun Section to implement the test road paving. The G309 Qinglan Line is the "Horizontal Five" line in the "Two Rings, Two Connections, Seven Verticals, and Twelve Horizontals" trunk road network planned in the "Thirteenth Five-Year Plan" of Zibo City. It passes through Linzi District, Zhangdian District and Zhoucun District in Zibo, with a route length of 66.1 km. The current G309 Qinglan Line Zhangdian District to Zhoucun District Gaozi Road intersection section is 21.2 km long, with four and six lanes in both directions. Among them, the city crossing mileage is 11.5 km (7 km in Zhangdian urban area and 4.5 km in Zhoucun urban area), and the city crossing section accounts for 54% of the total length of this section. The current transit vehicles and urban traffic are mixed, posing a significant safety risk, which not only restricts the function of the trunk road, but also has a significant impact on urban development, road environment, and public travel. The whole line adopts a two-way six-lane first-class highway standard, with a design speed of 80 km/h; the road base width of the overlapping Lutai Avenue section is 51 m, and the road base width of the remaining sections is 33 m/33.5 m (the bridge width is 32.5 m, and the culvert is the same width as the road base); the design flood frequency of the road base, bridge and culvert is 1/100; the design vehicle load level of the bridge and culvert is Highway-I level; the peak acceleration coefficient of the earthquake is 0.10, and traffic safety and along-line facilities are executed according to relevant regulations.

① The inorganic material modified solidified red mud test section is k453 + 700 ~ k453 + 850, k453 + 950 ~ k454 + 050 and k454 + 100 ~ k454 + 250 (segmented calculation 150 + 100 + 150), with a total length of 400 m, disposed in a single width.

② The test section of the composite stabilized red mud is located at the engineering project k454 + 160 ~ k454 + 360, with a length of 200 m and a road width of 31.5 m. Among them, the stabilized red mud structure is located on the right half of the road, and the conventional section is located on the left half of the road with the same pile number. The composition of the road surface structure of the two test sections is shown in Table 5.25.

5.4.2 Modified Red Mud Base Layer Construction Process

The modified solidified red mud base layer adopts the factory mixing method for construction, and its main process flow is: ① Pre-construction preparation → ② Surface treatment → ③ Inspection and acceptance → ④ Leveling and setting out the center stake and side stake line → ⑤ Centralized mixing at the mixing station → ⑥ Transportation of red mud mixture to the site → ⑦ Spreading of red mud by paver → ⑧ Stabilization → ⑨ Vibration compaction (strong vibration 2 times, weak vibration 2 times) → ⑩ Final pressure → ⑪ Quality inspection and acceptance → ⑫ Curing → ⑬ Construction of the next layer (return to the 5th step and repeat the construction until the specified filling height is reached).

The centralized factory mixed red mud mixture meets the following requirements:

① The moisture content of the modified solidified red mud should be controlled slightly higher than the optimum moisture content + 3%, and the moisture content

Table 5.25 Composition of the test road surface structure

Structural layer	Test section A		Test section B	
	Material type	Thickness/cm	Material type	Thickness/cm
Surface layer	SMA.13	4	SMA.13	4
Underlying layer	AC.20C	6	AC.20C	6
Upper base layer	Cement stabilized gravel	18	Cement stabilized gravel	18
Lower base layer	Cement stabilized gravel	18	Cement stabilized gravel	18
Bottom base layer*	Cement stabilized gravel	18	Composite stabilized red mud	18
Roadbed*	Inorganic stabilized red mud	20	Composite stabilized red mud	20
	Inorganic stabilized red mud	20	8% lime soil	20
	Inorganic stabilized red mud	20	6% lime soil	20
Subgrade	Slag replacement		Slag replacement	

of the mixture should not be less than the optimum moisture content when it is transported to the site for spreading and rolling;

② The red mud lumps should be uniformly crushed to a maximum size not exceeding 5 cm;

③ The dosage control of the composite stabilizing material is accurate, with an error not exceeding $\pm$ 0.5%;

④ The mixture is uniform, with no interlayers.

The qualified red mud mixture from the factory should be transported to the construction site as soon as possible. It should be covered during transportation to reduce moisture loss and spillage. The production capacity of the centralized mixing equipment and the paver should match each other, and continuous paving should be achieved. When the production capacity of the centralized mixing equipment is small, low-speed paving is used when paving the red mud mixture with a paver to reduce the situation where the paver is waiting for materials.

After the red mud mixture is paved, it is compacted in a timely manner following the principle of light before heavy. First, a light roller follows the paver for timely rolling, and then a medium and heavy vibrating roller or tire roller of 20 ~ 50 t is used to continue rolling and compacting. After the red mud subgrade is compacted, the compaction degree should be tested in time, and the sand filling method should be used for the compaction degree test. The compaction degree should meet the standard specified in the specification. If it does not meet the requirements, it should be re-compacted in time until it is qualified before proceeding to the next process.

The maintenance of the red mud subgrade meets the following requirements:

① For the modified and solidified red mud subgrade that meets the quality requirements after testing, cover curing or water spraying curing should be used. The number of water sprays per day should be determined according to the climate, and the surface of the red mud layer should be kept moist during the entire curing period;
② After the construction and rolling of the red mud mixture is completed, continuous construction can be carried out within 24 h or the upper layer of red mud mixture can be paved after 3 days of curing;
③ During the curing period, traffic should be closed or sections of the road where the upper layer of red mud cannot be paved immediately should be prohibited or restricted from heavy vehicle traffic. If it is necessary to open to traffic, the speed limit should be below 30 km/h.

1. Construction of inorganic stabilized red mud subgrade test road

The test section is 400 m long, K453 + 700 ~ K453 + 850, K453 + 950 ~ K454 + 050 and K454 + 100 ~ K454 + 250 (calculated in Sects. 150 + 100 + 150); inorganic stabilized red mud is used to treat the subgrade area, using 60 cm thick inorganic stabilized red mud, the determined ratio is: lime: fly ash: red mud (Shan Aluminum): phosphogypsum is 15:7:70:8, the construction of the inorganic stabilized red mud subgrade is divided into upper, middle, and lower layers, each layer is compacted to a thickness of 20 cm. Considering the construction quality, construction progress, and on-site construction capabilities, the lower layer mixture is paved with a grader after factory mixing; the middle and upper layers are first mixed with lime, fly ash, and phosphogypsum in the factory, and then transported to the site for paving with a paver, and then mixed with red mud for road mixing construction. The construction process is shown in Fig. 5.23, and the construction process is shown in Fig. 5.24.

2. Construction of composite stabilized red mud subgrade and subbase test road

The composite stabilized red mud test section is located at K454 + 160 ~ K454 + 360, with a total length of 200 m; the stabilized red mud is constructed in two layers, respectively replacing the 20 cm 10% lime soil and 18 cm cement stabilized gravel subbase in the original design. Factory mixing method is used for construction, the construction process is shown in Fig. 5.25, and the construction process is shown in Fig. 5.26.

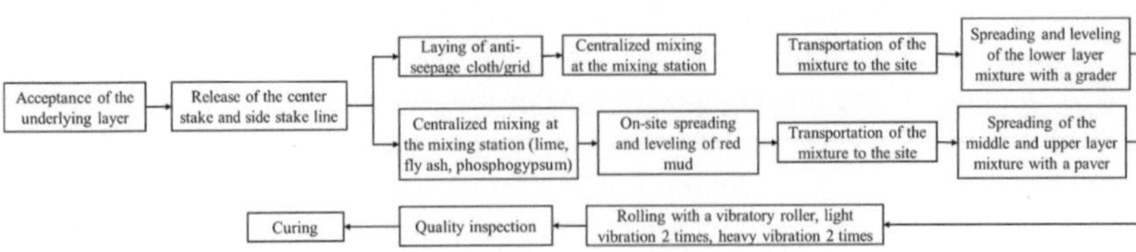

Fig. 5.23 On-site construction process diagram

(a) Waterproof geotextile layout (b) Transportation and unloading of red mud mixture

(c) Paving and leveling of the lower layer of red mud mixture (d) Leveling and rolling of the lower layer of red mud mixture

(e) Rough leveling of red mud by crawler bulldozer (f) Mixing of two ash stabilized red mud by road mixer

Fig. 5.24 Construction process of test section A

(g) Leveling of the upper layer of red mud mixture (h) Rolling of red mud mixture

(i) Curing of the upper layer of red mud mixture

Fig. 5.24 (continued)

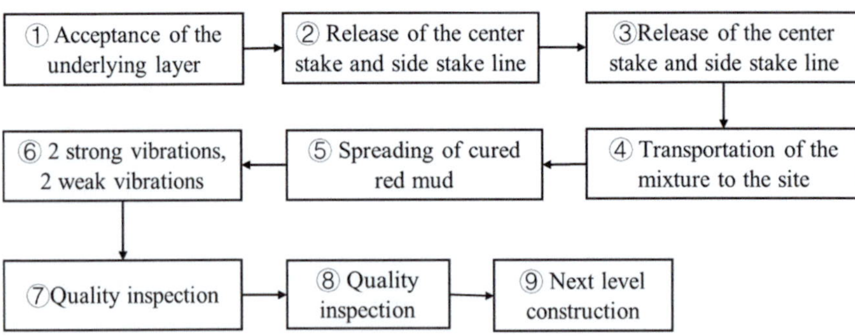

Fig. 5.25 Red mud subgrade construction flow chart

(a) Construction preparation

(b) Factory mixing

(c) Loading and transportation

(d) Unloading and paving

(e) Combined rolling

(f) Quality inspection

Fig. 5.26 Construction process of test section B

(g) Cover curing (h) On-site coring

Fig. 5.26 (continued)

5.4.3 Quality Control and Evaluation of Subbase on Site

1. Raw material preparation

Before construction, samples of red mud must be taken for natural moisture content, liquid plastic limit, CBR value, standard compaction, and unconfined compressive strength tests to fully grasp the index values of its main physical and mechanical properties and determine stable parameters. When the source of the ore or the production process changes, the tests must be redone.

2. Optimal moisture content control

The optimal moisture content and maximum dry density of stabilized red mud are determined through heavy-duty standard impact tests. After the red mud is spread, a moisture content test is conducted before proceeding to the next operation. The on-site measured moisture content is controlled to be 2 ~ 3 percentage points higher than the optimal moisture content. If the moisture content is insufficient, water is sprinkled in time to supplement it, and if the moisture content is too high, appropriate turning and drying are carried out.

3. Layer thickness

The control of the construction layer thickness directly affects the uniformity of the mixing of the stabilized material and the compaction quality. The compaction layer thickness of the fill is generally 25 ~ 30 cm, and the loose spreading thickness is 30 ~ 35 cm. The layer thickness should be uniform, the surface should be flat, and the pit digging method or level measurement should be used to measure the loose spreading thickness, and the allowable error should be controlled within ± 3 cm.

4. Mixing uniformity

According to the optimal dosage of the stabilized material design, the mixing station accurately controls the dosage of the composite stabilized material, and the error is not more than ± 0.5%. The stabilized red mud after mixing requires a uniform and consistent surface color, no ash lumps or ash strips, no flower surface and "plain soil" blocks, and the red mud particles are controlled to be less than 5 mm. The EDTA titration method is used to detect the dosage of the stabilized material, and the deviation of the measured content from the standard value does not exceed the control range of 0.5% ~ 1%.

5. Compaction time control

Due to the easy wind drying and water loss of the red mud surface, to ensure the uniformity within the spreading layer thickness range, the compaction process is completed within 4 h after the mixing is completed. After the mixing is uniform, the roller is quickly statically pressed to prevent water loss. Then, a 15 ~ 20 t vibrating roller is used for weak vibration once, strong vibration 3 ~ 4 times, and finally static pressure once. After the compaction is completed and the inspection is qualified, curing is carried out immediately. If multi-layer continuous filling construction technology is adopted, it is ensured that the construction layer can be compacted within the 4-h compaction time of the previous filling layer at the corresponding position.

6. Curing conditions and age

Moist curing is the key process for the formation of stabilized subgrade strength and the stabilization reaction of metal ions. After each layer of layered filling construction is compacted or the last layer of multi-layer continuous filling construction is compacted, felt is used to cover and sprinkle water for curing, and the curing period is not less than 2 days. During the curing period, the surface is kept moist, water is sprinkled at least 4 times a day in summer, and 20 cm thick soil is covered when the minimum temperature in winter is below 0 °C. Traffic is restricted during the curing period, the speed of the water truck is limited to not exceed 15 km/h, and heavy vehicles are prohibited.

(1) Detection method

The "point-to-point" corresponding method is used to arrange test points at intervals of 10 m for test road lengths of 100 ~ 200 m, and PFWD dynamic modulus, FWD deflection, and DCP detection are performed on conventional lime soil and stabilized red mud test sections, etc. The detection data is analyzed to evaluate the road performance of the stabilized red mud material.

1) PFWD dynamic deformation modulus

① Testing equipment

The Prima100 portable drop hammer deflection instrument is used, including loading equipment, load plate, and integrated data recording and collection instrument, as shown in Fig. 5.27. The loading device 10 kg drop hammer, free fall 90 cm to the steel spring bar and generates a peak 7.07 kN pulse width of 18 ms load pulse. The drop hammer is equipped with a transmission lock pin and guide column. To ensure a stable drop height for each measurement, the mechanical device is placed on top

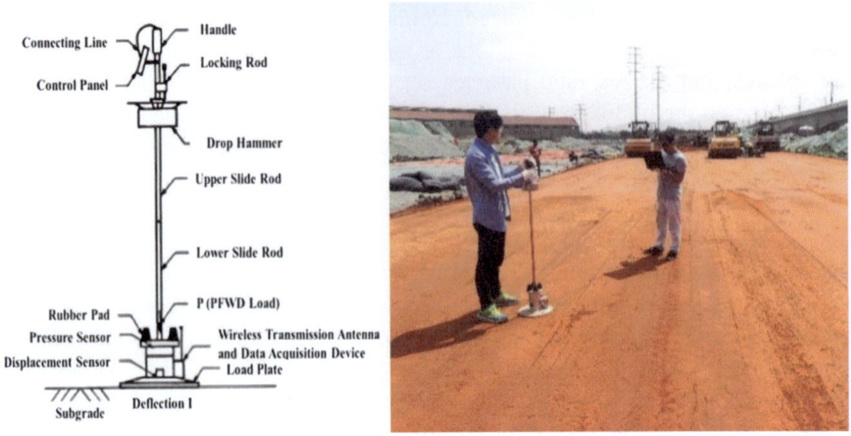

Fig. 5.27 Schematic of PFWD equipment construction and field testing

of the column to fix and release the drop hammer. The average pressure amplitude of the 30 cm diameter load plate is 0.1 MPa, two handles and a small ball bearing the concentrated load are placed on top of the load plate, and the accelerator in the protective sleeve is used for deformation measurement.

② Testing principle

In the PFWD testing process, potential energy is first converted into kinetic energy (drop hammer falling), and then potential energy is converted into kinetic energy (drop hammer impacts the load plate, the load plate compresses the soil base deformation), the load amplitude is controlled by the hammer weight and drop height, which is 0.1 MPa (7.07 kN). At the same time, the following assumptions are made when calculating the dynamic modulus:

(1) The elastic half-space Boussinesq theory is used, assuming that the soil base is an elastic half-space, and the calculation formula is as follows (5.5);

$$E_p = \frac{\pi}{4} \times \frac{2p\delta(1 - \mu^2)}{l} \tag{5.5}$$

where

E_p Dynamic deformation modulus measured by handheld drop hammer deflection meter;
p The circular uniform load received by the load plate is 0.1 MPa;
l The radius of the load plate is 0.15 m;
μ The Poisson's ratio of the soil is 0.35;
δ Represents the deformation (mm) under the bearing plate.

(2) The interaction with the soil foundation is ignored, there is no load sensor, and the load is obtained through the principle of energy conversion;

(3) The calculation parameters ignore the type of soil used in the soil foundation, and all use a Poisson's ratio of 0.35. This is a larger error for semi-rigid materials, but it is acceptable for the soil foundation.

(2) In-situ test results

1) PFWD dynamic deformation modulus

Figure 5.28 shows the dynamic deflection results measured by PFWD after the stable red mud test road is moisturized and cured for 7 days. Figure 5.29 is the dynamic modulus calculated from the deflection in Fig. 5.28. From the figure, the PFWD dynamic modulus in the up and down directions does not differ much, mainly concentrated between 300 and 600 MPa.

Fig. 5.28 Dynamic deflection results

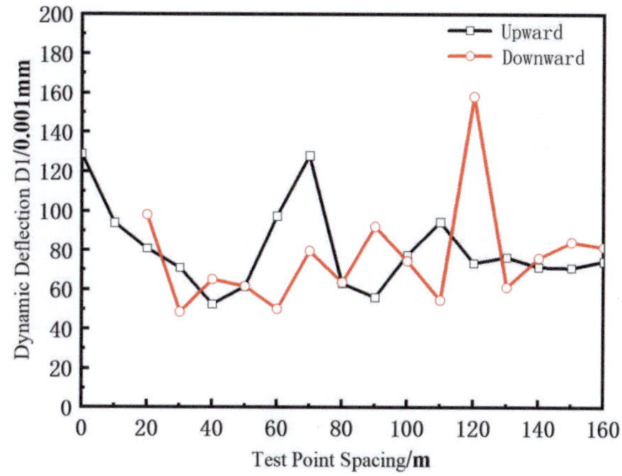

Fig. 5.29 Dynamic modulus E

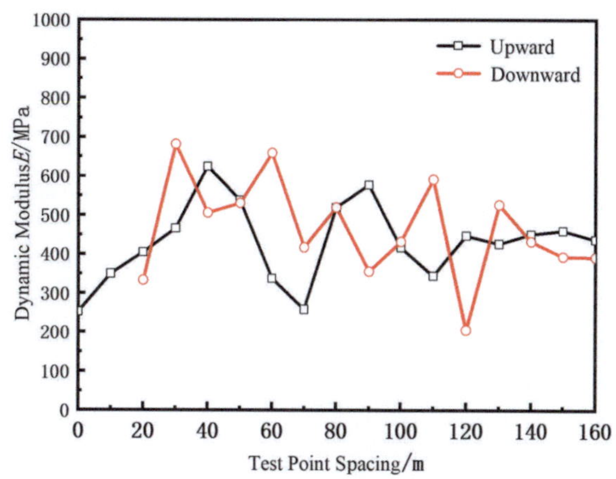

2) FWD deflection

After the curing period of the test road ends for 7 days, the conventional lime soil and stable red mud test section roadbeds are tested with FWD. The changes in the deflection of the load center point are shown in Fig. 5.30 and Table 5.26. Overall, the deflection of the stable red mud roadbed is slightly smaller than that of the conventional lime soil roadbed, meeting the roadbed design deflection requirements, and the bearing capacity of the red mud base is slightly better than that of the conventional lime soil subgrade.

Figure 5.31 and Table 5.27 are the deflection test conditions conducted on the right track red mud top (base layer) and left track lime soil top (roadbed) after cleaning the surface covering soil in 2018. The results show: It can be clearly seen that the deflection of the base layer of the stable red mud subgrade test road is slightly smaller than the deflection of the top of the conventional lime soil roadbed. Overall, the bearing capacity of the stable red mud subgrade is slightly better than that of lime soil, and it can obtain better bearing capacity support for the upper base structure than the lime soil subgrade.

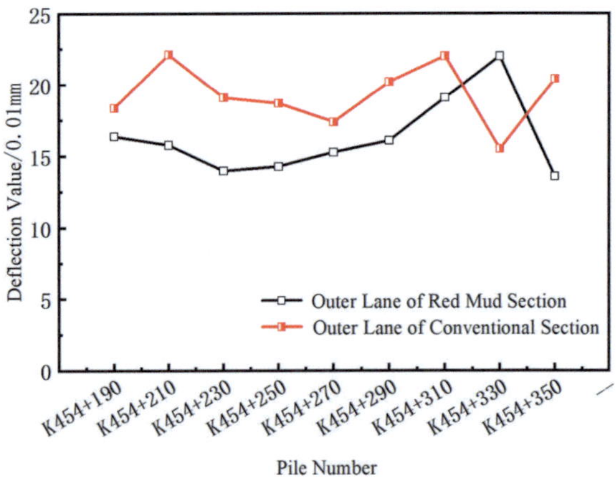

Fig. 5.30 Deflection of the inner and outer sides of the roadbed 7 days after the end of construction and curing

Table 5.26 Deflection of the inner and outer sides of the roadbed 7 days after the end of construction and curing

Test location	Right track (red mud top)	Left track (gray soil top)
Number of test points	9	9
Minimum deflection	31.3	38.3
Maximum deflection	41.8	42.5
Standard deviation of deflection	3.52	1.15
Average value l	35.1	40.1
β value	1.28	1.28
Representative value lr	39.6	41.5

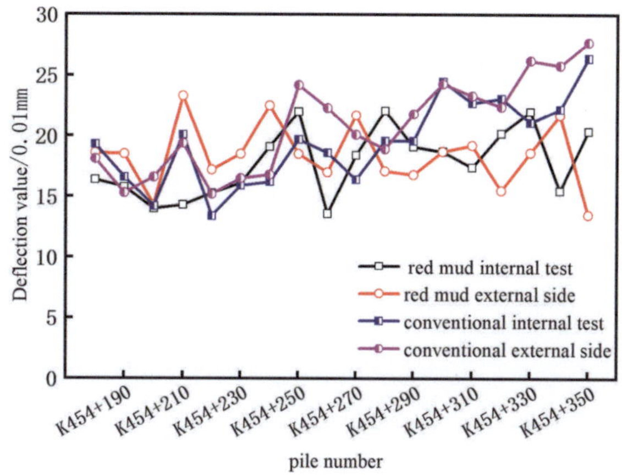

Fig. 5.31 Deflection of the top of the right track red mud base layer and the top of the left track lime soil roadbed

Table 5.27 Deflection of the top of the right track red mud base layer and the top of the left track lime soil roadbed

Test location	Right track (red mud top)	Left track (lime soil top)
Number of test points	36	36
Minimum deflection	13.5	13.4
Maximum deflection	23.3	27.7
Deflection standard deviation	2.69	3.71
Average value l	18.1	20.1
β value	1.28	1.28
Representative value l_r	21.5	24.9

Figure 5.32 and Table 5.28 show the deflection of the cement stabilized gravel base layer after construction and curing of the right (red mud) water stabilized gravel top and the left (conventional) water stabilized gravel top. The results show that the deflection of the water-stabilized base layer of the stabilized red mud base test road is slightly larger than that of the conventional lime soil subgrade water-stabilized base layer top. The reason for the analysis is that the conventional pavement has set up two layers of water-stabilized gravel base layer plus one layer of water-stabilized gravel bottom base layer, while the stabilized red mud structure has replaced the bottom base layer structure, but the overall deflection is not much different, which can meet the design deflection requirements and has good bearing capacity.

Figure 5.33 and Table 5.29 show the deflection of the right (red mud) asphalt pavement top and the left (conventional) asphalt pavement top. The results show that due to the different design of the "base layer + bottom base layer" structure of the pavement, the deflection of the top of the stabilized red mud base asphalt pavement surface layer is slightly larger than the deflection of the top of the conventional asphalt pavement surface layer. When the two are relatively close, it indicates that

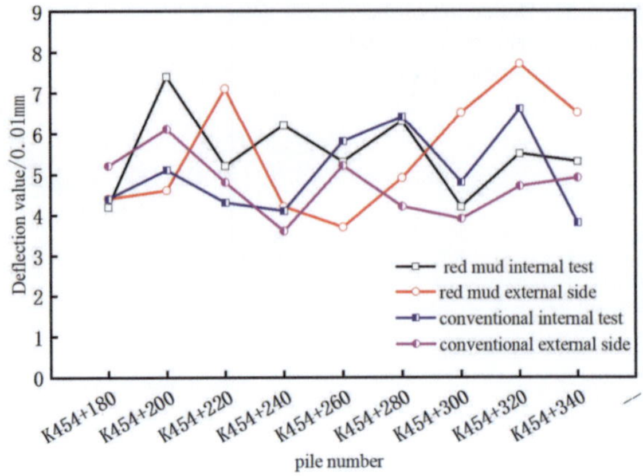

Fig. 5.32 Deflection of the right base top and left base top

Table 5.28 Deflection of the right base top and left base top

Test location	Right base top (red mud structure)	Left base top (conventional structure)
Number of measurement points	18	18
Minimum deflection	3.7	3.6
Maximum deflection	7.7	6.6
Deflection standard deviation	1.18	0.86
Average value l	5.5	4.9
β value	1.28	1.28
Representative value l_r	7.0	6.0

the overall bearing capacity of the test road pavement structure can meet the design and use requirements.

In summary, the overall strength and bearing capacity of the stable red mud replacing the traditional lime-stabilized soil subgrade and cement-stabilized gravel base are higher than that of the lime-stabilized soil. As shown in Fig. 5.34, under different pavement structure combinations, the overall deflection of the stable red mud structure is slightly higher than that of the conventional structure (with an additional layer of water-stabilized base), but it can meet the road design deflection and bearing requirements, thus achieving good economy.

3) DCP detection

DCP detection was carried out on the top of the stable red mud subgrade at the age of 3d and 7d, and the results of the rebound modulus, CBR value and UCC are shown in Figs. 5.35, 5.37 and 5.39. Tables 5.30, 5.31 and 5.32 show the data analysis results obtained through SPSS software. Figures 5.36, 5.38 and 5.40 show the normal distribution of the rebound modulus, CBR value, and UCC of the stable

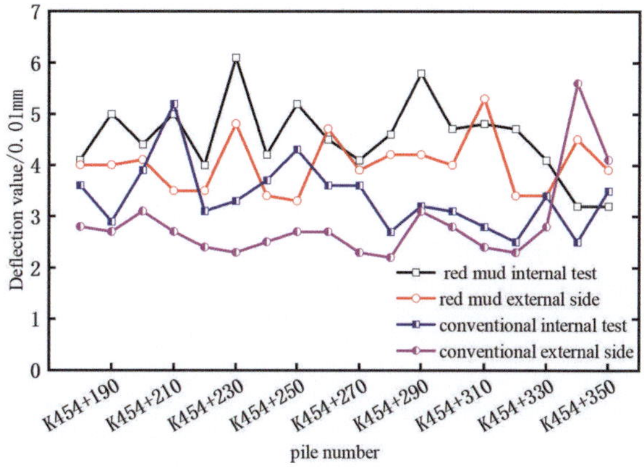

Fig. 5.33 Deflection of the right pavement top and left pavement top

Table 5.29 Deflection of the right pavement top and left pavement top

Test location	Right pavement top (red mud structure)	Left pavement top (conventional structure)
Number of test points	36	36
Minimum deflection	3.2	2.2
Maximum deflection	6.1	5.6
Standard deviation of deflection	0.69	0.77
Average value l	4.3	3.1
β value	1.28	1.28
Representative value l_r	5.2	4.1

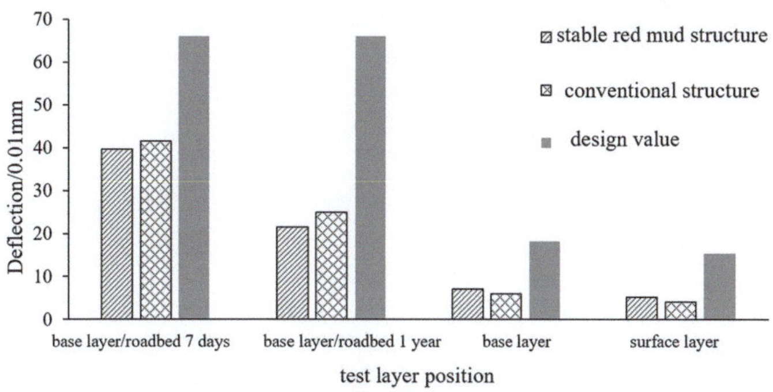

Fig. 5.34 Comparison of deflection between stable red mud and conventional structures at different layers

red mud subgrade at different ages. It can be seen from the figure that with the increase of the curing age, the rebound modulus, CBR value and UCC of the stable red mud subgrade all show an increasing trend, showing an increase in mean and median, and a decrease in variance and standard deviation, and the upper and lower limits of the three indicators increase within the 95% confidence interval.

This shows that as the curing age increases, the quality control indicators and design indicators of the stabilized red mud subgrade increase, the dispersion decreases, the overall uniformity of the subgrade is improved, and the performance of the subgrade is greatly improved.

Fig. 5.35 Rebound modulus

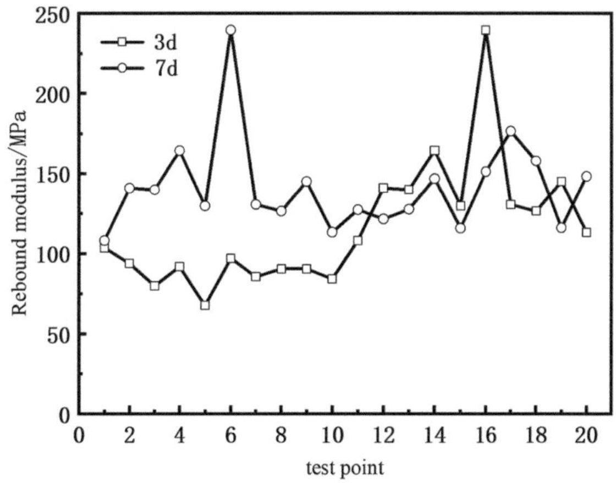

Table 5.30 Rebound modulus E/MPa at different ages

Age	Mean	Median	Variance	Standard deviation	Maximum	Minimum	95% confidence interval	
							Lower limit	Upper limit
3d	116.01	105.7	1511.395	38.88	239.67	67.62	97.82	134.2
7d	141.20	135.06	861.52	29.35	249.21	108.03	127.47	154.94

Table 5.31 CBR value/% at different ages

Age	Mean	Median	Variance	Standard deviation	Maximum	Minimum	95% confidence interval	
							Lower limit	Upper limit
3d	26.84	24.27	90.99	9.54	57.45	15.17	22.37	31.30
7d	32.96	31.426	52.75	7.26	61.23	24.83	29.56	36.35

Table 5.32 UCC at different ages/kPa

Age	Mean	Median	Variance	Standard Deviation	Maximum	Minimum	95% confidence interval	
							Lower limit	Upper limit
3d	2615.64	2381.95	9628.81	98.13	5341.12	1462.94	2191.71	3112.56
7d	3283.06	3122.01	5646.49	75.14	5821.44	2442.69	2922.905	3631.23

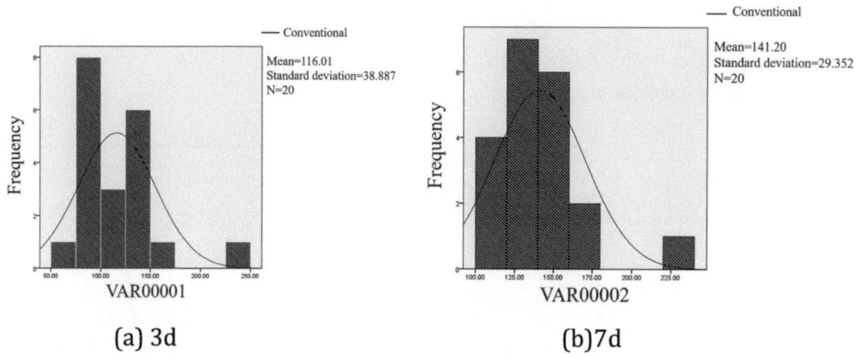

Fig. 5.36 Normal distribution of rebound modulus

Fig. 5.37 CBR value

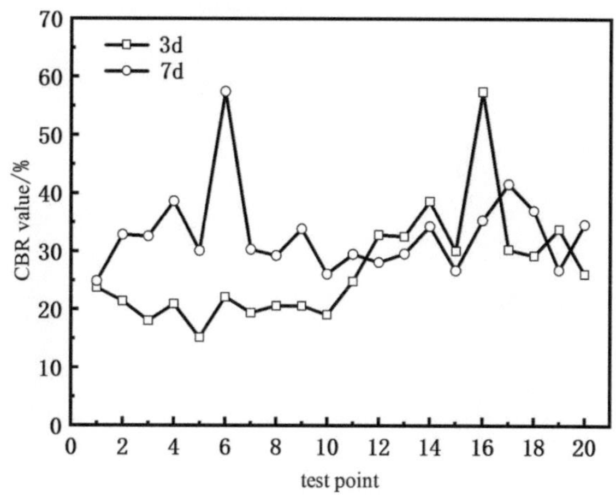

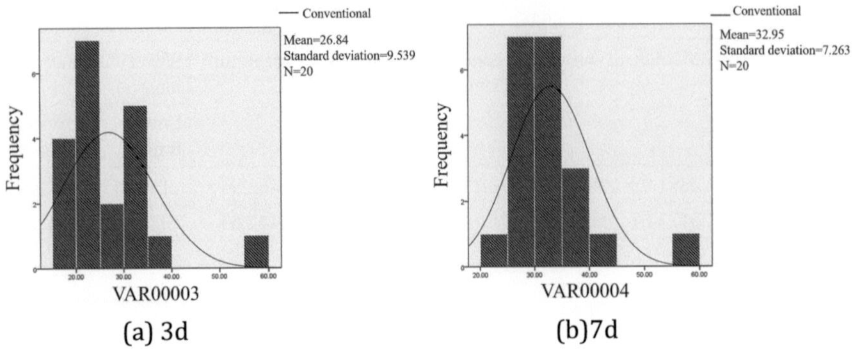

Fig. 5.38 Normal distribution of CBR

Fig. 5.39 UCC

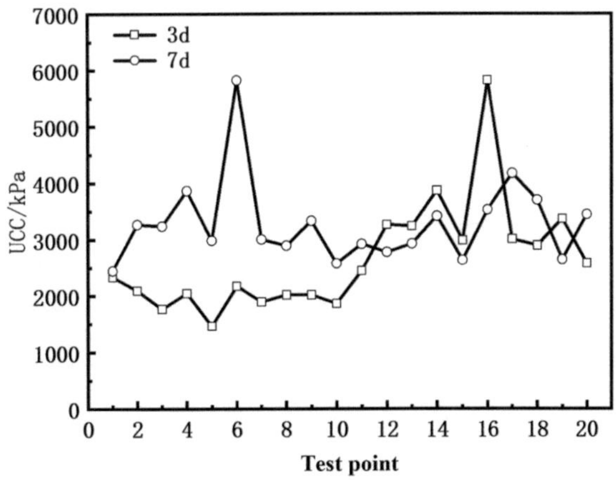

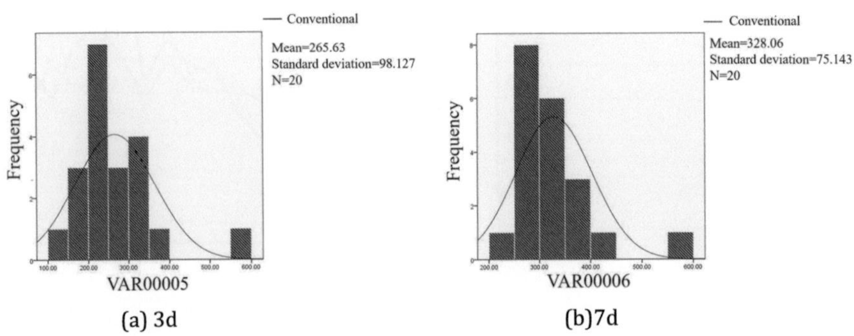

Fig. 5.40 Normal Distribution of UCC

5.5 Analysis of Pavement Structure Based on Modified Cured Red Mud Subgrade

Using stable red mud as roadbed and subgrade filling material, combined with the characteristics of stable red mud material, the structure of conventional lime soil asphalt pavement and red mud base asphalt pavement is optimized and analyzed; using elastic layer theory, the bending response, fatigue life and permanent deformation of conventional lime soil base and stable red mud base asphalt pavement structure are studied. The depth of rutting and the cumulative equivalent axle times of fatigue cracking are used as indicators to evaluate the performance of red mud base pavement structure.

5.5.1 Asphalt Pavement Structure Mechanical Calculation

1. Mechanical Analysis Model

The influence of material nonlinearity and anisotropy on the mechanical response state within each structural layer of the pavement is considered and analyzed differently at home and abroad. A large number of studies have shown that there is a certain difference between the horizontal stresses obtained under different types of material constitutive models (elastic linear and nonlinear, isotropic and anisotropic), and the deflection values calculated by the nonlinear model are closer to the measured values, but the vertical stresses within the corresponding layers are not much different. Based on this basic feature, considering the calculation cost, this project does not consider the material nonlinearity and anisotropy of the pavement structure layer and subgrade soil when calculating the stress level within the asphalt pavement structure layer, and still uses the layered elastic continuum mechanical analysis model under the action of double circular vertical uniform load, that is:

1) The pavement is a multi-layer elastic layered system, each layer is a homogeneous elastic material with isotropy, characterized by rebound modulus and Poisson's ratio;
2) Semi-infinite space, the subgrade is infinite in both horizontal and depth directions, the thickness of each structural layer of the pavement is finite, but infinite in the horizontal direction;
3) The surface load is a double circular vertical load, and the contact surface load pressure is uniformly distributed;
4) It is assumed that the contact conditions between each layer are completely continuous;
5) The self-weight of each layer is not considered.

Table 5.33 Test road A section

Structural layer	Material type	Thickness/cm	Rebound modulus/MPa	Poisson's ratio	Remarks
Surface layer	SMA-13	4	11,500	0.35	
Underlying layer	AC-20C	6	11,000	0.35	
Upper base layer	Water-stable gravel	18	20,000	0.25	
Lower base layer	Water-stable gravel	18	20,000	0.25	
Subbase layer	Water-stable gravel	18	14,000	0.35	
Roadbed	10% lime soil	20	3000	0.40	
Roadbed	6% ~ 8% lime soil	40	2000	0.40	Calculated as one layer
Subgrade	Raw soil	–	160	0.40	

2. Selection of Pavement Structure Parameters and Calculation Indicators

This part of the research mainly analyzes the pavement structure performance of red mud as roadbed filling material, compares and analyzes the influence of red mud subgrade and conventional lime soil subgrade on the mechanical response of each structural layer of the pavement, and provides theoretical support for the design of red mud base asphalt pavement structure. Tables 5.33, 5.34 and 5.35 respectively represent the forms and structural parameters of conventional lime soil subgrade pavement structure (test road A section), inorganic stabilized red mud base pavement structure (test road B section) and composite stabilized red mud base pavement structure (test road C section).

This section mainly focuses on three types of structures: conventional lime soil asphalt pavement, inorganic stabilized red mud asphalt pavement, and composite stabilized red mud asphalt pavement. It analyzes the bending response at different depths of the pavement structure under different axle load conditions, and analyzes the fatigue life and permanent deformation of the asphalt pavement structure.

5.5.2 *Pavement Structure Dynamic Response Analysis*

1. Bending response analysis

Figures 5.41, 5.42 and 5.43 respectively show the distribution cloud maps of pavement bending under standard axle load, 20% overload, and 60% overload conditions. As can be seen from the figures, as the distance from the center of the load increases, the load gradually decreases, and the bending gradually decreases; under standard

Table 5.34 Test road section B

Structural layer	Material type	Thickness/cm	Resilient modulus/MPa	Poisson's ratio	Remarks
Surface layer	SMA-13	4	11,500	0.35	
Under layer	AC-20C	6	11,000	0.35	
Upper base layer	Water-stable crushed Stone	18	20,000	0.25	
Lower base layer	Water-stable crushed stone	18	20,000	0.25	
Bottom base layer	Water-stable crushed stone	18	14,000	0.35	
Roadbed	Lime-stabilized red mud	20	4000	0.35	
Roadbed	Lime-stabilized red mud	40	3000	0.4	Calculated as one layer
Subgrade	Raw soil	–	160	0.40	

Table 5.35 Test road section C

Structural layer	Material type	Thickness/cm	Resilient modulus/MPa	Poisson's ratio	Remarks
Surface layer	SMA-13	4	11,500	0.35	
Lower layer	AC-20C	6	11,000	0.35	
Upper base layer	Water-stable gravel	18	20,000	0.25	
Lower base layer	Water-stable gravel	18	20,000	0.25	
Bottom base layer	Stabilized red mud	18	10,000	0.35	
Roadbed	Stabilized red mud	20	7000	0.35	
Roadbed	6% ~ 8% lime soil	40	2000	0.4	Calculated as one layer
Subgrade	Natural soil	\	160	0.40	

axle load conditions, the surface bending at the center of the load of the conventional lime soil subgrade pavement structure is about 8.65 mm/100, the surface bending at the center of the load of the inorganic stabilized red mud subgrade pavement structure is about 7.12 mm/100, and the surface bending at the center of the load of the composite material stabilized red mud subgrade pavement structure is about 8.76 mm/100. The bending of the stabilized red mud subgrade pavement structure is slightly smaller than that of the conventional lime soil subgrade pavement structure, indicating that the red mud subgrade pavement structure can meet the performance requirements of the conventional pavement structure under standard axle load; under 20% overload conditions, the surface bending at the center of the load of the conventional lime soil subgrade pavement structure is about 10.38 mm/100, the surface bending at the center of the load of the inorganic stabilized red mud subgrade pavement structure is about 8.50 mm/100, and the surface bending at the center of the load of the composite material stabilized red mud subgrade pavement structure is about 11.92 mm/100. Under 60% overload conditions, the surface bending at the center of the load of the conventional lime soil subgrade pavement structure is about 13.83 mm/100, the surface bending at the center of the load of the inorganic stabilized red mud subgrade pavement structure is about 9.93 mm/100, and the surface bending at the center of the load of the stabilized red mud subgrade pavement structure is about 15.89 mm/100. Under overload conditions, the surface bending of the stabilized red mud subgrade pavement structure significantly increases, which is significantly larger than that of the conventional lime soil subgrade pavement structure. The greater the degree of overload, the greater the increase in surface bending, indicating that the red mud subgrade is more prone to damage under overload conditions and should avoid overload situations as much as possible.

2. Fatigue Axle Time Analysis

According to the OD survey analysis, the traffic volume AADTT of large buses and trucks at the cross-section is 3560 vehicles/day, the annual traffic growth rate γ is 4.0%, the direction coefficient DDF is 55.0%, and the lane coefficient LDF is 50.0%.

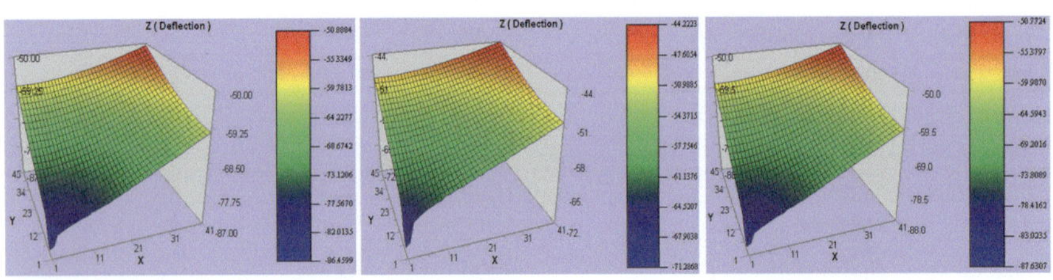

(a) Test road section A (b) Test road section B (c) Test road section C

Fig. 5.41 Distribution cloud map of pavement bending under standard axle load

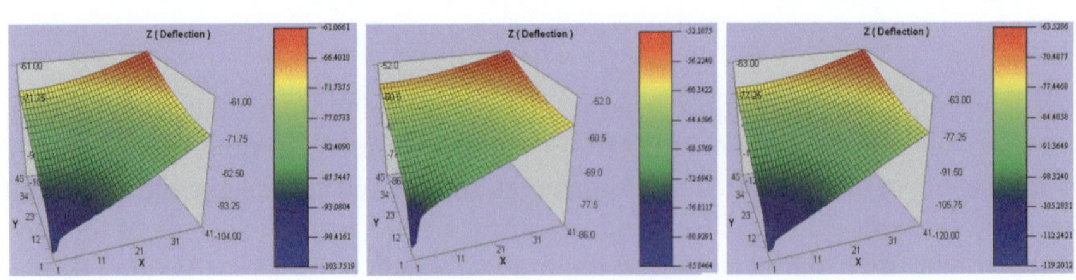

(a) Test road section A (b) Test road section B (c) Test road section C

Fig. 5.42 Distribution cloud map of pavement bending under 20% overload

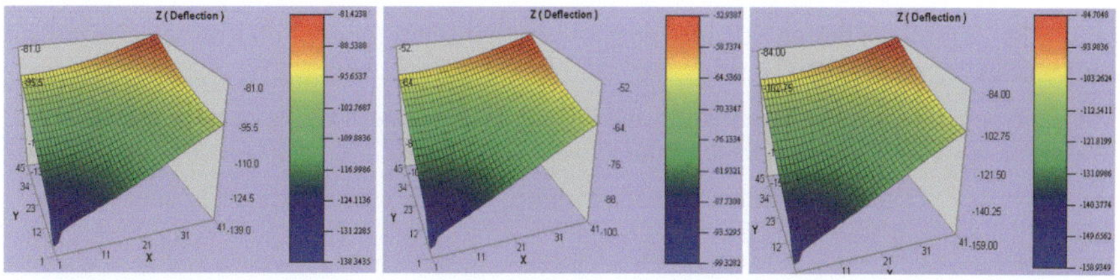

(a) Test road section A (b) Test road section B (c) Test road section C

Fig. 5.43 Distribution cloud map of pavement bending under 60% overload

According to the traffic history data, the design road is determined to be TTC1 class based on the local experience value of this TTC classification, and the vehicle type distribution coefficient is shown in Table 5.36.

Based on the survey and analysis of the vehicle load situation and historical data of the adjacent roads in the road network, the proportion of non-full load and full load for each type of vehicle is obtained, as shown in Table 5.37.

Table 5.38 shows the equivalent design axle load conversion coefficients for non-full load vehicles and full load vehicles under different design indicators.

Based on the material and structural parameters in Tables 5.33, 5.34 and 5.35 and the traffic parameters in Tables 5.36, 5.37 and 5.38, the rutting and semi-rigid base fatigue cracking indices of the asphalt layer of the red mud base and conventional lime soil subgrade test roads are analyzed. The calculation results are shown in Table 5.39. As can be seen from Table 5.39, within the design life (15 years), the rutting of the

Table 5.36 Vehicle type distribution coefficient

Vehicle type	Class 2	Class 3	Class 4	Class 5	Class 6	Class 7	Class 8	Class 9	Class 10	Class 11
Vehicle distribution coefficient (%)	6.4	15.3	1.4	0	11.9	3.1	16.3	20.4	25.2	0

Table 5.37 Proportion of non-full load vehicles and full load vehicles/%

Vehicle type	Class 2	Class 3	Class 4	Class 5	Class 6	Class 7	Class 8	Class 9	Class 10	Class 11
Non-full load vehicle proportion	85	90	65	75	55	70	45	60	55	65
Full load vehicle proportion	15	10	35	25	45	30	55	40	45	35

Table 5.38 Equivalent design axle load conversion coefficients for non-full load vehicles and full load vehicles

Vehicle type	Permanent deformation and fatigue cracking of asphalt mixture layer		Fatigue cracking of inorganic binder layer	
	Unloaded vehicle	Loaded vehicle	Unloaded vehicle	Loaded vehicle
Category 2	0.8	2.8	0.5	35.5
Category 3	0.4	4.1	1.3	314.2
Category 4	0.7	4.2	0.3	137.6
Category 5	0.6	6.3	0.6	72.9
Category 6	1.3	7.9	10.2	1505.7
Category 7	1.4	6	7.8	553
Category 8	1.4	6.7	16.4	713.5
Category 9	1.5	5.1	0.7	204.3
Category 10	2.4	7	37.8	426.8
Category 11	1.5	12.1	2.5	985.4

asphalt layer of both the red mud base pavement structure and the conventional lime soil subgrade pavement structure is 7.52 mm, which is less than the design value of 15 mm, meeting the pavement structure design requirements; the fatigue life of the semi-rigid base of both the red mud base pavement structure and the conventional lime soil subgrade pavement structure meets the design requirements, and the fatigue life of the semi-rigid base of the red mud base pavement structure is longer than that of the conventional lime soil subgrade pavement structure.

3. Permanent Deformation Estimation

The long-life asphalt pavement structure only needs to restore its functionality during the design period, while the red mud base pavement structure and the regular lime soil base pavement structure both need to repair ruts during the design period. Based on this, the development trend of rut depth with the years of traffic for two types of pavement structures was analyzed. The estimated results of asphalt layer rut depth are shown in Table 5.40 and Fig. 5.36. The surface layer of the regular lime soil asphalt pavement structure (test road A section), inorganic stabilized and composite stabilized red mud asphalt pavement structure (test road B section and C section) are all 4cmSMA-13 upper layer and 6cmAC-20C lower layer. The permanent deformation of the asphalt layer of the inorganic stabilized red mud, composite stabilized red mud asphalt pavement structure, and regular lime soil asphalt pavement structure increases with the increase of service years. Within the design life of 15 years, the asphalt layer ruts of the red mud base pavement structure and the regular lime soil base pavement structure are both 9.02 mm, which are less than the design value of 15 mm, meeting the pavement structure design requirements. When the service

Table 5.39 Verification index analysis results

Verification index	Pavement structure type	Design value	Calculated value	Does it meet the design requirements
Cumulative equivalent axle times of semi-rigid base fatigue cracking	Regular lime soil	1,632,830,418	25,395,548,783	Yes
	Inorganic stabilized red mud	1,632,830,418	19,892,385,845	Yes
	Composite stabilized red mud	1,632,830,418	26,449,622,529	Yes
Asphalt layer rut depth/mm	Regular lime soil	15	9.02	Yes
	Inorganic stabilized red mud	15	9.02	Yes
	Composite stabilized red mud	15	9.02	Yes
Road surface acceptance bend sink/0.01 mm	Regular lime soil	20.0	8.76	Yes
	Inorganic stabilized red mud	20.0	7.12	Yes
	Composite stabilized red mud	20.0	8.65	Yes
Roadbed top surface acceptance bend sink/0.01 mm	Regular lime soil	150	31.70	Yes
	Inorganic stabilized red mud	150	26.90	Yes
	Composite stabilized red mud	150	26.80	Yes

life reaches 20 years, the permanent deformation is 10.91 mm. The highway maintenance specification stipulates that 15 mm permanent deformation is the standard for judging whether to carry out rut repair maintenance. During the 20 years of use, neither the inorganic stabilized red mud nor the composite stabilized red mud asphalt pavement structure needs maintenance (Fig. 5.44).

5.6 Performance Evaluation of Red Mud Road Engineering

Generally, the focus of quality control in highway subgrade construction is on compaction, dosage, and moisture content. However, these three are physical state indicators of the embankment fill, and can only indirectly reflect the road performance of the fill. They often do not accurately reflect the mechanical properties and water stability of such embankment structures. The indoor CBR value of modified cured red mud is generally greater than 50%, significantly higher than the lower limit of the strength requirement for fine-grained road fill. Under this condition, whether it can be considered that reducing the compaction or density of the modified cured red mud fill can still meet the design requirements of fill strength performance (CBR

Table 5.40 Estimated results of asphalt layer permanent deformation

Wheel track depth/mm	6 years	7 years	8 years	9 years	10 years	11 years	12 years	13 years	14 years	15 years	20 years
Regular lime soil	5.31	5.77	6.21	6.64	7.06	7.46	7.86	8.25	8.64	9.02	10.91
Inorganic stabilized red mud	5.31	5.77	6.21	6.64	7.06	7.46	7.86	8.25	8.64	9.02	10.91
Composite Stabilized red mud	5.31	5.77	6.21	6.64	7.06	7.46	7.86	8.25	8.64	9.02	10.91

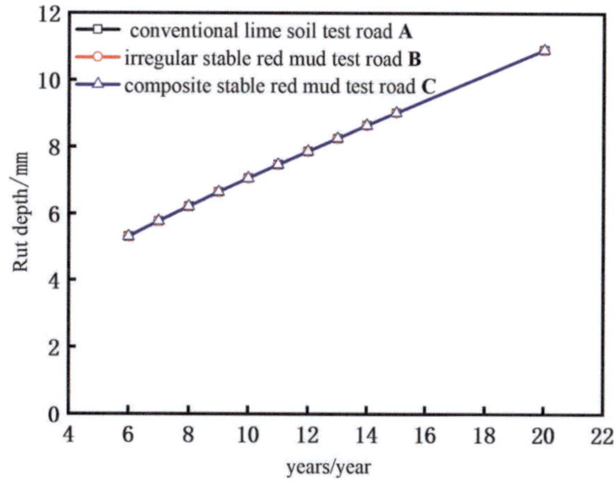

Fig. 5.44 Permanent deformation estimation results

value, strength stability test, field strength test, etc.) and subgrade stiffness (rebound modulus, bending settlement, etc.) is a topic that needs to be studied urgently. Therefore, the indoor and outdoor testing and evaluation of the mechanical properties of the modified cured red mud subgrade structure, and the accumulation and analysis of valid data from the modification effect test of the modified cured red mud, are more in line with engineering needs than traditional methods of quality control of compaction, effective modification component dosage, and moisture content. Direct testing of the road mechanical properties of the modified cured red mud subgrade structure (field rebound modulus, bending settlement, dynamic probing), and its application in the quality inspection and evaluation of the modified cured red mud road material subgrade, is of great engineering significance.

According to the subgrade field test, although the PFWD and FWD tests are accurate and convenient to use, they are relatively expensive. The field bearing plate test is low in cost but time-consuming, and the sand filling compaction test has a certain accuracy, but it has a certain destructiveness to the newly built subgrade and cannot be used for a large number of comparative tests. In order to find a more economical and reasonable test method on the basis of ensuring the accuracy of the test data, this section is based on the DCP field test, with the hammer penetration depth as the indicator (mm/blow), to analyze the relationship between this indicator and various quality control and design indicators, so as to provide a reference for quick and large-scale testing during the construction of the red mud subgrade.

To ensure the accuracy and representativeness of the test data, this experiment uses the penetration depth (mm/blow) within 20 cm after 7 days of water curing as the control indicator, and analyzes its relationship with the quality control and design control indicators under the 7d curing conditions. The specific data is shown in Table 5.41.

According to the data in the above table, a linear regression analysis was performed to analyze and compare the relationship between the hammer penetration DN (mm/

Table 5.41 Summary of test results

Test number	Penetration (mm/blow)	Rebound modulus (MPa)	Dynamic deformation modulus (MPa)	Bending (0.01 mm)	Compaction (%)
1	9.03	119.1	86	100.7	90
2	7.04	153.1	166	60.8	93
3	7.09	130.9	128.1	65.4	91
4	6.09	161.7	172.3	54.3	95
5	7.61	159.1	156.9	59.8	94
6	4.27	234.2	239.5	43.6	97
7	7.56	147.3	143.5	74.9	91
8	7.80	156.8	127.6	80.1	91
9	6.86	196.9	185	67.6	94
10	8.65	158.4	164.4	100.7	91
11	7.75	134.1	138.2	70.1	91
12	8.09	123.1	87.1	53.3	88
13	7.73	147.9	149.5	103.2	90
14	6.77	189.7	178	68.21	93
15	8.47	132.1	150.6	54.2	93
16	6.59	135.2	149.6	71.5	92
17	5.70	189.3	201.1	64.4	93
18	6.32	134.8	149.8	53.3	92
19	8.45	158.9	147.2	67.6	94
20	6.71	194.4	197.9	63.7	92

blow) and various indicators. The fitting results are shown in Figs. 5.45, 5.46, 5.47 and 5.48.

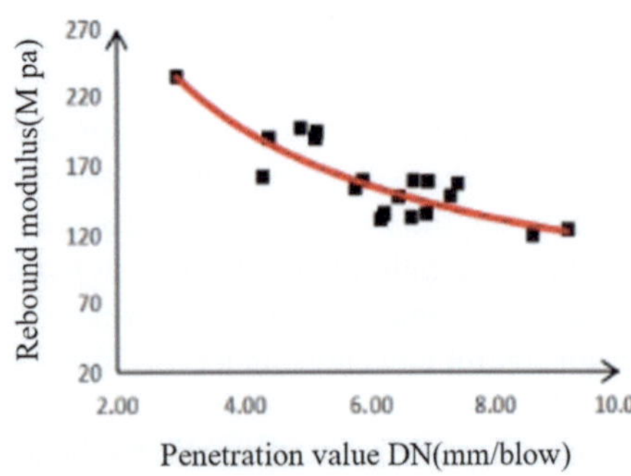

Fig. 5.45 Regression relationship between penetration value and rebound modulus

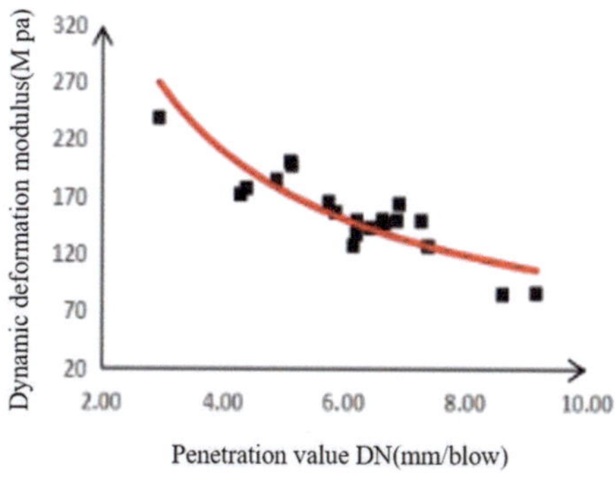

Fig. 5.46 Regression relationship between penetration value and dynamic deformation modulus

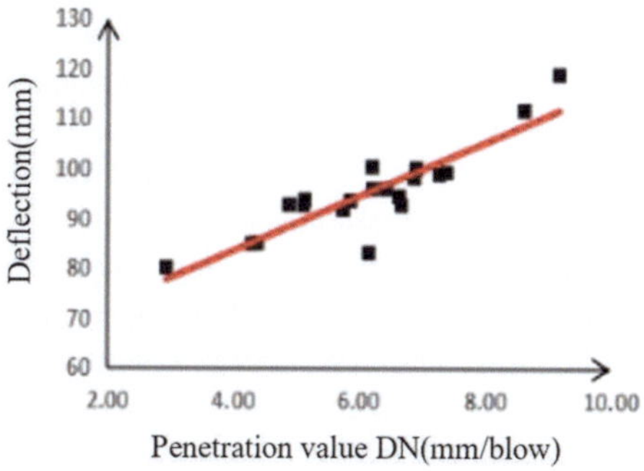

Fig. 5.47 Regression relationship between penetration value and deflection

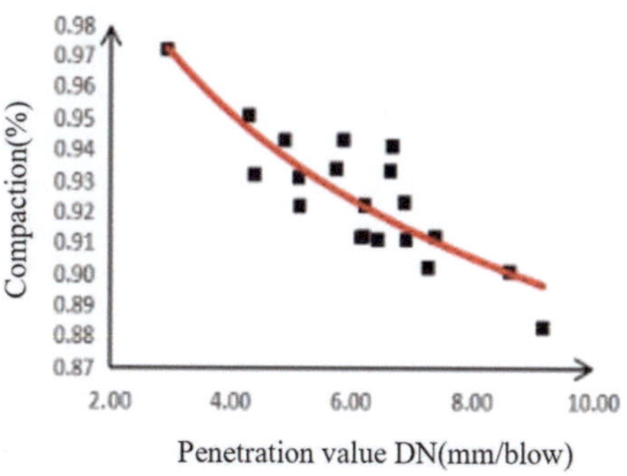

Fig. 5.48 Regression relationship between penetration value and compaction degree

According to the above analysis, the penetration DN (mm/blow) has a certain correlation with the field rebound modulus, dynamic deformation modulus, deflection value, and compaction degree, and the fitting degree R^2 are all above 0.7, as shown in the Table 5.42, which has a certain reliability. This indicates that using DCP as a method for quick testing of red mud subgrade not only has economic and convenient features, but also has a certain reliability, which has a certain reference value in the future construction process of red mud subgrade.

Table 5.42 Relationship between penetration value and various indicators

Indicator	Relationship formula	R^2
Penetration ~ rebound modulus	$E = 435.33DN^{-0.577}$	0.7023
Penetration ~ dynamic deformation modulus	$E = 647.6DN^{-0.813}$	0.7275
Penetration ~ deflection	$l = 5.452DN + 61.791$	0.7767
Penetration ~ compaction degree	$K = 1.0503DN^{-0.675}$	0.717

The evaluation of the road mechanical properties of stabilized red mud directly relates to the quality of the pavement structure layer, the operational quality after road construction, and the amount of maintenance work. The assessment of this issue in road construction is of great importance. For the highway engineering of stabilized and solidified red mud road materials, issues such as subgrade performance evaluation and the road mechanical performance testing and evaluation of stabilized and solidified red mud road materials used for embankment construction are addressed using PFWD dynamic modulus and FWD deflection detection, and a systematic field test analysis was conducted.

For the rebound modulus test of high modulus stabilized red mud subgrade, the appropriate test method and quality control standards should be selected according to the stress and deformation conditions of the subgrade. The peak impact load generated by PFWD ranges between 8 and 10 kN, and the corresponding bearing plate pressure is between 120 and 150 kPa, which is close to the pressure transferred to the top of the subgrade by the vehicle load. Therefore, the PFWD test load size has good consistency with the subgrade working conditions.

PFWD and FWD tests reflect the dynamic deflection and deflection basin generated under dynamic load, and the dynamic modulus characteristics of the subgrade under impact load can be calculated from this. Deflection is a macroscopic manifestation of the stress–strain properties of the subgrade, which can reflect the modulus and deformation characteristics of the subgrade. By comparison, it can be seen that there is a good power relationship between the deflection value and modulus of the stabilized red mud subgrade, as shown in Fig. 5.49 and Table 5.43.

Fig. 5.49 PFWD modulus E_d and FWD deflection/relationship

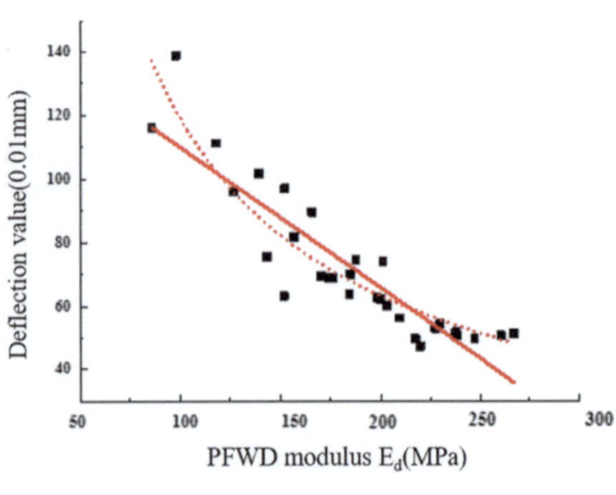

Table 5.43 Relationship between different modulus and deflection

Index	Fit type	Regression formula	R^2
$E_d \sim l$	Linear	$l = 0.0.4441\ E_d + 154.444$	0.7958
	Power	$l = 8280.45\ E_d{}^{0.92056}$	0.8363

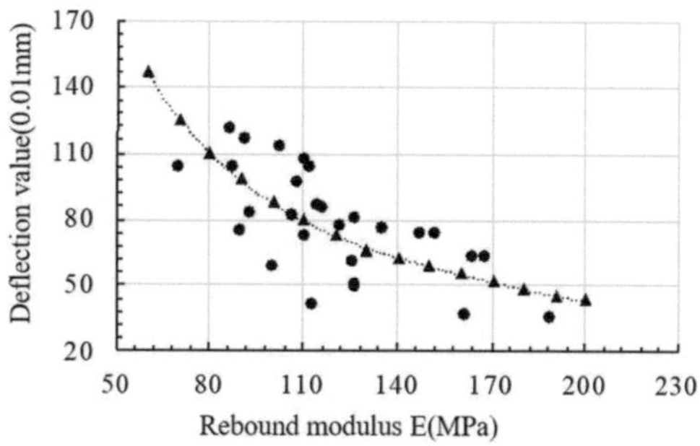

Fig. 5.50 Relationship between measured rebound modulus and deflection

Through the comparative testing of the PFWD dynamic deformation modulus and FWD dynamic deflection of the formed subgrade, there is a good correlation between the two test indicators. PFWD has the advantages of easy operation, high reliability, and test load close to the actual working conditions of the subgrade, suitable for construction process quality monitoring and evaluation. FWD has the advantages of high precision, high efficiency, and avoidance of manual reading errors, suitable for overall subgrade quality acceptance and evaluation.

In the detection of modified solidified red mud road materials test road, the parallel test of deflection and load plate rebound modulus can obtain the relationship between the measured rebound modulus and deflection, see Fig. 5.50. This relationship has important engineering guidance value for similar projects. Using the empirical relationship of measured regression, the rebound modulus control standards of different layers of the subgrade of modified solidified red mud road materials can be obtained.

5.7 Summary of This Chapter

(1) Through the red mud road demonstration project of the first phase of the road network project in Binzhou Beihai Vein Industrial Park, the field reliability and controllable pollution of modified solidified red mud road materials were verified, and the following conclusions were obtained:

① The original red mud has characteristics of high alkalinity, high liquid limit, and low strength and modulus. From the current highway design, acceptance and other norms or standards, the unconfined compressive strength, rebound modulus, California bearing ratio (CBR) and other indicators of the original red mud cannot meet the relevant requirements. Therefore, from an engineering perspective, red mud is not suitable for direct use as subgrade fill material for highways and municipal roads.

② The test results of modified solidified red mud have significantly improved compared to the original red mud. Its unconfined compressive strength and rebound modulus values have greatly increased, and can meet the road use requirements of relevant norms. The liquid limit and plastic limit are reduced, making the red mud have good on-site construction and ease of use.

③ According to the determination of the field bearing plate method, under the condition that the rebound modulus on the surface of the lime soil layer is less than 60 MPa; the rebound modulus on the top of the subgrade after the construction of the modified solidified red mud is greater than $E_0 >$ 130 MPa, far exceeding the requirements of design documents and relevant standards.

④ According to the results of DCP comprehensive dynamic probing, the CBR value of the modified solidified red mud subgrade is distributed between 25 and 60, and the measured rebound modulus of the subgrade surface is distributed between 100 and 250; the CBR value of the lower lime soil layer is distributed between 10 and 15, and the measured rebound modulus of the lime soil layer surface is distributed between 20 and 50 MPa. The CBR and corresponding layer top rebound modulus indicators obtained from the on-site DCP comprehensive dynamic probing are consistent, meeting the subgrade design control requirements.

(2) Relying on the implementation of the modified red mud subgrade demonstration project in the experimental section of the Jiqing Expressway reconstruction and expansion project, the reliability of the modified solidified red mud road materials for highway use and the feasibility of the construction process are verified:

① According to the engineering characteristics and process requirements of the modified solidified red mud, the on-site road mixing and centralized factory mixing construction methods were proposed, and the related technical requirements of key processes such as mixing, spreading, rolling, and curing during construction were pointed out.

② Through on-site core sampling for unconfined compressive strength tests, the unconfined compressive strength of the on-site core cut into pieces is 2.0 ~ 2.8 MPa, the on-site core variance is 0.246, and the standard deviation is 0.49621, indicating that the strength of the improved red mud mixed on the construction site can be guaranteed accordingly.

③ Through on-site bearing plate rebound modulus, PFWD, FWD, DCP detection, the results show: the subgrade's rebound modulus, dynamic rebound modulus, strength, CBR value are large and higher than the specification value, the deflection value is small, the deformation produced under the impact of vertical load is small, and with the increase of age, its strength, modulus all show a significant increase, and the deflection value decreases, indicating that with the increase of curing age, the strength, stiffness and other road performance of the subgrade have been greatly improved.

④ Through the sand filling method for on-site compaction test, the on-site compaction in the 95% confidence interval is 91% ~ 93%, indicating that the improved red mud has good road performance even when the compaction does not meet the specification requirements;

⑤ Through the regression analysis of DCP penetration DN with rebound modulus, dynamic deformation modulus, deflection value, and compaction, it is known that the penetration DN has a good correlation with the quality control indicators and design indicators of the red mud subgrade, and the fitting degree is all greater than 0.7, providing an economical and convenient way for the rapid detection of the modified solidified red mud subgrade.

(3) Relying on the implementation of the experimental road construction in the Zhoucun section of the G309 Qinglan line, the road performance of the stabilized red mud subgrade is studied:

① Aiming at the engineering characteristics and construction process requirements of the modified solidified red mud material, a continuous centralized factory mixing construction method for the red mud subgrade layer of the highway project was proposed, and specific technical requirements and quality control points for key processes such as mixing, spreading, rolling, and curing during construction were put forward.

② The modified solidified red mud test road was subjected to PFWD deformation modulus, FWD deflection, and DCP on-site testing. Through fixed-point testing, the correlation between the dynamic rebound modulus and deflection value (deformation) measured by PFWD and FWD was established, providing application reference for the design and quality control of the modified solidified red mud pavement structure. The impact force of the PFWD test drop hammer is close to the pressure transmitted to the top of the subgrade by the vehicle load, which can effectively evaluate the filling quality of the stabilized red mud subgrade.

③ According to the tracking test results of FWD different structural layers, the overall strength and bearing capacity of the modified solidified red mud top surface are higher than that of lime stabilized soil. The overall deflection of the modified solidified red mud pavement subgrade layer structure is basically the same as that of the conventional pavement structure, which can meet the requirements of road design deflection and bearing capacity, and the economy is obvious.

(4) A systematic study was carried out on the dynamic response of the modified solidified red mud subgrade or asphalt pavement structure under vehicle load, the distribution and attenuation characteristics of the deflection response of the

red mud base pavement structure were discussed, and the permanent deformation and fatigue life of the asphalt pavement structure were analyzed, and the following conclusions were drawn:

① Under the standard axle load, the deflection of the inorganic binder modified red mud road surface is about 7.12 mm/100, and the deflection of the composite material solidified red mud road surface is about 8.76 mm/100, which meets the performance requirements of the asphalt pavement structure. The deflection response of the inorganic stabilized red mud road surface is even slightly better than that of the conventional asphalt road surface. However, the overload has a large difference in the impact on different stabilized red mud base road structures. Among them, the composite stabilized red mud asphalt road structure, due to the lack of a layer of water-stable structure, is more damaged under overload conditions, and the increase in road surface deflection is significantly greater than that of the conventional lime soil asphalt road structure. The greater the degree of overload, the greater the increase in the deflection of the composite stabilized red mud road structure; while the inorganic stabilized red mud, due to the three layers of water stability and the high bearing capacity of the red mud subgrade, the impact of overload on it is significantly less than that of the conventional lime soil asphalt road structure.

② Within a design life of 15 years, the cumulative equivalent axle times of fatigue cracking of the semi-rigid base layer of inorganic stabilized red mud and composite stabilized red mud asphalt pavement structure are approximately 2.0 × 10 10 and 2.6 × 10 10 times, respectively, both meeting the design requirements.

③ Within a design life of 15 years, the asphalt layer ruts of the red mud base pavement structure and the conventional lime soil subgrade pavement structure are both 9.02 mm, both less than the design value of 15 mm, meeting the pavement structure design requirements.

④ The permanent deformation of the asphalt layer of the inorganic material modified red mud and composite stabilized red mud asphalt pavement structure, and the conventional lime soil asphalt pavement structure reaches 10.91 mm after 20 years of use. According to the highway maintenance specifications, the two types of stabilized red mud asphalt pavement structures do not need maintenance during the 20-year use period.

In summary, the red mud-based asphalt pavement structure should adopt the integrated collaborative design method of stabilized red mud materials for different application parts, that is, the subgrade part uses inorganic material modified red mud or composite stabilized red mud to replace conventional lime soil, and the base layer uses composite stabilized red mud materials, the overall service performance of the road structure is better.

Chapter 6
Exposure Scenarios and Risk Assessment of Modified Solidified Red Mud for Road Use

This chapter focuses on the use of modified red mud for highway construction, analyzes the exposure scenarios of modified red mud for road use, including exposure pathways and how to determine the simulation parameters of exposure scenarios, further studies the method for determining the exposure concentration of pollutants in modified red mud, and proposes the method for evaluating the impact of the process of using modified red mud on the soil. Based on the red mud from Chiping, Liaocheng as the base material, a specific simulation scenario for the use of modified solidified red mud for road use is established, and a risk assessment of the exposure scenario is conducted based on this, providing a reference for the environmental risk assessment of the actual engineering application of modified red mud.

6.1 Analysis of Exposure Scenarios for Modified Red Mud Road Use

The road structure of conventional material sections mainly consists of the following five parts from bottom to top:

①The bottom layer is the natural soil base; ② The ordinary material base layer, located above the natural soil base, is the bottom layer of the road structure, with a thickness of not less than 50 cm, the higher the highway grade, the greater the thickness; ③ Lime soil subgrade, the ordinary material base layer is lime soil subgrade, each layer is about 20 cm thick, generally 2 ~ 4 layers, with a total thickness of about 80 cm; ④ Cement stabilized gravel base layer, the lime soil subgrade is cement stabilized gravel base layer, each layer is 18 cm thick, generally 2 ~ 3 layers, with a total thickness of about 54 cm; ⑤ The surface layer is the pavement surface layer, generally asphalt or concrete pavement, with a thickness of about 15 cm, see Fig. 6.1.

The overall design of the road structure and the use of modified red mud during future construction have been carried out. In the road structure of the proposed

Y. Cheng and X. Zhu, *Utilization and Environmental Impact Assessment of Red Mud Resources*, https://doi.org/10.1007/978-981-96-7411-4_6

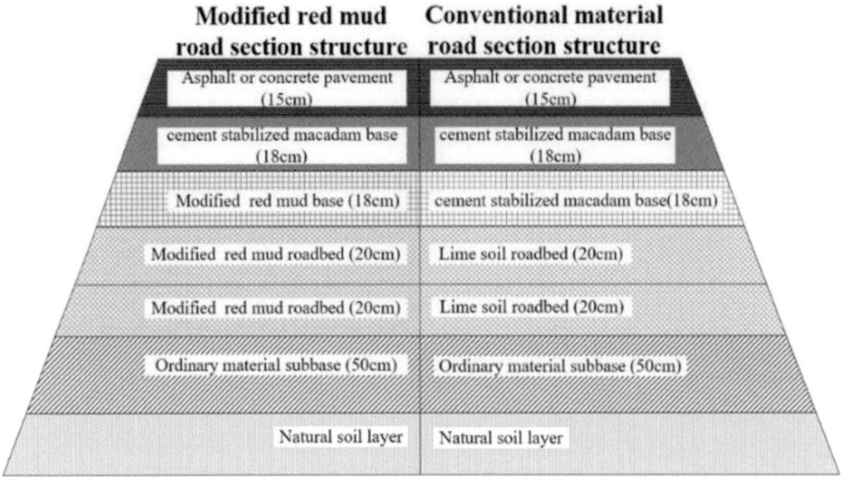

Fig. 6.1 Comparison of road structure schematic diagrams of modified red mud section and conventional material section

modified red mud section, modified red mud will replace the base fill material of two layers of pavement structure: ① Partially replace the cement stabilized gravel base layer. It is proposed to use the modified red mud base layer to replace the lower cement stabilized gravel base layer (18 cm). In the road structure of conventional sections, there are generally 2 ~ 3 layers of cement stabilized gravel base layers, and the modified solidified red mud base layer will replace the bottom layer of cement stabilized gravel layer. ② Replace all lime soil subgrades. Use modified red mud to replace all lime soil subgrades (each layer is 20 cm, a total of about 80 cm), the actual number of layers and thickness will be determined according to the on-site implementation situation, see Fig. 6.1.

6.1.1 Exposure Pathways for Modified Red Mud Road Use

① Exposure pathways during construction

The exposure pathways of pollutants during the use of modified solidified red mud for road use mainly include: inhalation of dust or volatile pollutants, accidental ingestion of red mud dust, contact exposure of dust falling on the skin, and drinking of pollutants leached from modified red mud into the water environment. During the laying process of modified red mud, it will be compacted layer by layer. In windy weather, there may be dust, which will be accidentally inhaled, ingested, or skin contacted by on-site workers. At the same time, the settled dust will also have a certain impact on the soil and groundwater environment of the settlement point. However, the risk can be controlled by taking dust prevention and suppression measures and doing a good job

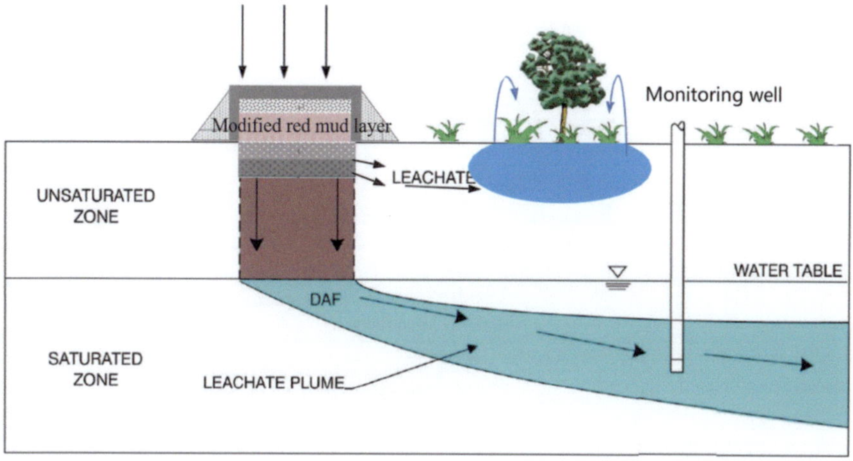

Fig. 6.2 Schematic diagram of exposure scenarios of red mud as subgrade filling material

in occupational health protection for construction workers during the construction process. Therefore, the environmental risk during the construction period will not be considered in the assessment.

② Exposure pathways during use

In the evaluation, the focus will be on the exposure pathways during the use of modified red mud for road construction, and there is no possibility of exposure to modified red mud dust through ingestion, inhalation or skin contact during the use process after completion. The main exposure pathway during the use of highways built with red mud as road material is: the leaching (filtration) action of precipitation causes harmful substances in the modified solidified red mud layer to dissolve, migrate and transform into environmental media, thereby polluting surface water or groundwater, as shown in Fig. 6.2.

③ Analysis of project exposure pathways

The main topography of Liaocheng City is the alluvial plain of the lower Yellow River, with relatively flat and open terrain; the main strata are Quaternary sedimentary strata. Except for the Ordovician limestone outcrops along the Yellow River in Dong'e in Liaocheng City, the rest are all covered by Quaternary strata of the Cenozoic era, with a thickness generally between 30 and 400 m, well-developed aquifers, and the aquifer groups composed of Quaternary loose sediments constitute the most important exploitable aquifers, with many aquifer layers but not large thickness; due to the flat terrain, the particles that make up the aquifer are fine, the permeability of the aquifer is weak, resulting in a gentle groundwater hydraulic gradient and extremely slow groundwater flow; the total amount of fresh groundwater resources in Liaocheng City is 90081 million m^3, the groundwater resources in the eastern Tuha River basin are abundant, and the groundwater resources in the western Majia River basin are

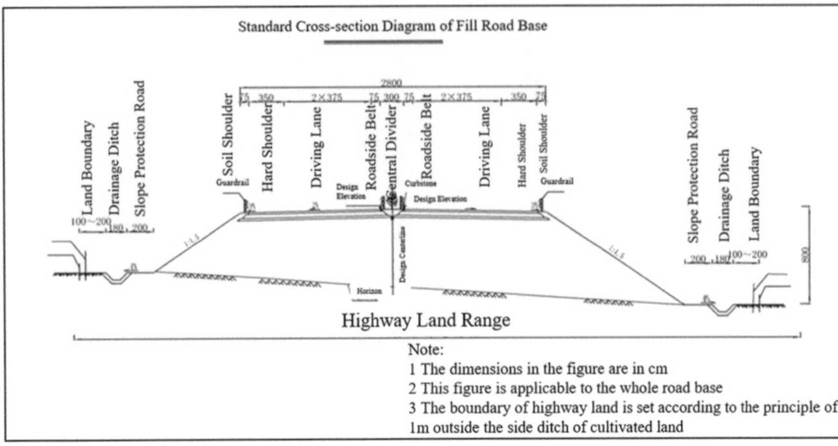

Fig. 6.3 Standard cross-section of fill subgrade highway

poor; from the vertical distribution of groundwater, there are three types: shallow groundwater (the bottom interface is about 60 m deep), medium-deep brine (buried at 60 ~ 180 m) is all brine, deep fresh water (buried below 180 m). The sources of groundwater recharge are atmospheric precipitation and Yellow River seepage. The overall groundwater quality in Liaocheng City is relatively poor, with more than half being Class V water; Class I and II waters are sporadically distributed in a few townships; Class III waters are distributed in Liangcun, Gaotang, the northern part of Guan County, the northwestern part of Dongchangfu District, and Wang Laoxiang in Chiping County, Liuji and Li Yang in Yanggu County. The highways in Liaocheng City are mostly fill-type highways, so this scenario is a fill-type subgrade highway based on modified red mud (standard cross-section see Fig. 6.3).

According to Article 4.1.5 of the "Highway Asphalt Pavement Design Code", waterproof and drainage measures should be taken for the pavement structure to prevent precipitation from infiltrating into the pavement structure layer. According to the design requirements of expressways, for fill-type subgrade expressways, the edge of the subgrade should be higher than the water accumulation height on both sides of the subgrade, and the effects of groundwater and capillary water should be considered so as not to affect the strength and stability of the subgrade. In addition, the internal drainage of the pavement structure should be connected with other relevant drainage systems of the highway, and should comply with the current "Highway Drainage Design Code" (JTG/T D33) relevant provisions. When designing highway drainage, the size of rainfall should be considered according to the topography and geological conditions along the highway, and the subgrade water should be led into nearby ditches through interception, drainage, and diversion.

According to the above analysis, there are two exposure pathways for this modified red mud road use.

① After the rainwater infiltrates into the pavement structure, it seeps vertically downwards, and the pollutants are released and migrate to the groundwater body due to the leaching (filtration) action; ② After the rainwater infiltrates into the pavement structure, it is laterally discharged from the subgrade structure, together with the pavement runoff and other road structure drainage, it enters the side ditch or road drainage system, and after entering the side ditch, it seeps into the groundwater.

6.1.2 Determination of Exposure Scenario Simulation Parameters

The factors that affect the release of pollutants in the exposure scenario mainly include: the release rate of the pollutants themselves, the pH value of the rainfall, and the amount of rainfall. Therefore, after determining the method of simulating the exposure scenario, specific simulation parameters need to be further determined.

① Simulation leaching method

According to the schematic diagram of the road structure of the modified red mud road, in the actual road subgrade filling process where the modified red mud is used as the subgrade material, it is generally layered, evenly compacted, and the compacted modified red mud base layer can be regarded as a dense block structure. The "Test of Diffusion Leaching Behavior of Inorganic Components in Block Waste or Building Materials" (i.e., NEN7375) method is used to understand the leaching characteristics of inorganic components in block materials in specific exposure scenarios, obtain the release amount and release mechanism of inorganic components in block materials, and evaluate the release behavior of pollutants in block materials in various specific exposure scenarios. Therefore, the NEN7375 leaching method can be used to simulate and study the long-term release behavior and change trend of pollutants in the modified red mud in the environment under actual road base application conditions.

② Determination of simulation parameters

In the actual application process of subgrade materials, it is necessary to combine relatively suitable experimental parameters to reflect the real situation. The parameters that need to be simulated in the leaching experiment mainly include the type and pH of the leaching agent, the liquid–solid ratio, and the leaching method. The following will determine each parameter reasonably in combination with the actual application scenario of modified red mud as subgrade filling material.

(1) Selection of Leaching Agent

The main factor affecting the release of pollutants in road materials is the action of rainfall. The nature of rainfall, the amount of rainfall, and the periodic characteristics of rainfall (intermittency) are the main external factors affecting the release of pollutants in road materials. When evaluating the environmental risk of using inorganic industrial slag waste as a resource for road materials, the release behavior of metal pollutants in road materials in the exposure environment and its influencing factors are key to determining its environmental safety. Among them, the pH of the exposure environment medium is one of the main factors affecting the release of metal pollutants. Combined with the analysis of rainfall components and rainwater pH in our country, such as the long-term acid rain type in our country is mainly sulfuric acid and nitric acid, the most unfavorable principle often considered is the leaching action of rainfall under acid rain conditions, and it is determined to carry out scenario simulation experiments. More often, a mixed acid leaching agent of sulfuric acid and nitric acid that is more conducive to the leaching of heavy metals is selected. However, when modified red mud is used as a road material, considering that the alkalinity of red mud is too high, the main pollution problem considered is not only heavy metals, but the more prominent environmental problem is the environmental impact that may occur when the pH is too high. If a mixed acid leaching agent of sulfuric acid and nitric acid is selected, it will play a neutralizing role and cannot reflect the most unfavorable principle. According to the "Summary of Environmental Quality in Liaocheng City" (2016–2019) released by the Liaocheng City Environmental Protection Bureau in recent years, the maximum range of rainfall pH in the urban area of Liaocheng City was 6.26 ~ 8.73 in 2016, and the minimum was 6.63 ~ 8.14 in 2019. The highest annual average value is 7.32 ~ 7.84, and no acid rain samples have been detected, as shown in Table 6.1. Therefore, under this specific scenario, considering the principle of more unfavorable conditions, pure water (pH = 8.6) is selected as the leaching agent.

(2) Determination of Liquid–Solid Ratio for Simulation Experiment

Generally, the liquid–solid ratio selected for the leaching experiment is determined by the actual application disposal scenario being simulated. The liquid–solid ratio of the actual application disposal scenario can be estimated by dividing the total volume of liquid in contact with the solid material over a certain period of time by the total mass of the application disposal material.

The road section using modified red mud simulated by the leaching experiment is an asphalt concrete pavement. According to the "Highway Engineering Technical Standards", the design service life of the asphalt surface layer of highways is 15 years, and the design reference period of the cement concrete surface layer of highways

Table 6.1 Statistics of rainfall pH monitoring data in Liaocheng City

Year	Rainfall pH range	pH average	Acid rain detection rate (%)	Acid rain pH limit
2019	6.63–8.14	7.32	0	≤ 5.60
2018	7.18–8.50	7.67	0	≤ 5.60
2017	7.04–8.51	7.54	0	≤ 5.60
2016	6.26–8.73	7.84	0	≤ 5.60

Data Source Liaocheng City Ecological Environment Bureau, Summary of Environmental Quality in Liaocheng City 2016–2019

is 30. Therefore, the service life of the proposed road surface is 15 or 30 years. Since the proportion of highways choosing asphalt concrete pavement is very high at present, the service life is temporarily set at 15 years in this study. The runoff coefficient of the asphalt concrete pavement is 0.95, and its infiltration coefficient is 0.05. According to relevant data, the average annual rainfall in Liaocheng City from 1962 to 2013 was 564.9 mm, the maximum annual rainfall was 977.2 mm, and the minimum was 187.2 mm, as shown in the Table 6.2, the average annual rainfall here is 565 mm. Based on the above analysis, the rainfall, infiltration coefficient, red mud block thickness and density, etc., are substituted into formula (6.1) to calculate the liquid–solid ratio. The volume of the leachate calculated from this can only partially submerge the test block. In order to facilitate the actual operation of the simulation experiment, referring to the NEN7375 method, the amount of leachate added in the simulation experiment should keep 2 cm of leachate above, below and around the modified red mud block. The actual water usage is estimated to be about 1.37 l, as shown in Fig. 6.4.

(3) Leaching method

In actual application scenarios, since the modified red mud is unlikely to remain in a fully wet or fully dry environment for a long time, but is in a state of alternating dry and wet with the change of natural environmental conditions, continuous leaching conditions are more conducive to the dissolution of pollutants in the test block than intermittent leaching. Therefore, according to the principle of assuming the most unfavorable conditions, continuous leaching is chosen as the leaching method for the simulation leaching experiment.

In summary, referring to the NEN7375 standard method, according to the determined simulation parameters of the exposure scenario, using pure water as the leaching solution, the modified red mud solidification test block is kept with 2 cm of leaching solution above, below and around, and continuously leached for 64 days.

Table 6.2 Characteristics of precipitation in each administrative region of Liaocheng City from 1962 to 2013

Administrative region	Annual precipitation/mm		
	Average	Maximum	Minimum
Dongchangfu District	580.4	954.2	308.5
Linqing	546.4	977.2	187.2
Guan County	553.8	892.7	280.6
Xin County	529.1	937.9	256.1
Yanggu	565.8	920.3	286.0
Dong'e	602.8	897.7	275.4
Chiping	588.9	942.1	308.5
Gaotang	552.6	954.0	287.1
City Average	564.9	917.2	312.7

Data source Wang Y et al (2015) Analysis of precipitation trend in Liaocheng City, Anhui Agricultural Science 43(26):207–210, 265

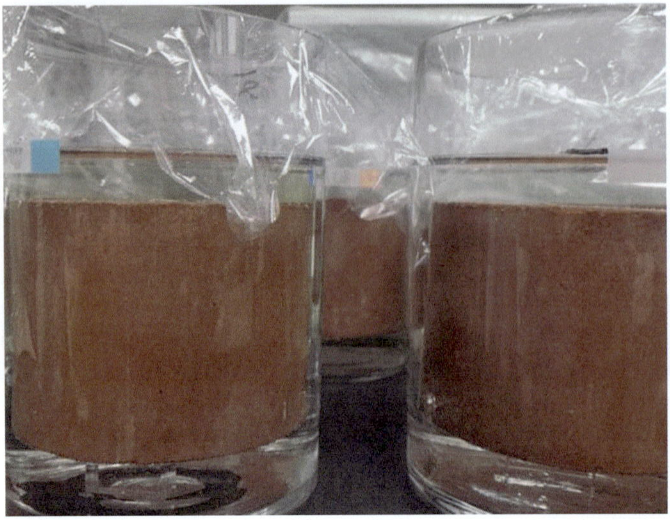

Fig. 6.4 Modified red mud block in the simulation experiment (there is 2 cm of leachate above, below and around the modified red mud block)

Later, according to the actual situation in the experiment, it was continued for a total of 92 days.

③ Leaching test method

The continuous leaching experiment in the NEN7375 standard method is divided into 8 stages to change the leaching solution. In the actual continuous soaking process of the modified red mud test block for road use, due to the high pH, in order to obtain an accurate trend, it is continued according to the actual pH situation. The experimental device is shown in Fig. 6.5, the leaching solution replacement time is shown in Table 6.3, and the actual experimental situation is shown in Figs. 6.6 and 6.7.

④ Accumulated release of pollutants

By analyzing and measuring the concentration of each characteristic pollutant in the leaching solution replaced in the above 8 stages of the NEN7375 experiment, the release mechanism and amount of pollutants in the modified red mud can be understood.

Calculation of heavy metal release in each leaching stage *I*:

$$E_i = \frac{C_i \times V}{f \times A} \tag{6.1}$$

where E_i is the heavy metal leaching amount of leaching stage *I*, mg/m^2; C_i is the heavy metal concentration in the leaching solution of leaching stage *I*, μg/L; *V* is the

Fig. 6.5 Schematic diagram of continuous leaching tank experiment (*Note* 1—modified red mud test block, 2—test block bracket, height 2 cm, total 2 pieces)

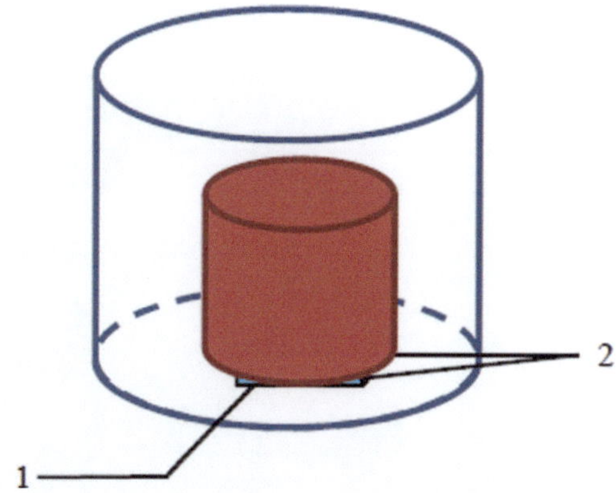

Table 6.3 Continuous leaching experiment design

Leaching stage	Leaching solution replacement time		Remarks
	Cumulative time/d	Interval time/d	
1	0.25	0	The regular leaching stage lasts for 64 days in total
2	1	0.7518	
3	2.25	1.2530	
4	4	42	
5	9	120	
6	16	168	
7	36	480	
8	64	672	
Extension stage	92	672	Due to high pH, to obtain accurate trend changes, the experiment was extended for 28 days

volume of the leaching solution, L; A is the surface area of the test block, m^2; f is the conversion factor, 1000 μg/mg.

Calculation of accumulated release ε_n:

$$\varepsilon_n = \sum_{i=1}^{n} E_i \quad n = 1, 2, \ldots, 8 \tag{6.2}$$

⑤ Prediction of long-term release of pollutants

Through the results of the leaching experiment and the study of the release mechanism, we can understand the release amount and rate of pollutants in the modified red mud test block. In order to grasp the release of pollutants from the modified red mud

Fig. 6.6 Site picture of the continuous leaching experiment of modified red mud (container mouth uncovered)

Fig. 6.7 Site picture of the continuous leaching experiment of modified red mud (container mouth covered)

during actual use, a pollutant release model needs to be established. The release model proposed by Crank, which is based on diffusion as the main release mechanism for inorganic components, is selected. This model is a one-dimensional diffusion model based on Fick's second law of diffusion and is widely used for predicting the release amount controlled by diffusion. The release model is as follows:

$$M_{area} = 2 \times \rho \times C_0 \left(\frac{D^{obs} \times t}{\pi} \right)^{1/2} \tag{6.3}$$

where M_{area}: t: Accumulated release of heavy metals in unburned brick test blocks per unit area over time, mg/m^2; C_0: Effective amount of heavy metals, mg/kg; ρ: Density of unburned brick test blocks, kg/m^3; D^{obs}: Diffusion coefficient, m^2/s; t: Leaching time, s.

The diffusion coefficient D_i^{obs} is calculated as follows:

$$D_i^{obs} = \pi \left[\frac{M_i}{2 \times \rho \times C_0 \left(\sqrt{t_i} - \sqrt{t_{i-1}} \right)} \right]^2 \quad i = 1, 2, \ldots, 8 \tag{6.4}$$

where D_i^{obs}: Diffusion coefficient of pollutant at stage I, m^2/s; M_I: Leaching amount of pollutant at stage I, mg/m^2; C_0: Effective release amount of pollutant, mg/kg; ρ: Density of unburned brick test blocks, kg/m^3; t_I: Time at the end of leaching stage I, s; t_{I-1}: Time at the end of leaching stage $I - 1$, s.

The overall diffusion coefficient of heavy metals during the leaching cycle D^{obs} is the arithmetic mean of the diffusion coefficients at each leaching stage.

$$D^{obs} = \frac{1}{n} \times \sum_{i=1}^{n} D_i^{obs} \tag{6.5}$$

According to formula (6.5), a prediction model for the release amount of characteristic pollutants in modified red mud test blocks is obtained, which can be used to predict the long-term release amount of characteristic pollutants in modified red mud.

6.2 Risk Assessment Method for Exposure Scenarios of Modified Red Mud for Road Use

6.2.1 Method for Determining the Exposure Concentration of Pollutants in Modified Red Mud

The key to environmental risk assessment is to determine the source release concentration and exposure concentration. There are two approaches in this assessment:

① Primary assessment based on leaching values

According to the "Groundwater Quality Standard" (GB/T18484-2017), groundwater is divided into five categories based on the groundwater quality status in China, health risks, etc. The third category is groundwater with medium chemical composition, mainly suitable for centralized drinking water sources and industrial and agricultural water use, i.e., drinking does not pose a health risk. Therefore, one risk assessment method is to compare the exposure concentration with the groundwater quality standard. If it exceeds the groundwater quality standard, it is considered risky; otherwise, it is not risky. To simplify the risk assessment, this study uses the leaching value of characteristic pollutants as the exposure concentration. If the leaching value is lower than the third-level standard value of groundwater quality, there is no risk; if the leaching value exceeds the third-level standard value of groundwater quality, further research is needed to determine whether there is a potential environmental risk.

② Risk assessment based on specific scenarios

This assessment is based on the exposure scenario analysis in the previous section and combined with environmental surveys. It calculates the leaching situation of pollutants in the red mud filling material layer through rainwater leaching (dissolution), and uses the Industrial Waste Evaluation Management (IWEM) model developed by the U.S. EPA to assess the leaching and environmental risk of modified red mud as a substitute for road base material during road use. IWEM is an evaluation model developed by the U.S. EPA's Office of Resource Conservation and Recovery (ORCR) to manage the impact of pollutant leaching after the resource utilization process of general industrial solid waste on groundwater. The initial version was developed in 2002, and it has now been upgraded to version 3.1. This system simulates the waste treatment units, structural fill bodies. In the scenario of road use, the process of pollutant release from waste, migration and transformation through the soil layer and groundwater saturation layer to the groundwater monitoring well is considered. This model can help managers briefly assess the suitability of general industrial solid waste for structural filling or road reuse with minimal data input, and propose corresponding suggestions or permit conditions. IWEM uses Monte Carlo analysis to handle the uncertainty brought about by default values and other non-user-specified modeling parameters.

In the Monte Carlo analysis, thousands of complete simulation processes (source module, destination and migration modeling) are run, taking input values from each iterative distribution, generating the probability distribution of the expected concentration of each waste component in groundwater. By improving different schemes, such as the choice of different lining layers for waste treatment units, IWEM will simulate the concentration of 90% percentile in groundwater and compare it with the maximum contaminant level (MCL) in groundwater or health-based numbers (HBN) provided by users to determine a more effective scheme. For waste road use situations, IWEM calculates the expected concentration distribution of pollutants in general industrial solid waste used as road-building materials to reach the groundwater monitoring well. IWEM adds up the 90% percentile exposure concentration of all such pollutants in the road structure to get the total 90% percentile groundwater exposure concentration, and compares it with the benchmark of the component.

IWEM includes a database of over 200 waste component chemical properties and MCLs, and allows for the addition of waste components or modification of component properties, such as soil–water partition coefficients (Kd) or degradation coefficients, in the database. It also allows for the customization of groundwater benchmark concentrations and exposure times.

IWEM consists of the following main components (or modules):

(1) User-friendly graphical interface;
(2) Pollutant source release module, simulating the release process from waste disposal units, road use, and structural filling;
(3) Migration transformation and destination model: IWEM evaluation system integrates the EPA CMTP migration transformation model (i.e., leachate migration and transformation composite model), which consists of a vadose zone model and a saturated layer model, and can predict the exposure amount of pollutants in groundwater downstream (at the extraction well). This model can simulate the process of pollutant release from the waste in the treatment unit (WMU) and its migration to groundwater through the vadose zone, including convection, hydraulic diffusion, linear/non-linear adsorption, first-order chain decay reactions, etc., and predict the exposure concentration of pollutant components in groundwater at downstream target sensitive points. This model can be used to judge the severity of site pollution and assist in determining the priority order of site risk control. The EPA CMTP model mainly consists of four modules: pollutant, pollution source, unsaturated zone, and saturated zone. The model assumes that pollutant migration in the unsaturated zone is mainly vertical transport, and after entering the saturated zone, it is three-dimensional migration and diffusion.
(4) A series of database parameters for waste components and specific site-related information. Waste road use evaluation: The road use evaluation module in IWEM is an independent component. The use of industrial solid waste as road-building material for a part of the road structure determines the pollutant release pattern. The output from the road use evaluation module includes the trend of pollutant release flux over time and the time trend of the concentration of each pollutant component in the waste. The output results are provided to EPACMTP, which simulates its migration and transformation in soil and groundwater, obtains the total exposure concentration of all components in 90% percentile groundwater, and is used to determine whether the groundwater is below or exceeds the quality benchmark provided by the user.

In the road assessment model of IWEM software, each part in Fig. 6.8, such as the road surface, shoulder, median strip, embankment, valley, etc., is assumed to be a relatively independent vertical unit from the surface to the saturated aquifer. In this assessment, only the road surface base material uses modified red mud, and the shoulder, valley slope, etc., do not use industrial waste materials. Therefore, only the vertical unit of the road surface part can be evaluated.

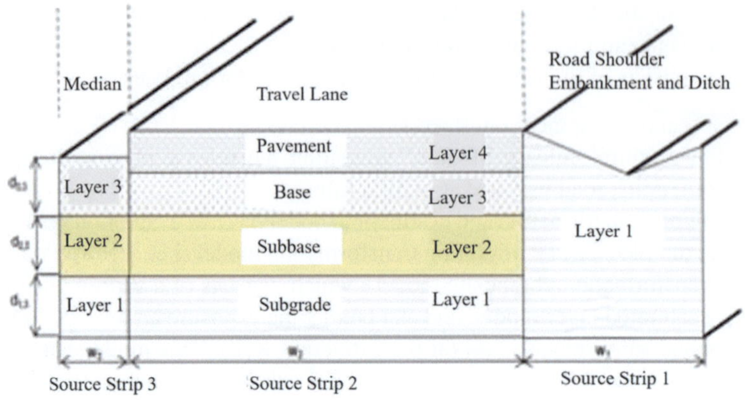

Fig. 6.8 Example of typical road structure suitable for IWEM and layer information required for model input

6.2.2 *Evaluation Method of Soil Impact in Modified Red Mud Road Use Process*

Liaocheng City governs six counties, one city, and two districts, with a total of 8.29 million mu of arable land and a population of 5.8 million, of which 4.28 million are agricultural population. The entire territory of Liaocheng is flat, with an altitude between 22.6 and 49.0 m, higher in the southwest and slightly lower in the northeast; it is mainly the Yellow River alluvial plain, with fertile land, and its natural conditions are very suitable for the cultivation of high-efficiency economic crops such as vegetables and edible fungi. It is one of the key areas for facility vegetable cultivation planned by the Ministry of Agriculture in the Yellow River, Huai River, and Bohai Sea. After many years of unified planning and development, six major industrial belts have been formed along the main highways. Since the pH of red mud remains relatively high after modification, considering the actual situation in Liaocheng, if there is agricultural land on both sides of the highway, the impact of the leachate on the soil pH during the process of modified red mud road use needs to be considered.

Earthworms, as representatives of soil invertebrates, play an important role in maintaining and improving soil structure, enhancing soil fertility, improving soil aeration, absorbing and degrading organic matter in the soil. The acute toxicity test of earthworms is a basic short-term test and one of the most basic ecological toxicity tests, which uses terrestrial organisms to use soil as a medium for transmitting pollutants. Therefore, this study chose earthworms as one of the subjects to assess the growth inhibition effect of modified red mud leachate on them.

Plant growth tests are used to evaluate the effects of solid or liquid compounds on the germination and initial growth stages of various plants planted in the soil. Seed germination is a critical period in the plant life cycle and is one of the most sensitive periods to external environmental factors. The germination rate is not only the most intuitive physical quantity to measure the germination ability of seeds, but also an important indicator of higher plant toxicology tests. Therefore, seed germination tests can be used to preliminarily assess the toxicological effects of pollutants on plants growing in the soil.

Therefore, this study intends to study the impact of pH on soil during the use of modified red mud through earthworm and plant growth tests.

1. Evaluation basis

ISO 11269-1: 2012 Soil quality—Plant germination and root elongation inhibition test.

GB/T 21809: 2008 Chemicals—Earthworm Acute Toxicity Test.

2. Test samples

Since modified red mud is mainly used as a subgrade material during highway construction in the jurisdiction of Liaocheng City, the simulation test abandons the practice of artificially preparing experimental soil and directly selects farmland soil collected from Chiping County, Liaocheng City. The soil samples used were collected from the east side of the village committee of Beizhangzhuang Village, Hutun Town, Chiping County, Liaocheng City, Shandong Province, 1.5 km away, near the Tuha River, as shown in Fig. 6.9.

3. Test organisms

(1) Earthworms

Eisenia foefide, often chosen for earthworm acute toxicity tests, is not a typical soil species, but its sensitivity to chemicals is similar to that of earthworms that actually inhabit the soil. The life cycle of Eisenia foefide is relatively short, requiring 3–4 weeks to hatch at 20 °C, and can mature in 7–8 weeks. It is a good test species, so this earthworm acute toxicity test chose Eisenia foefide. The test Eisenia foefide was purchased online from Beijing Kunlong Breeding Center in Changping District, Beijing.

Before the test, the test earthworms were cultured in unpolluted soil rich in organic matter. After receiving the earthworms sent by the breeding center, they were first

Fig. 6.9 Simulation soil sample collection point

placed in a light incubator and cultured under the same light and temperature conditions as the experiment for 1 week, with no abnormalities found; then, the earthworm incubator was moved to the laboratory for indoor culture until the last batch of experiments began, the test earthworms were all cultured under natural room temperature and light in the laboratory.

(2) Test crop—Pak Choi

The test crop for the plant growth test was Pak Choi, which is sensitive to germination environment and common in Liaocheng area. The change of environmental factors on seed germination and seedling production is evaluated through crop seed germination tests. The test crop seeds were purchased from Tianlifang Seed Industry on Taobao, and the pre-test found that the germination rate of the purchased Pak Choi seeds was 100%.

4. Earthworm survival rate test

(1) Test preparation

The earthworm survival rate test was conducted in a glass beaker. The glass beaker was soaked in acid, rinsed with tap water, and then rinsed with deionized water before and after use. Before the test, select mature earthworms with obvious clitellum, healthy and active, and similar size (visually estimated natural length is about 4 ~ 5 mm, weight is about 300 ~ 500 mg) for standby. The earthworm survival rate test was conducted in a light incubator, with the experimental conditions in the box: temperature 20 °C, humidity 80%, light 1333 lx (intermittent light, i.e., 12 h light, 12 h dark). Check the status of the artificial light incubator before the test to ensure normal operating conditions.

(2) Preliminary test

The same simulated experimental conditions as the formal test are used, including: test soil, test earthworms, water-soil ratio, etc. The difference is that in the preliminary test, deionized water is used instead of the leachate of the modified red mud solidified block to adjust the water-soil ratio to 1:4. Weigh 400 g of test soil into a glass beaker, then add 100 mL of deionized water, let it sit indoors for more than 1 h, and manually mix evenly.

After placing 10 earthworms in each beaker, seal it with plastic wrap, and then use tape to seal the connection between the plastic wrap and the glass beaker. Use a toothpick to evenly poke about 30 holes in the plastic wrap at the mouth of the beaker to facilitate the breathing of the earthworms. Place the beaker with holes into the light incubator with adjusted experimental conditions, and cultivate under the same temperature, humidity, and light experimental conditions as the formal experiment (i.e., temperature of 20 °C, humidity of 80%, 12 h light and 12 h dark intermittent light conditions). Spray a small amount of water in the incubator according to the actual situation to maintain humidity.

After 7 and 14 days of cultivation, the soil in the beaker is gently poured into a shallow dish, the earthworms are picked out and placed in a glass petri dish, and the survival of the earthworms and their reaction to mechanical stimulation at the front and tail are examined. After 7 days of exposure, about 5 g of nutrient soil is added to the soil and an appropriate amount of water is sprayed in, mixed evenly, and then the soil and earthworms are put back into the original beaker, repeating the steps of sealing, sealing, poking holes, and cultivating in the incubator. After 14 days of exposure, the experiment is terminated, the earthworms are checked and counted, and the soil and earthworms are no longer reused. The preliminary experiment is repeated three times. In the preliminary experiment, the survival rate of earthworms after 7 and 14 days of exposure was 100%.

(3) Formal Experiment

In the test soil, the experiment of adding leachate from different treatment modified red mud solidification blocks at different leaching time periods is called modified red mud leachate treatment; the experiment of adding deionized water is the control treatment. The experimental process and conditions of the control treatment are the same as those of the preliminary experiment.

The process of the modified red mud leachate treatment experiment is as follows: weigh 400 g of test soil and put it into a glass beaker, then add 100 mL of leachate collected from different treatment modified red mud solidification blocks at different time periods, and let it stand for more than 1 h and mix manually. Then select healthy, obvious clitellum, similar size and quality adult earthworms and put them into the beaker, each parallel beaker of different treatments contains 10 test earthworms. The sealing of the beaker, the sealing of the connection, and the poking of holes at the mouth of the beaker are the same as in the preliminary experiment and the control experiment. The glass beaker treated as described above is placed in a light incubator for cultivation, and the cultivation experiment temperature, humidity, and light conditions are the same as in the preliminary experiment and the control experiment. The earthworms are taken out after 7 and 14 days of exposure, and the steps of the preliminary experiment and the control experiment are repeated to check the survival and mechanical response of the earthworms, and corresponding records are made. Each treatment is repeated three times.

(4) Counting

Count the survival of the test earthworms after 7 and 14 days of exposure in the counting experiment.

Quality control

All earthworms are purchased from the same batch of the same variety, under the same cultivation conditions;

Select healthy, active earthworms with obvious clitellum and similar size and quality before the experiment;

In the pre-cultivation experiment, the survival rate of earthworms after 7 and 14 days of exposure is more than 80%.

The glass beakers used in each experiment must be strictly cleaned, using a three-step cleaning process of acid washing—tap water rinsing—tap water rinsing to avoid cross-contamination.

(5) Data Processing

Calculate the average and standard deviation of the number of surviving earthworms in each treatment, and calculate the survival rate of earthworms after 7 and 14 days.

5. Seed Germination and Root Elongation Test

(1) Test Preparation

All the petri dishes used in the plant germination test are disposable plastic petri dishes, with a diameter of about 15 cm. Each batch of tests uses new petri dishes, and the used petri dishes are discarded directly and not reused.

Before the start of the experiment, the seeds need to be sorted, selecting fuller bok choy seeds, eliminating as much as possible those that are visibly not full or incomplete, without excluding individual misjudgments. Select a number of grains that is at least 1.5 times the required number for this batch of experiments.

The experiment is conducted in a light incubator, using natural light, and the incubator temperature is controlled at 25 ± 1 °C. Check the status of the incubator before the experiment to ensure that the operating conditions are normal, especially the temperature control conditions.

(2) Preliminary Experiment

The same simulation experiment conditions as the formal experiment were adopted, including: experimental soil, test crop seeds, soil–water ratio, etc. The difference is that in the preparatory experiment, deionized water was used to replace the leachate of the modified red mud solidification block to adjust the soil–water ratio to 1:5. The petri dishes with deionized water were placed indoors for 1–2 h and manually mixed evenly. The sorted seeds were sown in disposable plastic petri dishes, 10 in each group, with three parallels. After 4 days of cultivation in a light incubator, the germination rate was counted, and the germination rate was 100%.

(3) Formal experiment

In the test soil, the experiment treatment of adding the leachate of the modified red mud solidification block at different leaching time periods is called the modified red mud leachate treatment; the addition of deionized water is the control treatment. 75 g of test soil was spread in a disposable plastic petri dish. In the red mud leachate treatment experiment, after each leaching time period, the leachate of the modified red mud solidification block was added to the test soil in the plastic petri dish;

correspondingly, an appropriate amount of deionized water was added to the petri dish of the control treatment experiment. The soil–water ratio of different treatments was adjusted to 1:5, and the petri dishes with leachate or deionized water were placed indoors for 1–2 h and manually mixed evenly. Among the pre-selected crop seeds, 10 seeds of uniform size were planted in the test soil in the disposable plastic petri dish, the petri dish cover was covered, and it was placed in the light incubator. The incubator temperature was controlled at 25 ± 1 °C. After 4 days of cultivation under natural light, the experiment was stopped, the germinated seeds were counted, and the root length was measured. Each leachate of the modified red mud solidification block at different time periods was repeated three times (10 seeds $\times$ 3 = 30 seeds).

For the leachate of the modified red mud solidification block at different leaching time periods, if the pH of the leachate of the modified red mud solidification block shows a downward trend, the experiment will stop.

(4) Counting and measuring

Counting and measuring: Count the germinated seeds in each group of the formal experiment and measure their root length.

Quality control: 1) In a single experiment, only seeds from the same batch in the same year or season that have not been treated can be used, that is, they have not been treated with sterilants or insect repellents. Try to ensure that the size, saturation, and particle size grade are at the same level. The moisture content of the test seeds is less than 10%, and they are stored at room temperature. 2) The seed germination rate is greater than 80%. 3) The utensils used in each experiment must be clean and free of contamination.

(5) Data processing

Calculate the average value and standard deviation of the germination rate and root elongation of each treatment, and calculate the inhibition rate of germination and root elongation.

6. Evaluation standards

Refer to the method recommended by the International Organization for Standardization (ISO), use the acute mortality rate (%) of earthworms to suggest the following division of biological toxicity levels, see Table 6.4 for details.

Table 6.4 Earthworm biological toxicity level division standard

Toxicity level	Mortality (%)
Extremely toxic	≥ 80
High	80–50
Medium	50–30
Low	30–10
Slightly toxic or non-toxic	≤ 10

Table 6.5 Biological toxicity level division standard

Toxicity level	Germination inhibition rate (%)	Root elongation inhibition Rate (%)
Extremely toxic	≥ 80	≥ 90
High	80–50	90–80
Medium	50–30	80–50
Low	30–10	50–30
Slightly toxic or non-toxic	≤ 10	≤ 30

Refer to the method recommended by the International Organization for Standardization (ISO), use the seed germination rate, root elongation inhibition rate (%), to suggest the division of biological toxicity levels, see Table 6.5 for details.

6.3 Environmental Risk Assessment of Modified Red Mud for Road Use

6.3.1 Sample Collection

Seven original red mud samples were collected from the red mud dump at different production times, spanning from 2010 to May 2019. Detailed information about the samples can be found in Table 6.6. The seven red mud samples collected from the red mud dump were modified by adding 8% and 12% Heinchem modification materials, respectively, and processed into modified red mud powder (see Table 6.6 and Fig. 6.10) for evaluation. Both the red mud and modified red mud powder samples were naturally air-dried at room temperature in the laboratory and stored for future use.

Table 6.6 Overview of red mud and modified red mud sample information

Serial number	Sample number	Production time	Modified red mud sample number	
			8% Modification material added	12% Modification material added
1	XFCN1	Before 2010 (wet discharge)	XFHY1-8	XFHY1-12
2	XFCN2	2010	XFHY2-8	XFHY2-12
3	XFCN3	2014	XFHY3-8	XFHY3-12
4	XFCN4	2016	XFHY4-8	XFHY4-12
5	XFCN5	2017	XFHY5-8	XFHY5-12
6	XFCN6	June 2018	XFHY6-8	XFHY6-12
7	XFCN7	May 2019	XFHY7-8	XFHY7-12

Fig. 6.10 Modified red mud powder (left) and schematic diagram of modified red mud solidified test block (right)

In actual operation, the modified red mud is compacted layer by layer during the roadbed filling process, so the entire roadbed is considered as a large block. According to the actual road construction conditions and actual paving compaction situation, modified red mud test blocks were prepared according to the "Highway Geotechnical Test Specification" (JTG E40-2007), and the block density was controlled by a compaction degree of 96%. After the specimens were prepared, they were cured at constant temperature and humidity (temperature 20 ± 2 °C, relative humidity 95%) for 7 days, for the study of pollutant release patterns during the use of modified red mud roads (see Fig. 6.10, the right picture is a cylindrical test block).

6.3.2 Identification of Characteristic Pollutants in Red Mud and Modified Red Mud

1. Main Physical and Chemical Characteristics and Pollution Features of Red Mud

Red mud is an extremely fine solid waste produced from bauxite during the production of alumina. Its particle size is at the micron level, with a large specific surface area and porous structure. Its specific gravity is 2.7 ~ 2.9, and its natural density is 1412 ~ 1511 kg/m^3, specific surface area 64.09 ~ 186.9 m^2/g, and porosity ratio 2.53 ~ 2.95. The main components of Bayer process red mud are shown in Table 6.7.

As can be seen from the table, the main chemical components of red mud are Al_2O_3, Fe_2O_3, SiO_2, CaO and Na_2O, etc., with slight differences in content; red mud also contains a small amount of organic matter and valuable elements, as well as a small amount of natural radioactive material. The largest source of organic matter in the Bayer process for producing alumina is bauxite, with an organic carbon

Table 6.7 Main chemical components of Bayer process red mud (%)

Component	Al_2O_3	SiO_2	CaO	Fe_2O_3	Na_2O	TiO_2	Others	Remarks
A certain company	27.36	23.21	14.90	11.08	8.63	6.25	8.57	Source 1
Domestic typical	18.31	7.63	18.21	26.38	7.53	2.53	19.41	Source 1
Australia	24.0	18.8	5.26	28.5	10.26	3.11		Source 2
India	17–22	4–16	1–5	35–54	3–6	3–19	10–15	Source 2
United States	18.4	8.5	7.73	35.5	6.1	6.31		Source 2
World Average	6–35	2–20	1–41	35–57	1–10	3–26	10–22	Source 1

Source 1 "Li P (2019) Study on alkali removal and iron recovery from bayer process red mud" in Tables 1.2 and 1.3; *Source 2* Sutar H et al. (2014). Progress of red mud utilization: an overview. Am Chem Sci J 4(3):255–279

content of about 0.1 ~ 0.6%, mainly humic acid and bitumen. Bitumen, insoluble in alkali, is discharged with red mud, while humic acid reacts with sodium hydroxide during high-pressure leaching to form sodium humate, which enters the solution and eventually forms soluble sodium oxalate, sodium carbonate, and other low molecular weight sodium salts.

The phase composition of Bayer process red mud was analyzed by X-ray diffraction, and it was found to be mainly composed of calcite, hematite, limonite, $Ca(OH)_2$ and –C_2S, as shown in Fig. 6.11.

Red mud, as a strongly alkaline solid waste, has high pH and contains fluorine and metal elements, which are environmental issues of concern. The main pollutants in red mud include:

(1) Alkali.

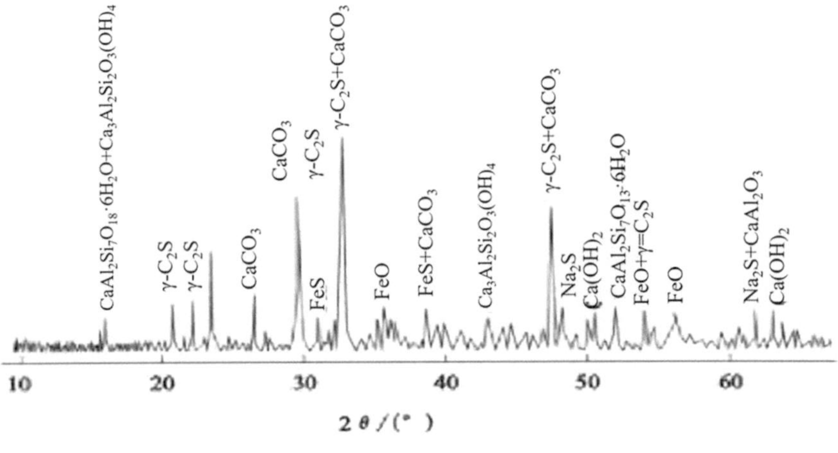

Fig. 6.11 X-ray diffraction pattern of Bayer process red mud

Red mud has a high pH due to the presence of a large amount of strong alkali, generally between 10.3 and 11.3, or even higher. It is generally believed that an alkali content of 30 ~ 400 mg/L is suitable for public water sources. Near red mud dumps without seepage prevention measures, high-alkaline wastewater seeps into the ground or enters surface water, causing the water body's pH to rise, leading to situations where the total hardness and pH of groundwater exceed the Class III groundwater quality standard, with groundwater pH reaching 11.2 and the highest total hardness approaching 1600 mg/L, more than twice the standard.

There are mainly two types of alkali substances in Bayer process red mud: one is soluble alkali, mainly including residual NaOH, Na_2CO_3, soluble sodium aluminate, etc.; the other is insoluble alkali, which is the insoluble hydrated sodium aluminosilicate formed by the reaction of SiO_2 in bauxite with sodium aluminate solution, its chemical formula can generally be expressed as $Na_2O \cdot Al_2O_3 \cdot xSiO_2 \cdot nH_2O$, mainly in the form of cancrinite. Among them, the soluble alkali content in red mud accounts for about 54% of the total alkali, and the insoluble alkali accounts for about 46% of the total alkali.

(2) Fluoride

Fluoride is another major pollutant in red mud, often discharged with red mud to the dump. The standard limit for fluoride in drinking water in our country is 1.0 mg/L, and the Class III standard limit for fluoride in groundwater quality standards is 1.0 mg/L. Some studies have reported that the fluoride content in red mud is 4.89 ~ 8.6 mg/L, and the fluoride content in the associated liquid of red mud can reach 11.5 ~ 26.7 mg/L. If red mud is not properly disposed of, the fluoride in it may contaminate groundwater.

(3) Other pollutants

In addition to high alkali and fluoride, red mud also contains a number of major inorganic non-metallic and metallic elements. He Yi (2013) analyzed the total amount of inorganic pollutants in 24 red mud samples from 8 companies (see Table 6.8) and the leaching content (see Table 6.9). Sun Zhaoyun (2017) analyzed the toxic and harmful substances in red mud samples from Zibo and Binzhou (see Tables 6.9 and 6.10). The leaching content of lead (Pb), arsenic (As), selenium (Se), mercury (Hg), and hexavalent chromium Cr^{6+} in red mud is higher than the Class III limit of groundwater quality standards.

2. Metal pollutants in Chiping Xinfachem red mud and modified red mud.

The results of metal content in red mnud raw samples and modified red mud samples are shown in Tables 6.11 and 6.12.

(1) Main metal pollutants in red mud

From Table 6.11, it can be seen that except for individual elements in individual samples, the content of most metal pollutants in red mud raw samples produced at

Table 6.8 Analysis results of major inorganic elements in red mud samples (wt%)

Element	C	O	Na	Mg	Al	Si	S	Cl	K	Ca	Ti	Fe
Company A	5.77	38.6	6.35		9.29	11	0.18			1.15	4.16	23.5
	28.25	37.67	4.54		6.45	4.03	0.25			0.68	2.33	15.8
	7.75	36.28	6.89		9.79	8.18	0.3			0.96	4.09	25.76
Company B	2.08	41.54	0.22	0.2	4.21	5.04				36.69	1.3	8.72
	2.48	40.75	0.18	–	3.64	4.62				38.28	1.53	8.52
	2.35	39.71	0.1	0.15	3.94	5.05				38.33	1.29	9.08
Company C	4.93	40.92	5.62	0.53	8.11	7.81	1.27		1.44	18.6	2.83	7.94
	3.81	40.68	6.21	0.52	9.18	7.94	1.22		1.84	17.13	2.97	8.5
	5.54	42.00	5.94	0.53	9.49	8.21	1.05		1.66	14.17	3.63	7.78
D Company	5.95	42.99	16.08	0.18	9.81	8.41	0.38	0.15	0.32	8.12	2.22	5.39
	3.3	42.74	9.00	0.19	11.91	10.88	0.28	0.15	0.50	11.34	2.84	6.87
	4.58	41.76	8.78	0.20	11.93	10.37	0.34	0.17	0.56	11.54	2.85	6.92
E Company	4.65	38.63	9.84		13.66	10.53	0.41		0.43	0.61	1.25	19.99
	8.48	40.34	9.58		12.45	10.23	0.31		0.46	0.49	0.87	16.79
	4.45	39.42	9.78		13.58	11.18	0.26		0.45	0.75	0.94	19.19
F Company	4.66	42.99	6.63	0.88	12.55	10.91	0.58		1.89	10.27	2.27	6.37
	5.39	44.92	6.48	0.96	12.18	10.3	0.45		1.65	10.34	2.36	4.97
	3.43	42.68	6.54	0.86	12.86	11.47	0.43		2.36	9.72	2.63	7.02
G Company	6.4	45.78	0.15		23.47	15.93	0.38		2.78		2.1	3.01
	5.61	46.74	0.3		23.51	15.92	0.34		2.59		2.23	2.76
	5.56	47.74	0.17		24.69	14.92	0.18		2.65		2.18	1.91
H Company	5.13	36.85	5.84		8.16	5.73	0.35			10.25	4.55	23.14
	3.67	37.41	5.52		9.34	6.07	0.41			11.10	4.94	21.54
	3.18	35.31	5.8		8.38	5.42	0.41			9.89	4.88	26.73

Data source: He Y et al (2013) Analysis of the current situation of red mud environmental management in China. Environ Sustain Dev 6:43–46

different time periods is quite close. Taking the two samples XFCN6 and XFCN7 produced in June 2018 and May 2019 as examples, the content of each metal pollutant is very close, indicating that the grade of raw ore and production process in the production of alumina this year are relatively stable.

(2) Main metal pollutants in modified red mud

From Table 6.12, compared with the level of metal pollutant content in red mud raw samples, after adding modified materials, the calcium content in modified red mud has increased, and except for the metal content of individual samples will change slightly, the content level of most metal pollutants in the rest of the modified red mud has not changed significantly. After statistical analysis, there is no significant difference in the content of metal pollutants between the modified red mud samples with 8%

Table 6.9 Test results of toxic and harmful substances in red mud

Serial number	Project	Unit	Zibo red mud	Binzhou red mud	Analysis method
1	Cyanide	mg/kg	0.40	0.40	CJ/T 221-2005
2	Total mercury	mg/kg	0.10	0.11	CJ/T 221-2005
3	Lead	mg/kg	17.9	28.8	GB/T 17141-1997
4	Cadmium	mg/kg	0.52	0.37	GB/T 17141-1997
5	Total chromium	mg/kg	639	698	CJ/T 221-2005
6	Copper	mg/kg	102	53	CJ/T 221-2005
7	Zinc	mg/kg	84	67	CJ/T 221-2005
8	Beryllium	mg/kg	0.84	0.74	CJ/T 221-2005
9	Barium	mg/kg	54	58	CJ/T 221-2005
10	Nickel	mg/kg	76	48	CJ/T 221-2005
11	Arsenic	mg/kg	20.7	24.6	CJ/T 221-2005
12	Cobalt	mg/kg	16.7	13.5	CJ/T 221-2005
13	Selenium Se	mg/kg	0.24	0.32	CJ/T 221-2005
14	Molybdenum	mg/kg	7.8	9.2	CJ/T 221-2005
15	Silver	mg/kg	0.28	0.36	CJ/T 221-2005
16	Antimony	mg/kg	0.56	1.35	CJ/T 221-2005
17	Vanadium	mg/kg	596	879	CJ/T 221-2005
18	Manganese	mg/kg	718	607	CJ/T 221-2005
19	Hexavalent chromium Cr^{6+}	mg/kg	41.5	37.2	CJ/T 221-2005

Data source Sun Z (2017) Research on engineering technology and environmental impact of Bayer process red mud filling road base. Master's thesis of Shandong University

Table 6.10 Partial red mud leaching toxicity test results

Element	Unit	Company A*	Company B*	Company C*	Company D*	Company E*	Company F*	Company G*	Company H*	Zibo**	Binzhou**	Groundwater quality Standard Class III limit value	Hazardous waste leaching toxicity Identification standard limit value
Copper Cu	mg/L	0.04	0.02	0.02	0.02	0.02	0.02	0.02	0.02	0.01	0.01	1	100
Lead Pb	mg/L	0.1	0.4	0.1	0.1	0.1	0.1	0.1	0.1	0.02	0.02	0.05	5
Zinc Zn	mg/L	0.005	0.058	0.005	0.005	0.005	0.005	0.005	0.005	0.01	0.01	1	100
Total chromium Cr	mg/L	0.05	0.05	0.05	0.05	0.05	0.05	0.05	0.19	0.39	0.20	–	15
Cadmium Cd	mg/L	0.005	0.005	0.005	0.005	0.005	0.005	0.005	0.005	0.004	0.004	0.01	1
Beryllium Be	mg/L	0.005	0.005	0.005	0.005	0.005	0.005	0.005	0.005	0.001	0.001	0.0002	0.02
Barium Ba	mg/L	0.1	28	0.1	0.1	0.1	7.7	0.1	0.1	0.01	0.01	1.0	100
Nickel Ni	mg/L	0.04	0.04	0.04	0.04	0.04	0.04	0.04	0.04	0.02	0.02	0.05	5
Arsenic As	mg/L	0.188	0.0044	0.0566	0.439	0.0542	0.004	0.0938	0.767	0.0097	0.0087	0.05	5
Selenium Se	mg/L	0.0303	0.0002	0.0266	0.014	0.0356	0.0037	0.0013	0.132	0.01	0.0073	0.01	1

(continued)

Table 6.10 (continued)

Element	Unit	Company A*	Company B*	Company C*	Company D*	Company E*	Company F*	Company G*	Company H*	Zibo**	Binzhou**	Groundwater quality Standard Class III limit value	Hazardous waste leaching toxicity Identification standard limit value
Silver Ag	mg/L	0.01	0.02	0.01	0.01	0.01	0.01	0.01	0.01	0.01	0.01	–	5
Mercury Hg	mg/L	0.0034	0.0014	0.0083	0.0021	0.0049	0.0002	0.0013	0.0104	0.00005	0.00005	0.001	0.1
Hexavalent chromium Cr^{6+}	mg/L	0.019	0.004	0.004	0.004	0.004	0.004	0.004	0.004	0.342	0.148	0.05	5
Vanadium	–	–	–	–	–	–	–	–	–	0.70	1.10	–	–
Antimony	–	–	–	–	–	–	–	–	–	0.02	0.02	0.005	–
Manganese	–	–	–	–	–	–	–	–	–	0.01	0.01	0.1	–
Cyanide CN^--	mg/L	0.001	0.001	0.001	0.001	0.001	0.001	0.001	0.001	–	–	–	–
Fluoride	mg/L	5.95	0.166	2.65	5.25	2.75	0.8	2.3	0.46	16.1	10.6	1.0	100
Methylmercury	ng/L	10	10	10	10	10	10	10	10	–	–	–	–
Ethylmercury	ng/L	20	20	20	20	20	20	20	20	–	–	–	–

**Data source* He Y et al (2013) Analysis of the current situation of red mud environmental management in China. Environ Sustain Dev 6:43–46
***Data source* Sun Z (2017) Research on the engineering technology and environmental impact of Bayer process red mud filling road base. Master's thesis of Shandong University

Table 6.11 Metal pollutant content in red mud

Sample number and production time		Be mg/kg	Na g/kg	Mg g/kg	Al mg/kg	K g/kg	Ca g/kg	Ti mg/kg	V mg/kg	Cr mg/kg	Mn mg/kg	Fe g/kg	Co mg/kg	Ni mg/kg
XFCN1	Before 2010 (Wet Discharge)	0.82	66.3	0.11	9710.5	0.66	1.8	16,927.6	414.5	593.3	379.5	143.6	11.3	42.3
XFCN 2	2010	18.6	55.5	0.80	18,268.7	2.50	7.3	13,770.5	440.1	484.1	592.0	118.8	28.1	61.2
XFCN 3	2014	1.61	37.8	0.14	13,231.6	0.80	1.8	15,874.9	845.1	353.5	665.9	210.4	12.9	32.2
XFCN 4	2016	0.88	52.8	0.31	30,745.2	0.47	6.0	13,388.0	426.3	509.8	437.7	124.6	10.3	33.8
XFCN 5	2017	0.76	64.7	0.10	11,677.3	0.30	2.5	16,934.0	429.5	487.0	320.2	136.1	13.6	34.8
XFCN 6	June 2018	0.70	41.0	0.52	64,686.0	0.52	8.8	4054.7	349.8	452.0	383.7	100.1	7.30	22.5
XFCN 7	May 2019	0.77	45.4	0.45	52,162.3	0.71	7.9	7771.1	498.3	365.2	300.3	106.3	7.70	18.8
Sample number generation time		Cu mg/kg	Zn mg/kg	As mg/kg	Se mg/kg	Mo mg/kg	Ag mg/kg	Cd mg/kg	Sn mg/kg	Sb mg/kg	Ba mg/kg	Tl mg/kg	Pb mg/kg	
XFCN1	Before 2010	50.3	57.9	22.5	2.5	8.3	0.43	1.8	11.4	3.3	28.2	0.06	34.1	
XFCN 2	2010	60.3	94.1	49.5	20.4	23.5	19.3	22.4	13.4	23.9	70.7	17.2	58.5	
XFCN 3	2014	100.1	93.9	15.5	1.0	3.7	0.33	1.3	9.5	3.6	31.3	0.08	16.9	
XFCN 4	2016	37.3	136.6	23.9	3.3	7.1	0.36	2.1	10.7	3.2	44.6	0.07	42.2	
XFCN 5	2017	29.3	45.1	33.3	1.4	5.1	0.43	1.7	12.8	3.5	24.9	0.05	38.3	
XFCN 6	June 2018	27.3	60.2	9.7	1.8	0.6	0.30	1.7	1.2	0.2	53.0	0.07	40.8	
XFCN 7	May 2019	20.8	37.2	30.6	3.8	6.7	0.35	1.7	8.0	2.1	60.8	0.09	41.0	

Table 6.12 Metal pollutant content in modified red mud

Sample number generation time		Be mg/kg	Na g/kg	Mg g/kg	Al mg/kg	K g/kg	Ca g/kg	Ti mg/kg	V mg/kg	Cr mg/kg	Mn mg/kg	Fe g/kg	Co mg/kg	Ni mg/kg
XFHY1-8	Before 2010	0.77	54.1	0.28	18,418.8	0.66	19.9	12,347.1	366.3	511.0	365.6	112.5	8.7	31.8
XFHY1-12		0.75	48.1	0.37	38,479.5	0.69	25.4	11,049.6	357.1	500.1	389.9	101.1	7.6	29.6
XFHY2-8	2010	0.99	44.5	0.34	39,610.9	1.14	22.7	9838.5	378.3	404.9	546.0	91.6	8.8	36.9
XFHY2-12		0.93	42.3	0.87	52,961.0	1.03	29.9	8620.7	346.8	380.1	503.0	82.9	8.0	32.1
XFHY3-8	2014	1.31	24.6	0.49	35,701.1	0.62	19.7	8774.9	701.7	299.3	564.4	127.4	7.7	19.6
XFHY3-12		1.25	24.4	0.37	39,418.7	0.62	21.4	9483.3	685.6	302.3	601.5	127.5	8.0	20.4
XFHY4-8	2016	0.91	40.9	0.37	36,497.8	0.52	20.8	12,142.9	390.7	451.3	407.2	113.6	9.0	29.8
XFHY4-12		0.90	38.4	0.52	29,903.2	0.64	29.1	3803.2	313.7	437.7	424.5	104.1	8.8	27.1
XFHY5-8	2017	0.70	44.2	0.73	59,571.3	0.49	21.7	4970.4	109.8	343.2	325.3	87.0	8.6	22.6
XFHY5-12		0.73	43.7	0.48	46,229.3	0.49	27.0	10,791.9	362.7	431.3	330.6	95.3	8.8	26.2
XFHY6-8	June 2018	0.73	41.7	0.33	21,486.5	0.78	27.9	11,209.5	479.9	448.5	456.6	128.8	9.4	27.0
XFHY6-12		0.70	36.5	0.78	59,291.7	0.69	32.8	9154.6	417.7	400.7	365.3	101.4	7.4	22.7
XFHY7-8	May 2019	0.90	38.3	0.44	27,203.5	0.95	28.2	8611.4	479.4	344.6	316.7	115.4	8.7	21.4
XFHY7-12		0.86	34.5	0.67	48,987.1	0.86	31.2	7485.5	454.3	325.1	359.0	95.7	7.1	23.4

Sample number and production time		Cu mg/kg	Zn mg/kg	As mg/kg	Se mg/kg	Mo mg/kg	Ag mg/kg	Cd mg/kg	Sn mg/kg	Sb mg/kg	Ba mg/kg	Tl mg/kg	Pb mg/kg
XFHY1-8	Before 2010	39.5	52.6	17.8	2.4	7.6	0.45	1.7	10.0	3.1	44.4	0.09	34.3
XFHY1-12		35.6	91.1	25.0	2.6	7.5	0.43	1.7	10.9	3.5	52.1	0.11	34.4
XFHY2-8	2010	32.0	74.6	28.8	2.3	5.4	0.33	3.1	12.9	6.0	64.1	0.11	38.6

(continued)

Table 6.12 (continued)

Sample number and production time		Cu mg/kg	Zn mg/kg	As mg/kg	Se mg/kg	Mo mg/kg	Ag mg/kg	Cd mg/kg	Sn mg/kg	Sb mg/kg	Ba mg/kg	Tl mg/kg	Pb mg/kg
XFHY2-12		27.8	56.3	29.1	2.6	5.8	0.32	2.8	12.9	4.2	75.9	0.13	37.4
XFHY3-8	2014	61.4	81.3	11.1	1.2	2.9	0.28	1.2	10.1	2.7	41.0	0.05	16.4
XFHY3-12		61.9	65.9	13.6	1.3	3.1	0.29	1.1	10.1	3.3	40.6	0.09	17.7
XFHY4-8	2016	64.1	87.7	29.7	2.7	5.6	0.30	2.0	13.3	3.2	54.3	0.07	40.7
XFHY4-12		28.6	56.1	9.5	2.4	0.8	0.34	1.9	3.4	0.4	61.2	0.06	37.5
XFHY5-8	2017	20.0	154.1	7.4	1.0	0.9	0.18	0.8	3.7	0.6	48.0	0.06	44.6
XFHY5-12		21.2	67.8	26.1	1.8	4.5	0.32	1.6	11.9	3.2	40.2	0.05	38.5
XFHY6-8	June 2018	31.7	82.0	27.4	2.5	6.7	0.30	1.6	10.5	3.0	53.3	0.07	38.2
XFHY6-12		26.6	46.4	36.4	2.7	6.8	0.28	1.8	15.3	3.9	58.4	0.07	35.9
XFHY7-8	May 2019	22.9	76.0	31.4	3.4	6.8	0.35	1.6	9.9	3.4	69.7	0.09	40.4
XFHY7-12		20.8	67.2	36.6	3.4	6.8	0.30	1.5	11.0	3.0	77.3	0.09	38.9

Heinchem modified material and 12% Heinchem modified material. Therefore, the addition of 8% and 12% modified materials has little effect on the total metal content in the red mud raw samples.

The metal concentrations in the leachate of the original and modified red mud are shown in Tables 6.13 and 6.14.

(1) Leaching content of major metal pollutants in red mud

From the Table 6.13, it can be seen that the leaching content of most metals in the original red mud is lower than the "Groundwater Quality Standard" (GB/T14848-2017), and all or most of the aluminum (Al), chromium (Cr), arsenic (As), selenium (Se), and molybdenum (Mo) in the original red mud are higher than the third level standard of groundwater quality. At the same time, due to the high leaching content of vanadium (V), and there is no limit in GB/T14848-2017, the limit of 0.05 mg/L in the "Surface Water Environmental Quality Standard" (GB3838-2002) is used as a reference, and it is found that the leaching content of vanadium (V) is higher than the third level standard of surface water quality. There are 3 samples of calcium (Ca) and two samples of iron (Fe) that have leaching contents higher than the third level standard of groundwater.

(2) Leaching content of main metal pollutants in modified red mud

From Table 6.14, it can be seen that the leaching content of most metals in the modified red mud is lower than the third level standard of GB/T14848-2017, and all or most of the aluminum (Al), calcium (Ca), chromium (Cr), arsenic (As), selenium (Se), and molybdenum (Mo) in the original red mud are higher than the third level standard of groundwater quality, while the leaching content of vanadium (V) is higher than the third level standard of surface water quality. For the exceeded Al, Ca, V, Cr, As, Se, Mo, statistical analysis shows that the leaching content of V, As, Se, Mo in the modified red mud is significantly lower than that in the original red mud ($p = 0.00 < 0.05$), while the leaching content of Ca, Al, Cr in the modified red mud is significantly higher than that in the original red mud. The difference in the leaching content of V, As, Se, Mo, Ca, Al, Cr in the modified red mud with 8 and 12% modified materials is not significant. It can be seen that the modified material has a significant solidification effect on V, As, Se, Mo in the red mud, although the solidification effect of adding 12% modified material is slightly better than that of adding 8% modified material, but the difference is not significant.

3. Fluoride in the original and modified red mud from Chiping Xinfah

The leaching content of fluoride (F) in the original and modified red mud from Chiping Xinfah is shown in Table 6.15.

As can be seen from Table 6.15, the leaching content of F in the original XFCN2 red mud is relatively low, at 4.42 mg/L, while the F leaching concentration in the other 6 red mud samples ranges from 15.5 to 42.9 mg/L. After adding 8% modified material, the leaching content of F in the modified red mud ranges from 1.39 to 4.78 mg/L. After adding 12% modified material, the leaching content of F in the modified red mud is 1.15 to 4.61 mg/L. Statistical analysis shows that the F in the

Table 6.13 Metal leaching content in red mud

Sample number generation time		Be μg/L	Na mg/L	Mg mg/L	Al μg/L	K mg/L	Ca mg/L	Ti μg/L	V μg/L	Cr μg/L	Mn μg/L	Fe mg/L	Co μg/L	Ni μg/L
XFCN1	Before 2010 (wet discharge)	0.01	–	30	15,432.5	12,790	290	8.74	1468	18	0.57	70	0.64	1.33
XFCN 2	2010	0.02	–	60	12,212.5	2740	750	33.98	734.3	70.65	1.95	470	0.15	2.63
XFCN 3	2014	0	–	30	55,025	8380	510	11.51	5704	140.3	0.52	90	2.46	1.79
XFCN 4	2016	0	–	10	19,670	7920	240	4.23	1912	180	0.35	10	1.41	3.67
XFCN 5	2017	0	–	10	39,250	4350	590	12.17	2915	53.34	0.56	120	0.85	1.26
XFCN 6	June 2018	0.01	–	10	37,900	6940	260	3.77	7069	236.6	0.34	10	1.55	1.93
XFCN 7	May 2019	0.02	–	100	7802.5	4580	300	13.25	1542	516.6	2.33	290	3.22	3.05
GB14848-III		2	200		0.02		450			Cr^{6+} 50	100	300	50	20
GB3838-III		2						100	50	Cr^{6+} 50	100	300	100	20

Sample number generation time		Cu μg/L	Zn μg/L	As μg/L	Se μg/L	Mo μg/L	Ag μg/L	Cd μg/L	Sn μg/L	Sb μg/L	Ba μg/L	Tl μg/L	Pb μg/L
XFCN1	Before 2010 (Wet discharge)	53.43	− 1.06	217.5	280.9	470.8	0.07	− 0.06	0.13	2.44	0.9	0.01	0.01
XFCN 2	2010	30.67	4.63	46.39	18.79	55.46	0.48	− 0.1	0.11	1.02	2.28	0.01	0.19
XFCN 3	2014	98.92	− 0.48	951.1	478.7	789.4	0.08	− 0.06	0.17	2.01	0.43	0.01	0.2
XFCN 4	2016	88.07	− 0.53	275.1	451.5	492.6	0.01	− 0.1	0.18	2.34	0.08	0.01	− 0.05
XFCN 5	2017	40.56	− 0.02	326.9	178	263.9	− 0.03	− 0.01	0.13	2.23	0.29	0	0.04
XFCN 6	June 2018	95.71	− 1.2	981.1	271.7	403.9	− 0.02	− 0.08	0.55	2.22	− 0.06	0.01	− 0.08
XFCN 7	May 2019	45.69	0.87	163.3	260.5	274.5	− 0.01	− 0.15	0.18	2.09	2.98	0.02	0.19
GB14848-III		1000	1000	10	10	70	50	5		5	700	0.1	10
GB3838-III		1000	1000	50	10	70		5		5	700	0.1	50

Table 6.14 Leaching content of metals in modified red mud

Sample number	Generation time	Be μg/L	Na mg/L	Mg mg/L	A1 μg/L	K mg/L	Ca mg/L	Ti μg/L	V μg/L	Cr μg/L	Mn μg/L	Fe mg/L	Co μg/L	Ni ug/L
XF HY 1-8	Before 2010 (wet discharge)	0.01	–	0	58,925	6580	700	6.72	928.1	446.4	0.43	50	8	4.21
XF HY 1-12		0	–	10	59,825	8040	1110	4.31	545.1	893.5	0.33	30	3.27	3.65
XF HY 2-8	2010	0	–	0	32,700	3410	2610	3.19	626.8	331.2	0.33	20	1.63	0.99
XF HY 2-12		0.01	–	10	50,850	6690	11,680	3.92	291.4	787.1	0.3	40	0.78	0.88
XF HY 3-8	2014	0.01	–	10	38,275	5900	1940	4.94	971.9	186	0.33	40	0.67	0.47
XF HY 3-12		0	–	0	45,625	5910	1440	5.91	834.4	171.5	0.4	50	0.69	0.5
XF HY 4-8	2016	0	–	0	48,525	4470	850	5.83	1235	295	0.43	40	3.45	2.16
XF HY 4-12		0.01	–	0	71,300	6620	1870	7.83	687.1	366	0.48	60	3.66	2.36
XF HY 5-8	2017	0	–	0	60,650	3930	1030	8.6	1718	171	0.39	70	4.49	2.98
XF HY 5-12		0.01	–	0	52,175	6220	2610	5.56	947.5	202	0.34	40	5.45	3.29
XF HY 6-8	June 2018	0.01	–	0	110,575	5350	4590	8.85	420.1	254.8	0.53	90	1.18	0.78
XF HY 6-12		0	–	0	110,150	7490	1600	4.97	253.9	302.9	0.4	50	1.22	0.81
XF HY 7-8	May 2019	0	–	0	66,500	5750	3490	2.39	1184	166.7	0.34	20	2.08	1.61
XF HY 7-12		0.01	–	0	80,525	6620	5310	5.99	697.6	147.6	0.44	60	2.16	1.78
GB14848-III		2	200		20		450			Cr^{6+} 50	100	300	50	20
GB3838-III		2						100	50	Cr^{6+} 50	100	300	100	20

Sample number	Generation time	Cu μg/L	Zn μg/L	As μg/L	Se μg/L	Mo μg/L	Ag μg/L	Cd μg/L	Sn μg/L	Sb μg/L	Ba μg/L	Tl μg/L	Pb μg/L
XF HY 1-8	Before 2010	64.59	0.44	107.8	96.1	226.8	− 0.02	− 0.13	0.28	2.09	0.1	0.01	0.04
XF HY 1-12		45.67	− 1.2	69.8	79.7	219.5	0.07	− 0.09	0.17	1.09	− 0.12	0.01	10.04

(continued)

Table 6.14 (continued)

Sample number	Generation time	Cu μg/L	Zn μg/L	As μg/L	Se μg/L	Mo μg/L	Ag μg/L	Cd μg/L	Sn μg/L	Sb μg/L	Ba μg/L	Tl μg/L	Pb μg/L
XF HY 2-8	2010	25.88	0.6	45.2	86.9	97.3	− 0.04	− 0.09	0.13	1.06	− 0.01	0.01	0.18
XF HY 2-12		27.71	− 0.46	34.4	106.6	142.7	10.04	10.08	0.17	0.77	0.15	0.01	10.04
XF HY 3-8	2014	24.28	− 0.08	32.5	9.0	30.4	− 0.04	0.07	0.23	0.78	0.05	0.01	0.02
XF HY 3-12		28.34	− 0.11	27.0	7.9	27.2	10.02	10.04	0.15	0.69	0.05	0.01	0.41
XF HY 4-8	2016	55.01	− 0.16	106.2	132.2	169.8	− 0.03	− 0.08	0.21	1.48	0.11	0.01	10.04
XF HY 4-12		50.55	− 0.38	64.5	125.9	156.3	0	− 0.12	0.19	1.11	− 0.09	0.02	0.01
XF HY 5-8	2017	42.22	0.73	87.3	47.9	94.9	10.0A	0.07	0.23	1.45	0.01	0.01	0.07
XF HY 5-12		43.17	1.14	58.8	43.4	87.9	− 0.04	− 0.1	0.22	1.13	− 0.01	0.01	− 0.03
XF HY 6-8	June 2018	32.61	− 0.76	34.3	50.7	128.8	− 0.01	− 0.08	0.21	0.89	− 0.06	0.05	0.01
XF HY 6-12		35.02	− 1.1	31.2	52.3	136.9	0	− 0.12	0.18	0.49	0.18	0.03	0
XF HY 7-8	May 2019	48.32	− 0.36	82.63	98.46	262.1	10.04	10.04	0.22	0.97	− 0.08	0.04	10.00
XF HY 7-12		51.85	0.79	576	69.8	191.9	− 0.02	10.04	0.21	0.65	10.04	0.03	0.02
GB14848– III		1000	1000	10	10	70	50	5		5	700	0.1	10
GB3838-III		1000	1000	50	10	70		5		5	700	0.1	0

Table 6.15 Leaching content of F in red mud and modified red mud

Sample number	Red mud (mg/L)	8% Modified material (mg/L)	12% Modified material (mg/L)
XFCN1	15.5	2.45	1.81
XFCN2	4.42	3.01	3.10
XFCN3	42.9	1.39	1.15
XFCN4	35.4	4.60	4.61
XFCN5	20.7	3.53	1.64
XFCN6	36.1	3.31	1.38
XFCN7	20.3	4.78	4.05

Note The fluoride content in red mud is significantly higher than that in modified red mud

original red mud is significantly higher than that in the modified red mud; however, the difference in F between the modified red mud with 8% modified material added and that with 12% modified material added is not significant. It can be seen that the modified red mud has a certain solidification effect on F.

4. pH in Chiping Xinfah modified red mud

As mentioned earlier, the alkalinity of red mud is divided into soluble and insoluble. During the leaching of bauxite, NaOH reacts with the aluminum-containing substances in the bauxite to form soluble substances such as sodium aluminate and sodium silicate. The residual NaOH during the leaching process and the NaOH in the red mud during the storage process react with CO_2 in the air to form Na_2CO_3, which forms the main soluble alkali part in Bayer process red mud, namely NaOH residual liquid, Na_2CO_3 and Na_2SiO etc.

The pH value of the modified red mud powder after adding 8 and 12% modified materials was analyzed using the GB/T1555.12 method, and the results are shown in Table 6.20. The results show that the pH range of the modified red mud samples after adding 8% modified material is 11.82 ~ 12.07; correspondingly, the pH range of the modified red mud samples after adding 12% modified material is 11.81 ~ 12.04. Statistical analysis shows that there is no significant difference in pH between the two corresponding treatments, indicating that there is no significant difference in pH between the modified red mud with 8% modified material added and that with 12% modified material added.

5. Determination of characteristic pollutants in Chiping Xinfah modified red mud

According to the previous analysis, it is known that in the modified red mud with 8% and 12% modified materials added, the leaching content of Ca and Al is higher than the third level standard of groundwater quality, and also higher than the original red mud sample. This is due to the Ca and Al contained in the added modified materials. Considering that the total hardness of the four types of water standards that meet the requirements for industrial and agricultural water use in the "Groundwater Quality Standard" is $\leq$ 650 mg/L, it is recommended to ensure that the leaching content of Ca in the modified red mud after modification treatment is $\leq$ 650 mg/L when selecting modified materials. Al is a common element in nature and is also common in daily life. Although the third-level standard limit for Al in the "Groundwater Quality Standard" is 0.2 mg/L, Al is a general chemical indicator, which means that exceeding this value will affect the taste of drinking rather than endangering health. The "Food Additive Use Standard" (GB2760-2014) allows a residual limit of 100 mg/kg for aluminum in food. At the same time, although the iron (Fe) leaching content of 2 out of 7 original red mud samples exceeds the standard value of 300 g/L, the iron (Fe) leaching content in the modified red mud samples is far below the standard value of 300 g/L. Based on the above analysis, the probability of health hazards caused by the leaching content of Ca, Fe, and Al in the risk assessment process of modified red mud road use is very small and can be temporarily ignored. The metal pollutants that need to be focused on are V, Cr, As, Se, Mo. According to the previous analysis, although the content of F in the modified red mud is significantly lower than that in the original red mud, it is still higher than the third-level standard limit of groundwater quality (1.0 mg/L). Therefore, it still needs to be focused on in the risk assessment process of modified red mud road use. According to the previous analysis, the pH of the modified red mud is all above 11.8, and the potential environmental impact of the pH in the process of modified red mud road use needs to be considered in the future. The characteristic pollutants that need to be focused on in the risk assessment process of modified red mud road use are pH, F, V, Cr, As, Se, Mo. The addition of modified materials has little effect on the total metal content in the red mud, but has a greater impact on its leaching content. The risk assessment of modified red mud road use can focus on V, Cr, As, Se, Mo and other metal elements.

The study on the leaching content of F and characteristic metal pollutants in the modified red mud shows that the modified red mud has a significant solidification effect on F, V, As, Se, Mo. However, there is no significant difference between pH, F, V, Cr, As, Se, Mo in the modified red mud after adding 8% and 12% modified materials. The characteristic pollutants that need to be focused on in the long-term risk assessment process of modified red mud road use are pH, F, V, Cr, As, Se, Mo (Table 6.16).

6.3.3 *Environmental Impact of Metal Pollutants in Modified Red Mud*

As mentioned earlier, the characteristic pollutants that should be focused on in the long-term environmental risk assessment of modified red mud road use are pH, F, V, Cr, As, Se, Mo. Given the strong alkalinity of red mud, the release mechanisms of F, V, Cr, As, Se, Mo are not quite the same as the mechanism causing pH changes, so pH will be discussed separately later.

Table 6.16 pH in modified red mud

Sample number	Red mud production time	8% Modified material (mg/L)	12% Modified material (mg/L)
XFCN1	Before 2010 (wet discharge)	11.89	12.00
XFCN2	2010	11.89	11.81
XFCN3	2014	12.07	12.04
XFCN4	2016	11.98	11.96
XFCN5	2017	11.86	11.99
XFCN6	June 2018	11.85	11.95
XFCN7	May 2019	11.82	11.98

1. Release pattern of characteristic pollutants in modified red mud samples

Based on the exposure scenario analysis of modified red mud road use, according to the simulated leaching assessment method (NEN7375) introduced earlier, 8 different stages of leachate were obtained, and the content of characteristic pollutants in each stage of leachate was detected. According to the method introduced earlier, the cumulative release amount was calculated, and the cumulative leaching amount curve of pollutants was obtained, see Figs. 6.12, 6.13, 6.14, 6.15, 6.16 and 6.17.

From the above cumulative release curve, it can be seen that as the leaching time increases, the cumulative release amount of each pollutant shows an upward trend, basically releasing faster in the early stage, and slightly slowing down after five leaching stages (9 days), with the cumulative release amount of some elements almost showing a linear upward trend.

2. Long-term release prediction of pollutants

According to the release situation of pollutants in the leaching stage of modified red mud samples and the established prediction model of the release amount of

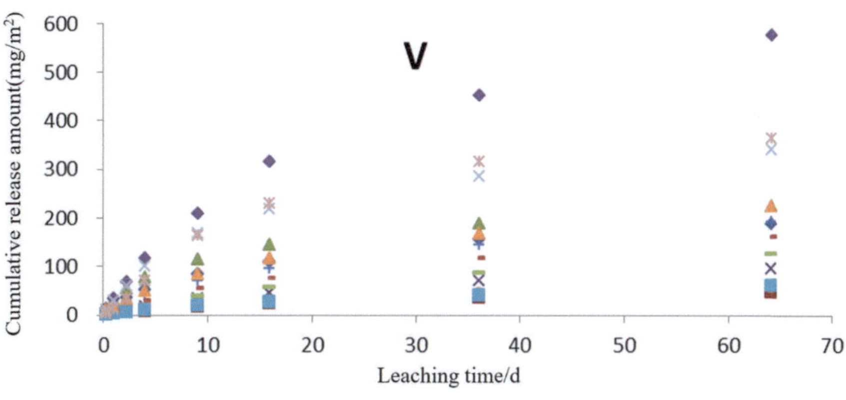

Fig. 6.12 Cumulative release curve of vanadium in modified red mud samples

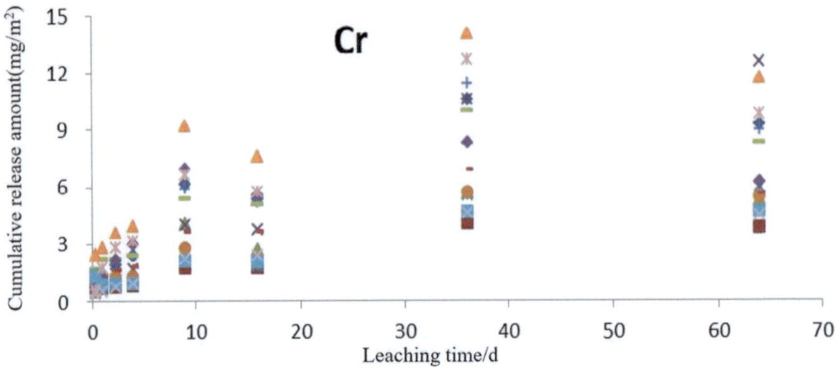

Fig. 6.13 Cumulative release curve of chromium in modified red mud samples

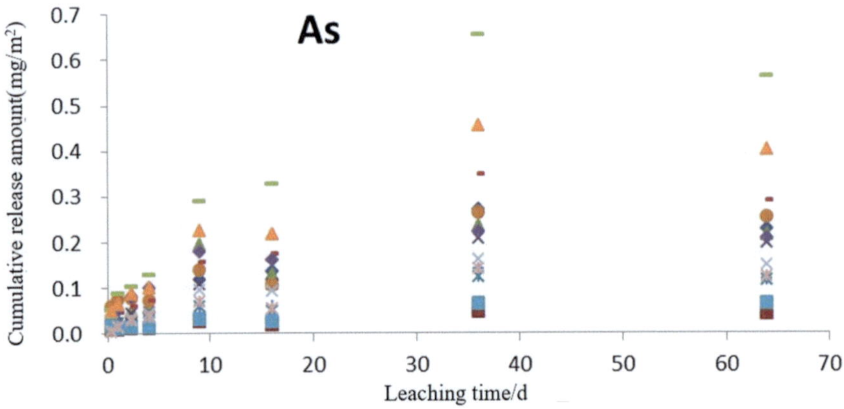

Fig. 6.14 Cumulative release curve of arsenic in modified red mud samples

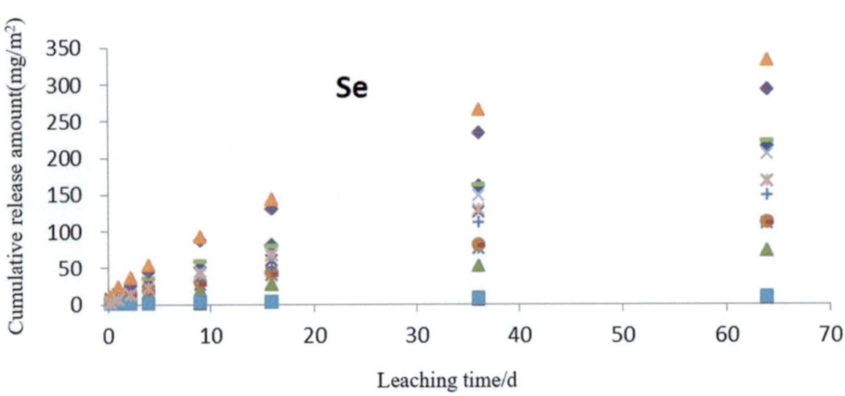

Fig. 6.15 Cumulative release curve of selenium in modified red mud samples

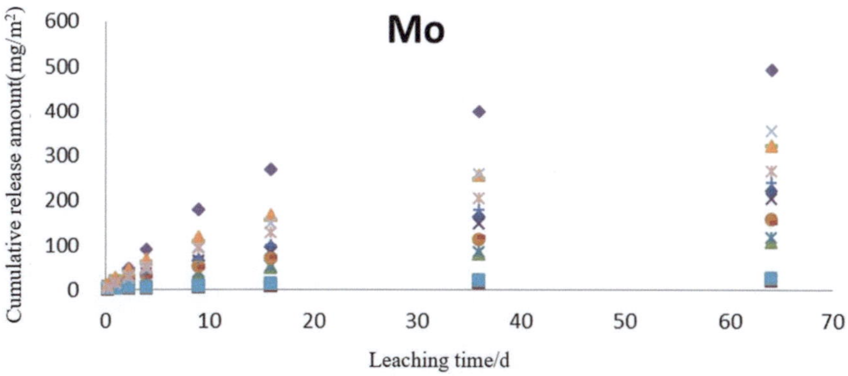

Fig. 6.16 Cumulative release curve of molybdenum in modified red mud samples

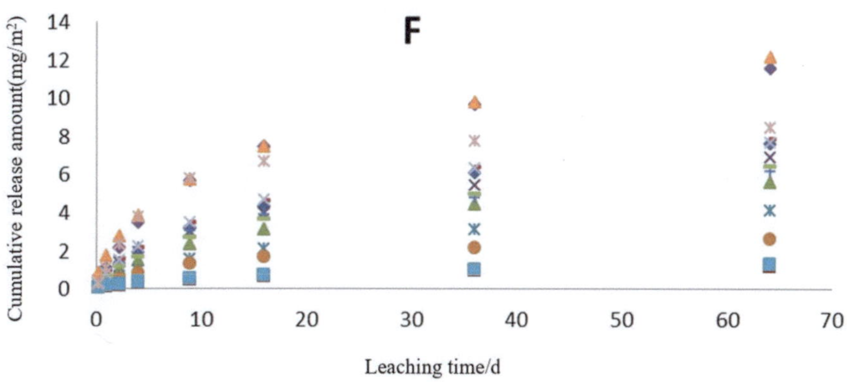

Fig. 6.17 Cumulative release curve of fluorine in modified red mud samples

characteristic pollutants in modified red mud samples, the long-term release amount of characteristic pollutants in modified red mud is predicted, and its change curve with time is shown in Figs. 6.18, 6.19, 6.20, 6.21, 6.22 and 6.23. Combining the cumulative release rate with the predicted cumulative release amount in the figure, it can be seen that the time required for the complete release of characteristic pollutants in different modified red muds varies.

3. Risk assessment of characteristic pollutants

First level: Preliminary assessment based on leaching amount

The first level of risk assessment is to compare the leaching content of characteristic pollutants in modified red mud with the third level standard limit of "Groundwater Quality Standard" (GB/T14848-2017). If the leaching value is lower than the third level standard of groundwater quality, there is no health risk after the pollutants are released and migrate to the groundwater body for residents to drink; if the leaching

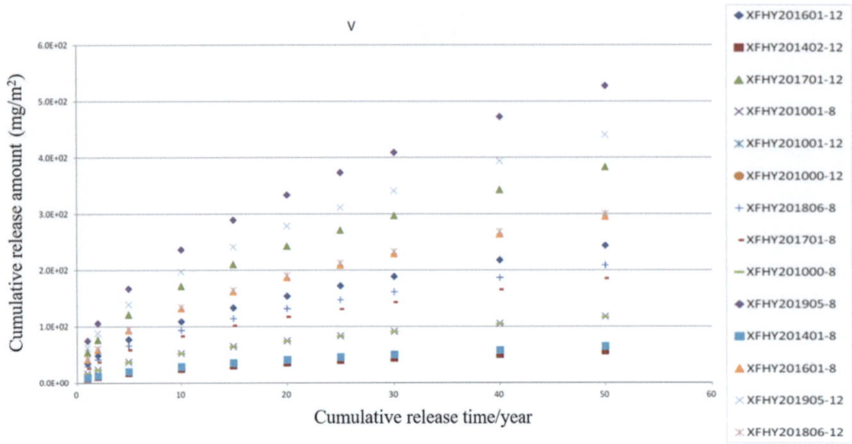

Fig. 6.18 Long-term cumulative release prediction of V in modified red mud blocks

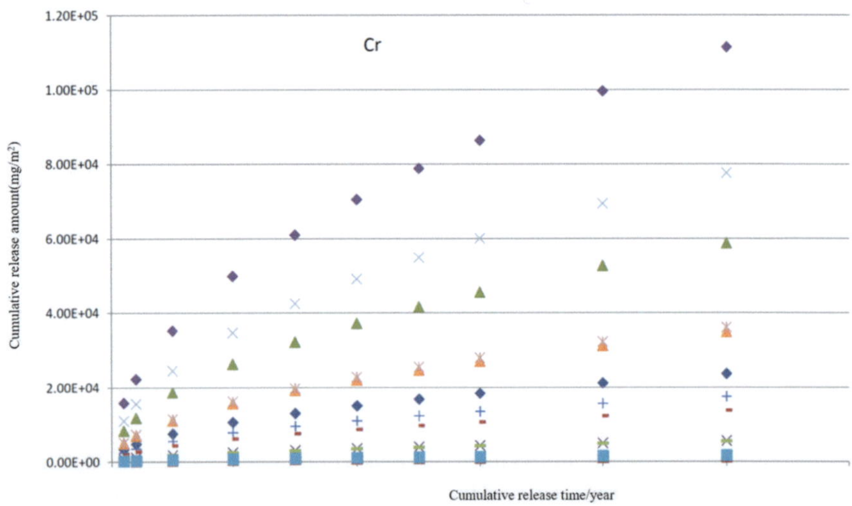

Fig. 6.19 Long-term cumulative release prediction of Cr in modified red mud blocks

value exceeds the groundwater environmental quality, further research is needed to determine whether there is a risk. The leaching contents of F, V, Cr, As, Se, Mo in modified red mud have been screened and determined to exceed the third level standard of groundwater quality, and the environmental risks of F, V, Cr, As, Se, Mo in the long-term road use process of modified red mud need to be further assessed in combination with specific exposure scenarios.

Second level: Risk assessment based on specific exposure scenarios

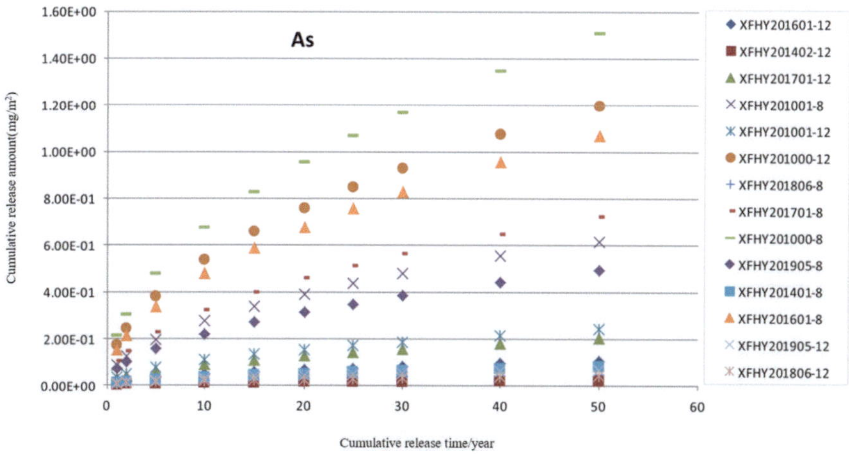

Fig. 6.20 Long-term cumulative release prediction of As in modified red mud blocks

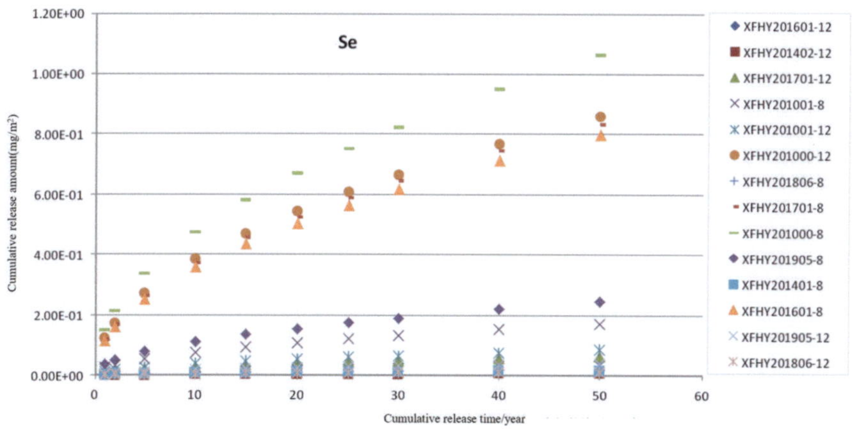

Fig. 6.21 Long-term cumulative release prediction of Se in modified red mud blocks

According to the aforementioned analysis, the risk assessment of the use of modified red mud roads mainly considers the exposure scenarios during the use process after completion, mainly: (1) After the rain infiltrates into the road surface structure, it seeps vertically downwards, and the pollutants are released and migrate to the groundwater body due to leaching (filtration); or (2) After the rain infiltrates into the road surface structure, it is discharged laterally from the road base structure, together with the road surface runoff and other road structure drainage, into the side ditch or road drainage system. Therefore, this assessment mainly considers the potential environmental risks to groundwater during the road use process after the completion of the modified red mud paving.

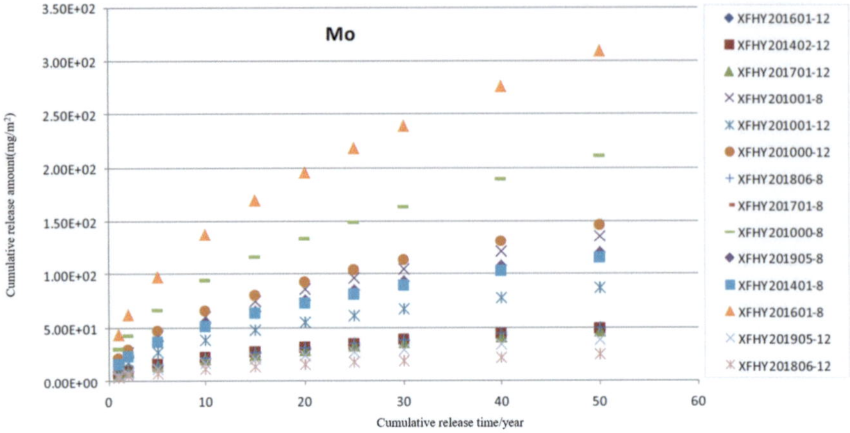

Fig. 6.22 Long-term cumulative release prediction of Mo in modified red mud blocks

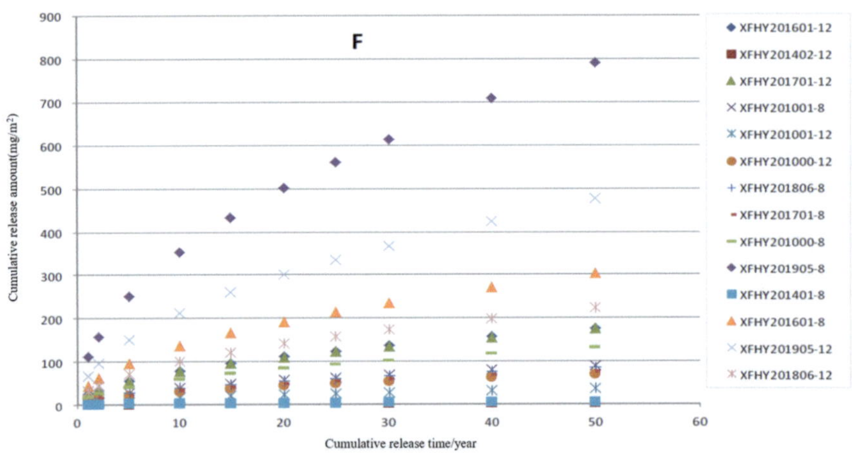

Fig. 6.23 Long-term cumulative release prediction of F in modified red mud blocks

This study uses the waste road use (roadway) evaluation module in the Industrial Waste Evaluation Model (IWEM) developed by the US Environmental Protection Agency (EPA) for evaluation. The working principle of this evaluation module is to evaluate the release flux of pollutants from industrial solid waste used as alternative materials for road construction, and integrate EPACMTP to simulate the migration and transformation of each pollutant component in soil and groundwater, obtain the overall exposure concentration of all components at the 90% percentile of groundwater, and compare it with the groundwater baseline value to obtain a suitability evaluation.

(1) Model assumptions

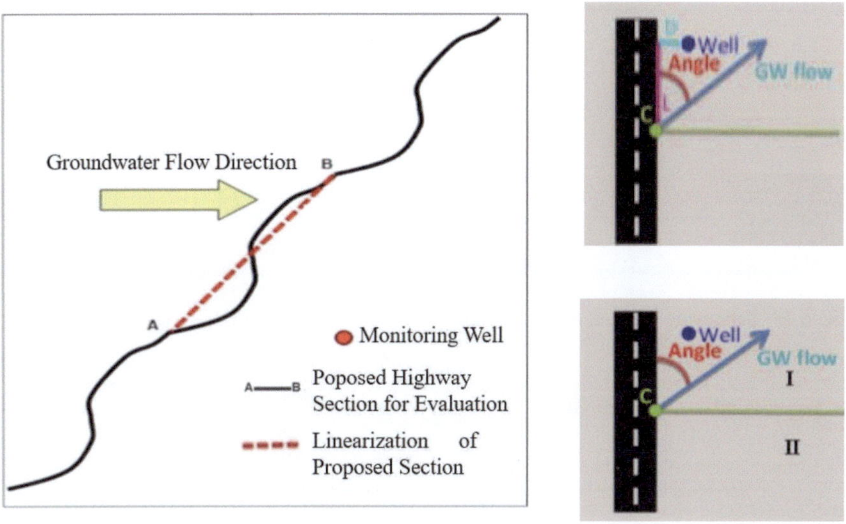

Fig. 6.24 Schematic diagram of the road section to be assessed

a. Linearization assumption of the assessment section

Highways are basically curved, and rarely straight. In actual assessment, the object to be assessed (highway section) needs to be linearized into a straight section. According to the position of the monitoring well, the angle between the groundwater flow direction and the centerline of the section, the assessment is divided into the following situations: (1) The monitoring well is located in area I, and the angle between the centerline and the groundwater flow direction is acute, right, or obtuse; (2) The monitoring well is located in area II, and the angle between the centerline and the groundwater flow direction is acute, right, or obtuse, as shown in Fig. 6.24.

b. Road striping assumption

When using IWEM3.1 for red mud road use assessment, the first thing to follow is the assumption conditions of the model, that is, the road surface, central isolation belt, shoulder, slope embankment, valley, etc. are regarded as relatively independent strip units (drip) in the vertical direction. In this study, the road cross-sectional structure was simplified, only considering the road surface (only considering the road surface part filled with modified red mud), slope embankment and valley, without considering the central isolation belt; at the same time, the valley, slope embankment and modified red mud filled road surface were regarded as relatively independent units in the vertical direction, as shown in Fig. 6.25.

At the same time, the different structural functional levels of each independent unit are regarded as a level (layer) in the assessment model. Since the slope embankment and valley parts do not use waste, only the road surface structure will be used, so here only the layered schematic diagram of modified red mud road use is drawn, with a total of 5 layers, as shown in Fig. 6.26.

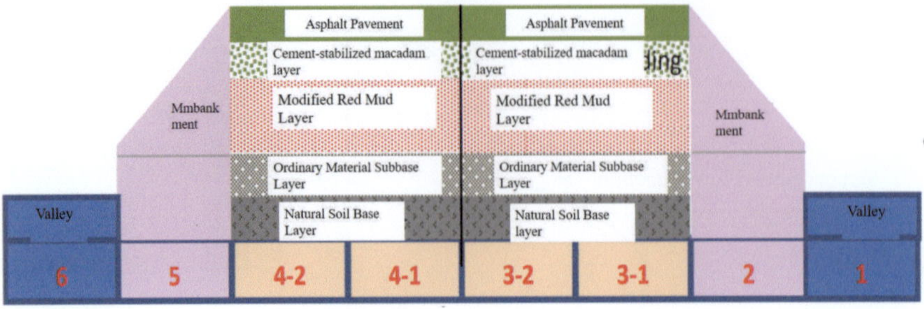

Fig. 6.25 Schematic diagram of the modified red mud road section strip

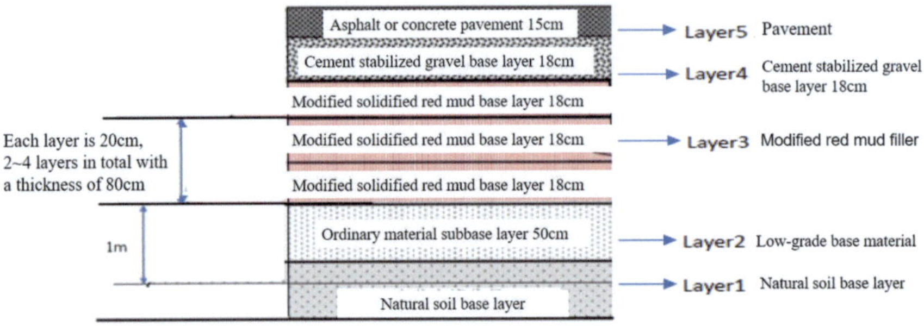

Fig. 6.26 Schematic diagram of the IWEM assessment layer of the road section using modified red mud

(2) Selection of key parameters.

a. Geometric location of the road

The model first needs to determine the length and location of the road section to be evaluated. In this evaluation, it is assumed that the object of evaluation is 0.5 km or 1 km; then the location of the monitoring well (in area I or II, see Fig. 6.27), the shortest straight-line distance D between the monitoring well and the edge of the road, the straight-line distance L between the center of the road and the vertical line between the monitoring well and the road, and the angle between the direction of groundwater flow and the centerline of the road, hereinafter referred to as D, L and the angle. The scenario where the monitoring well is in area II and the angle between the direction of groundwater flow and the centerline of the road is obtuse, and the scenario where the monitoring well is in area I and the angle between the direction

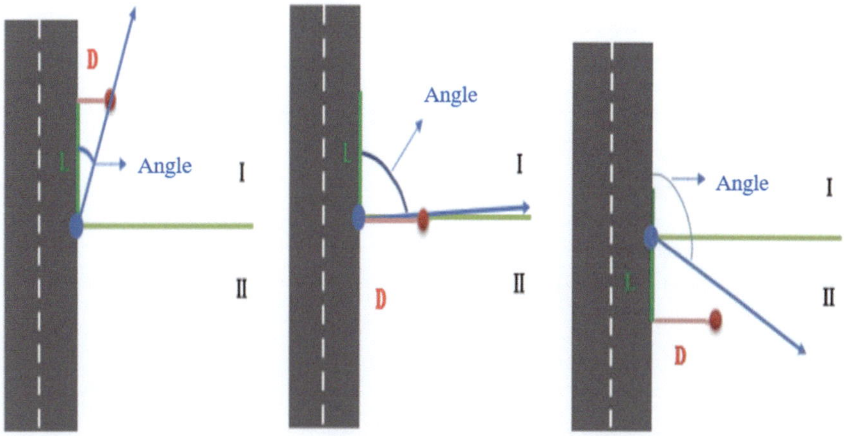

Fig. 6.27 Schematic diagram of the relationship between the road section to be evaluated, the direction of groundwater flow and the position of the monitoring well

of groundwater flow and the centerline of the road is acute, are similar. Therefore, only the situation in area I is considered in this evaluation.

b. Determination of strip parameters

Determine the strips and layers of the road section to be evaluated. If it is necessary to determine the width and number of layers of each strip of the road using modified red mud, see Table 6.17 for details.

Secondly, determine the specific parameters of each layer of each strip unit, generally set parameters from the bottom up. In the road section using modified red mud, because the density and permeability coefficient of the natural soil base layer and the ordinary material base layer are not much different, they are considered as one layer when determining the parameters, with a thickness of 1 m. The specific parameters of each strip of the road section to be evaluated are shown in Table 6.18.

c. Rainfall and other parameters

The annual rainfall in Liaocheng City is 565 mm, and the evaporation is 1800 mm, so the climate region selection is between 0.4 and 0.61 m for rainfall and more than

Table 6.17 Width and number of layers of the strips of the road section to be evaluated

Strip	Road function layer	Width/(m)	Number of layers
1	Gully	0.5	1
2	Embankment	0.45	2
3	Road surface	7.5	4
4	Road surface	7.5	4
5	Embankment	0.45	2
6	Gully	0.5	1

Table 6.18 Layer parameters of each strip of the road section to be evaluated

Strip number	Strip type	Layer	Layer type	Thickness (cm)	Permeability coefficient (m/year)	Density (g/cm^3)
1 Gully	Gully	1	Base layer	0.6	0.45	1.85
2 Embankment	Embankment	1	Subbase layer	0.6	0.36	1.85
	Embankment	2	Base layer	0.5	0.34	2.2
3 Road surface	Road surface	1	Natural soil base layer Ordinary material base layer	1	0.02	1.85
	Road surface	2	Red mud fill layer	0.6	0.03	1.88
	Road surface	3	Cement stabilized gravel layer	0.18	0.051	2.25
	Road surface	4	Asphalt concrete pavement	0.15	0.085	2.35
4 Road surface	Road surface	1	Natural soil base layer Ordinary material base layer	1	0.02	1.85
	Road surface	2	Red mud fill layer	0.6	0.03	1.88
	Road surface	3	Cement stabilized gravel layer	0.18	0.051	2.25
	Road surface	4	Asphalt concrete pavement	0.15	0.085	2.35
5 Embankment	Embankment	1	Base layer	0.5	0.36	1.85
	Embankment	2	Base layer	0.6	0.45	2.2
6 Gully	Gully	1	Base layer	0.6	0.45	1.85

1.5 m for evaporation. Other runoff coefficients and soil coefficients mainly refer to the specific situation of Liaocheng City or the default parameter data of the system.

(3) Pollutant parameters

The selection of pollutant content parameters is based on the highest value in the total input measured value, but the total size has a small impact on the exposure concentration. The factor that affects the prediction result of the exposure concentration is the leaching content of the pollutant. The leaching content of pollutants is mainly based on measured values, supplemented by estimated values.

a. Evaluation criteri

The system comes with and user-defined, choose the third-level standard of groundwater quality. Obtain the exposure concentration of the 90th percentile through simulation calculation, and compare it with the limit value of the third-level standard of groundwater quality. If it is lower than this standard value, the risk is acceptable, otherwise, the risk is unacceptable.

b. Evaluation situation of different exposure scenarios

According to the "Shandong Province Highway Administration Regulations", highways on both sides have side ditches (slope top interception ditches, slope foot protection roads, isolation fences, etc.), and their land use range is not less than one meter from the outer edge of the side ditches. According to the "Shandong Province Expressway Regulations", the land use on both sides of the expressway is fifty meters each, the sides of the overpass ramps, expressway connection lines, and around the toll stations are within a hundred meters, which is the expressway construction control area. Except for permanent buildings in the expressway control area, it should be greened in principle. At the same time, considering that there are shoulder ditches and valleys besides the driving lane of the expressway, therefore, during the assessment, choose different scenarios at distances of 1 ~ 10 m (including 1 m and 5 m) and 50 m from the expressway depending on the situation.

Randomly select D = 1 m, D = 5 m, D = 10 m, D = 50 m and other different situations for evaluation. Assuming that the modified red mud road section is 0.5 km or 1 km, after multiple adjustments, choose the more unfavorable angle (10 ~ 30) between the groundwater flow direction and the road, and determine that the environmental risk brought by the modified red mud used to build the road in the selected scenario is acceptable. The allowable leaching content and exposure concentration of characteristic pollutants are shown in Tables 6.19 and 6.20.

As can be seen from Table 6.19, in the specific scenarios listed in the table, D = 5 m, L = 5 m, and an angle of 10 is a relatively unfavorable scenario. In this scenario, when the leaching content of V, Cr^{6+}, Cr^{3+}, As^{3+}, Se^{4+}, Se^{6+}, Mo, F in the leachate of modified red mud is less than 0.15 mg/L, 0.1 mg/L, 0.2 mg/L, 0.012 mg/L, 0.012 mg/L, 0.012 mg/L, 0.075 mg/L and 1.2 mg/L respectively, the environmental risk of the long-term road use process after the designed highway construction can be accepted. No limit requirement for As^{5+} is proposed, because when the allowable leaching content of As^{5+} is 100 mg/L and 200 mg/L, the exposure concentration calculated by the IWEM simulation is far lower than the third-level standard of groundwater, and the highest total arsenic leaching content in the 14 evaluated modified red mud samples is 0.108 mg/L, therefore, it can be considered that as long as the leaching content of As^{3+} is less than 0.012 mg/L, its exposure concentration in groundwater is lower than the third-level standard of groundwater, and there is definitely no risk issue with the total arsenic leaching content. The same is true in other scenarios,

Table 6.19 Allowable leaching amount during modified red mud road use (D10m)

Scenario	D = 1; L = 1				D = 1; L = 5 D = 5; L = 5					D = 10				
	Leaching amount mg/L	0.5 km 10°	1 km 10°	1 km 15°	0.5 km 10°	Leaching amount mg/L	1 km 10°	1 km 18°	0.5 km 10°	Leaching amount mg/L	1 km L = 10, 10°	0.5 km L = 10,15°	0.5 km L = 10, 30° 30	0.5 km L = 1,90° 90
		Exposure concentration mg/L	Exposure concentration mg/L	Exposure concentration mg/L	Exposure concentration mg/L		Exposure concentration mg/L	Exposure concentration mg/L	Exposure concentration mg/L		Exposure concentration mg/L	Exposure concentration mg/L	Exposure concentration mg/L	Exposure concentration mg/L
V	0.16	0.0495	0.0494	0.0338	0.0467	0.15	0.0409	0.0409	0.0413	0.18	0.0475	0.0352	0.0193	0.0093
As^{3+}	0.014	0.0096	0.01	0.075	0.009	0.012	0.0078	0.0051	0.0078	0.015	0.0094	0.0068	0.004	0.002
Cr^{6+}	0.1	0.0499	0.0486	0.0352	0.0449	0.1	0.0458	0.0304	0.0446	0.12	0.0499	0.0363	0.0217	0.0119
Cr^{3+}	0.2	0.0296	0.0292	0.0184	0.026	0.2	0.0273	0.0159	0.0263	0.4	0.0944	0.0367	0.0174	0.0078
Se^{34+}	0.014	0.0094	0.0095	0.0073	0.0091	0.012	0.0074	0.005	0.0077	0.016	0.0099	0.0075	0.0044	0.0023
Se^{6+}	0.014	0.0094	0.0096	0.0071	0.009	0.012	0.0077	0.005	0.0077	0.016	0.0099	0.0074	0.0043	0.0023
F	1.4	0.9325	0.9578	0.741	0.8651	1.2	0.7787	0.5099	0.7719	1.4	0.8318	0.6627	0.3682	0.195
Mo	0.08	0.0538	0.0533	0.0415	0.0507	0.075	0.0486	0.0317	0.0494	0.08	0.0494	0.0476	0.0258	0.0132
As^{5+}	200	0.0037	3.60E−03	0.0025	0.0027	100	3.33E−04	1.08E−04	2.83E−04	200	0.0021	0.0017	0.0088	5.38E−04

Table 6.20 Allowable leaching amount when using modified red mud for roads (assuming the road section of the scenario is 1 km, D = 50 m)

Scenario	L = 50, angle 15°		L = 50, angle 20°		L = 50, angle 18°		L = 50, angle 10°		L = 50, angle 5°	
	Leaching concentration mg/L	Exposure concentration mg/L	Leaching concentration mg/L	Exposure concentration mg/L	Leaching concentration mg/L	Leaching concentration mg/L	Exposure concentration mg/L	Exposure concentration mg/L	Leaching concentration mg/L	Exposure concentration mg/L
V	0.3	0.0481	0.3	0.0369	0.3	0.0424	0.2	0.0346	0.2	0.0277
As^{3+}	0.026	0.0094	0.026	0.0075	0.026	0.008	0.02	0.0098	0.02	0.0084
Cr^{6+}	0.2	0.047	0.2	0.0372	0.2	0.0405	0.16	0.0463	0.16	0.0384
Cr^{3+}	0.4	0.0404	0.4	0.0311	0.4	0.0411	0.3	0.0422	0.3	0.0308
Se^{34+}	0.02	0.0073	0.02	0.0058	0.02	0.0063	0.02	0.0098	0.02	0.0084
Se^{6+}	0.022	0.008	0.022	0.0065	0.022	0.0068	0.02	0.0097	0.02	0.0084
F	2.2	0.7967	2.2	0.6345	2.2	0.6949	2	0.09662	2	0.8402
Mo	0.18	0.0661	0.18	0.052	0.18	0.0576	0.12	0.0582	0.12	0.0504
As^{5+}	20	5.52e−5	20	4.16E−5	200	3.22E−4	300	5.95E−4	300	3.74E−5

Scenario	L = 50, angle 60°		L = 5, angle 10°		L = 5, angle 10°		L = 10, angle 10°	
	Leaching concentration mg/L	Exposure concentration mg/L	Leaching concentration mg/L	Exposure concentration mg/L	Leaching concentration mg/L	Exposure concentration mg/L	Leaching concentration mg/L	Exposure concentration mg/L
V	0.2	0.0076	0.16	0.0411	0.15	0.00371	0.15	0.00362
As^{3+}	0.02	0.0026	0.016	0.0097	0.015	0.0089	0.015	0.0089
Cr^{6+}	0.16	0.0011	0.12	0.0479	0.12	0.0479	0.12	0.0475
Cr^{3+}	0.3	0.009	0.26	0.0416	0.25	0.0405	0.25	0.0363
Se^{34+}	0.02	0.0026	0.016	0.0097	0.015	0.0092	0.015	0.0088
Se^{6+}	0.02	0.0025	0.016	0.0097	0.015	0.009	0.015	0.0088
F	2	0.02562	1.6	0.9508	1.5	0.915	1.5	0.8666
Mo	0.12	0.015	0.11	0.0651	0.1	0.0602	0.1	0.0596
As^{5+}	300	9.99E−4	500	5.22E−04	300	0.0022	300	0.0019

that is, as long as As^{3+} meets the allowable leaching content requirement, the total arsenic meets the allowable leaching content requirement, here only considers the most unfavorable As^{3+} leaching content, there is no need to consider the leaching content of As^{5+}.

As can be seen from Table 6.20, taking a 1-km modified red mud road section as the evaluation object, under the scenario of D = 50, by adjusting the angle between the groundwater flow direction and the road centerline and the distance L from the road center point to the monitoring well and the vertical connection of the road multiple times, the relatively unfavorable scenario is determined to be the scenario when D = 50 m, L = 5, and the angle is 10. In this scenario, there are two groups of V, Cr^{6+}, Cr^{3+}, As^{3+}, Se^{4+}, Se^{6+}, Mo, F leaching content in the modified red mud leachate, and the simulated 90% percentile exposure concentration is lower than the third level standard of groundwater quality. Here, a more conservative group is selected as the allowable leaching amount, that is, the V, Cr^{6+}, Cr^{3+}, As^{3+}, Se^{4+}, Se^{6+}, Mo, F leaching content in the modified red mud leachate are respectively lower than 0.15 mg/L, 0.12 mg/L, 0.25 mg/L, 0.015 mg/L, 0.015 mg/L, 0.015 mg/L, 0.1 mg/L and 1.5 mg/L. In its long-term road use process, the risk of groundwater intake 50 m away from both sides of the highway is acceptable. In the above-discussed scenarios, among the 14 evaluated modified red mud samples, only individual samples of individual indicators meet the above leaching requirements. It is necessary to further optimize the modified materials and modification process, so that the leaching content of characteristic pollutants in the modified red mud can be reduced to an allowable range before it can be used on the road.

(4) Special considerations for drinking water surface water source areas

In China's "Groundwater Quality Standard" (GB/T14848-2017), there is no limit for vanadium, but the "Surface Water Quality Standard" gives a specific project standard limit for centralized drinking water surface water sources, where vanadium is 0.05 mg/L. Through the aforementioned assessment, it was found that due to the high leaching content of vanadium in modified red mud, and the relatively low leaching content of other characteristic pollutants, if the environmental risk of vanadium in modified red mud is considered, it will limit the road resource utilization of modified red mud; if it is hoped that 12% modified material of modified red mud is applicable, it is necessary to determine the IWEM simulation of modified red mud road use when the exposure concentration is lower than the surface water standard (0.05 mg/L) scenario and vanadium leaching content. In the evaluated 12% modified material of modified red mud, the highest leaching content of vanadium is 947.5 μg/L, the input value of the IWEM simulation of modified red mud road use specific scenario exposure concentration is taken as 0.95 mg/L, through multiple adjustments of the vertical distance D from the monitoring well to the edge of the road lane, the distance L from the road center point to the vertical connection of the monitoring well and the road, the angle between the groundwater flow direction and the road, the allowable leaching concentration of the characteristic pollutants of the environmental risk brought by the road construction of the modified red mud used in different selected scenarios, and its exposure concentration reaching the monitoring well are shown in Table 6.21.

As can be seen from the table, in the long-term use process after the use of 12% modified material of modified red mud road, the environmental risk of vanadium may be acceptable at least 300 m outside the highway, within 300 m the exposure concentration of vanadium pollutants will exceed 0.05 mg/L, in fact, the possibility of allowing such a scenario to exist is relatively small. It is recommended that enterprises further optimize the modified materials and modification process, reduce the leaching content of vanadium and its characteristic pollutants, and ensure the stability of the performance of modified red mud.

Table 6.21 Leaching concentration and exposure concentration of vanadium in different scenarios when using modified red mud road

Kilometers (km)	D	L	Angle between groundwater flow and road	Leaching concentration mg/L	Exposure concentration mg/L
1	160	50	15	0.95	0.041
1	160	100	15	0.95	0.1146
1	180	150	15	0.95	0.0287
1	185	150	15	0.95	0.0644
1	185	180	15	0.95	0.0688
1	190	190	15	0.95	0.0558
1	200	200	15	0.95	0.0372
1	200	210	15	0.95	0.0426
1	200	220	15	0.95	0.0485
1	200	250	15	0.95	0.0682
1	200	300	15	0.95	0.0838
1	200	310	15	0.95	0.0851
1	200	320	15	0.95	0.085
1	200	340	15	0.95	0.0849
1	200	350	15	0.95	0.0836
1	200	360	15	0.95	0.0835
1	220	310	15	0.95	0.0479
1	220	310	45	0.95	0.0389
1	220	310	20	0.95	0.0794
1	225	310	20	0.95	0.0787
1	240	310	20	0.95	0.0794
1	240	310	15	0.95	0.0495
1	280	310	20	0.95	0.0566
1	300	310	20	0.95	0.0381

Table 6.22 pH of leachate from modified red mud blocks at different test stages of the NEN7375 leaching method

Sample number	pH of modified red mud	pH of leachate from modified red mud blocks[a]									
		Batch	I	II	III	IV	V	VI	VII	VIII	Extended period
		Leaching time/d	0.25	1	2.25	4	9	16	36	64	92
		Interval time/d	0	1	1.25	2.25	5	7	20	28	28
XFHY1-8	11.89	skla	11.62	12.06	11.22	11.39	11.70	1235	12.63	12.40	11.67
XFHY1-12	12.00	skib	11.64	12.06	11.18	11.31	11.79	12.14	12.62	12.38	11.89
XFHY2-8	11.89	sk2a	157	11.97	11.20	1141	11.56	12.31	12.84	12.65	11.79
XFHY2-12	11.81	sk2b	11.51	11.95	11.12	11.43	11.68	12.02	12.74	12.06	11.81
XFHY3-8	12.07	sk3a	11.72	12.06	11.18	11.40	11.79	12.14	12.74	12.29	11.78
XFHY3-12	12.04	sk3b	11.59	12.04	11.33	11.43	11.66	12.35	13	12.81	12.09
XFHY4-8	11.98	sk4a	11.43	12.06	11.28	11.49	11.63	12.24	12.39	12.49	11.75
XFHY4-12	11.96	sk4b	11.17	11.79	11.08	11.45	11.70	12.29	12.63	12.30	11.72
XFHY5-8	11.86	skSa	11.40	11.96	11.12	11.24	11.74	12.23	12.42	12.42	11.6
XFHY5-12	11.99	sk5b	11.01	11.76	10.98	11.19	11.60	12.12	12.56	12.33	11.69
XFHY6-8	11.85	sk6a	11.04	11.88	11.13	11.28	11.84	11.86	12.64	11.96	11.7
XFHY6-12	11.95	sk6b	10.30	11.14	10.60	11.11	11.62	12.15	12.89	12.00	11.85
XFHY7-8	11.82[a]	sk7a	10.36	11.00	10.56	10.70	11.17	11.82	11.79	11.95	11.04
XFHY7-12 1	11.98[a]	sk7b	10.29	11.20	11.05	10.44	11.42	1181	12.48	12.36	11.61

[a]According to single factor ANOVA and multiple analysis, except for the significant differences between XFHY7-8 and other treatments with 8% modified red mud, XFHY7-12 and other treatments with 12% modified red mud, and XFHY3-12 and XFHY6-12, the pH differences between other treatments are not significant

6.3.4 Environmental Impact of pH During the Use of Modified Red Mud Road

Environmental Impact of pH in the Process of Modified Red Mud Road Use

1. pH value and its trend in modified red mud

After the completion of the 8-stage experiment simulating the release of pollutants in the modified red mud during road use by the NEN7375 leaching method in this evaluation, the pH of the leachate is still relatively high (between 11.95 and 12.81), so it is extended for 28 to 92 days based on the original basis. The pH results of different leaching stages in the conventional test and the extended test are shown in Table 6.22.

According to the pH and statistical analysis results of the leachate of the modified red mud block at different test stages in Table 6.22, the red mud produced at different times has different pH changes at different leaching stages. For red mud samples produced at the same time, the difference in the pH of the leachate of the modified red mud block after two treatments with 8% and 12% modified materials is not significant at different leaching test stages. For the modified red mud treated with 8% modified material, the pH of the leachate of the modified red mud block with shorter production time is lower, and the pH of the leachate of the modified red mud block with longer production/storage time is higher. The pH of the leachate of the modified red mud block produced in May 2019 is significantly lower than the pH of the leachate of the modified red mud block produced at other six different times. In addition, the pH differences between the leachate of the modified red mud blocks produced at other six different times are not significant. For the modified red mud treated with 12% modified material, similar to the 8% modified material, the pH of the leachate of the modified red mud block with shorter production time is lower, and the pH of the leachate of the modified red mud block with longer production/storage time is higher. The pH of the leachate of the modified red mud block produced in May 2019 is significantly lower than the pH of the leachate of the modified red mud block produced at other six different times. At the same time, the pH of the leachate of the modified red mud block produced in January 2014 is also significantly higher than the pH of the leachate of the modified red mud block produced in June 2018. In addition, the pH differences between the leachate of the modified red mud blocks produced at other different times are not significant.

The pH value changes of the leachate of the modified red mud block at different leaching experiment stages are shown in Figs. 6.28 and 6.29.

As can be seen from the figure, the pH change trends of the leachate of the modified red mud block treated with 8% and 12% modified materials are similar, with the following basic rules: a peak appears on the first day, about 11.0 ~ 12.06; then the pH immediately drops, then continues to rise and reaches a peak again at 36 days, after which the pH of the leachate of the modified red mud block shows a slow decreasing trend. At the 7th leaching stage when the pH reaches the second peak (cumulative leaching time 36 days), the pH of the leachate of the modified red mud

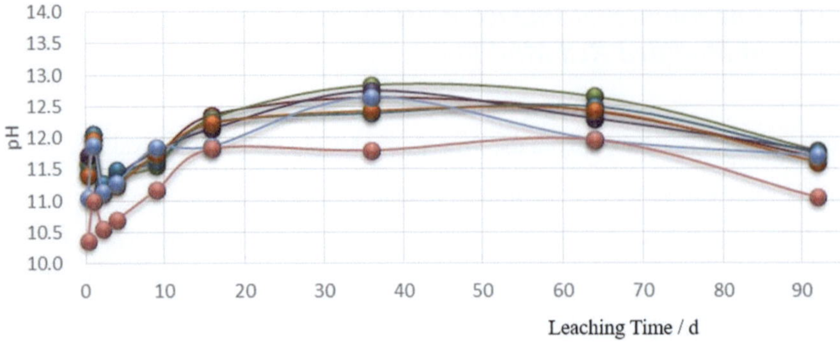

Fig. 6.28 pH of the leachate of the modified red mud block with 8% modified material by NEN7375 method

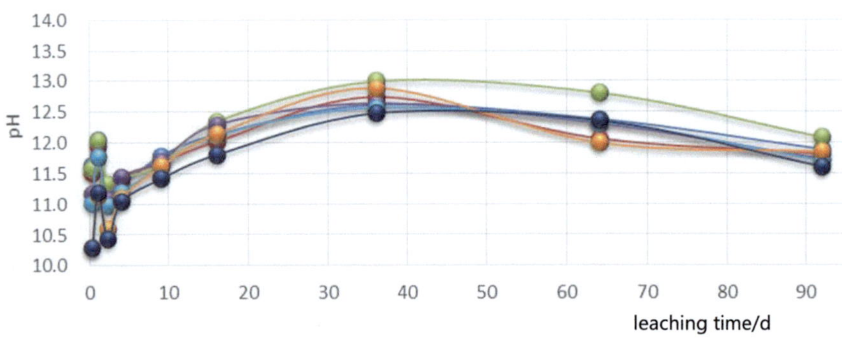

Fig. 6.29 pH of the leachate of the modified red mud block with 12% modified material by NEN7375 method

block is basically above 12.3, reaching a maximum of 13, with only XFHY201905-8 (red mud produced in May 2019, with 8% modified material) block leachate pH of 11.79. Overall, whether adding 8% modified material or 12% modified material, the pH of the leachate of the modified red mud block shows a slow decreasing trend after 36 days.

2. Prediction of hydrogen ion concentration release during the use of red mud road.

The cumulative release of hydrogen ion H^+ is calculated using the method in Sect. 6.2, and the cumulative release curves of H^+ are shown in Figs. 6.30 and 6.31.

As shown in the figure, during the leaching process of the modified red mud test blocks, with the increase of leaching time, the cumulative release of H^+ initially shows an upward trend, and quickly reaches a release equilibrium after the initial release, that is, it reaches equilibrium after the 4th stage (cumulative leaching time of 4 days), and the cumulative release of H^+ does not change much thereafter.

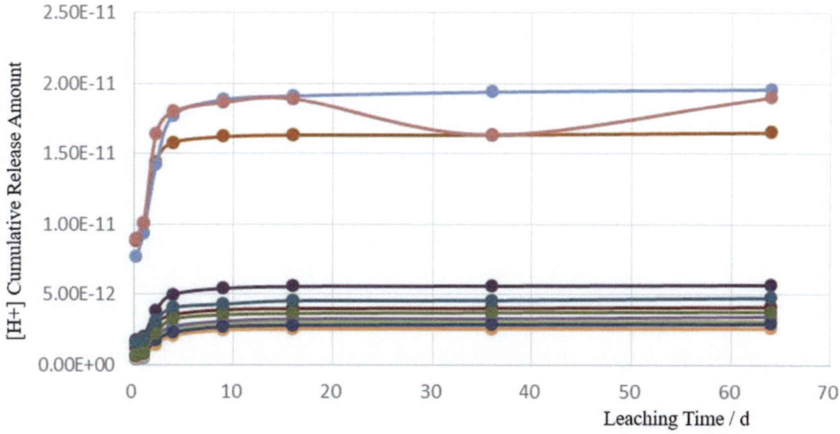

Fig. 6.30 Cumulative hydrogen ion release curve of modified red mud test blocks at different leaching stages (64 days)

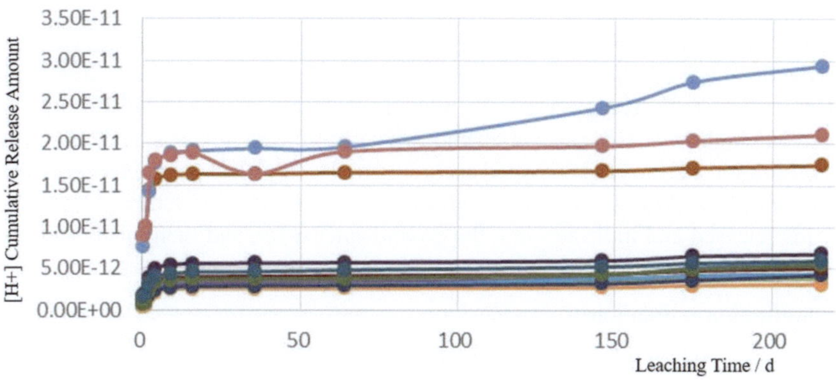

Fig. 6.31 Cumulative hydrogen ion release curve of modified red mud test blocks at different leaching stages (214 days)

3. Prediction of pH impact during the use of red mud road

Assuming the proposed highway is a two-way four-lane road, with two lanes in each direction, each lane is 3.75 m, then the lateral distance of the red mud substitute road base is 15 m. The schematic diagram is shown in Fig. 6.32. According to the "Highway Drainage Design Code" (JTG/T-2012), the infiltration coefficient of asphalt pavement is 0.15 $m^3/(d\ m^2)$. According to the "Highway Drainage Design Code" (JTG/T-2012), the average annual rainfall in Liaocheng area is about 565 mm, there is no need to design a road surface drainage system, and those that do not

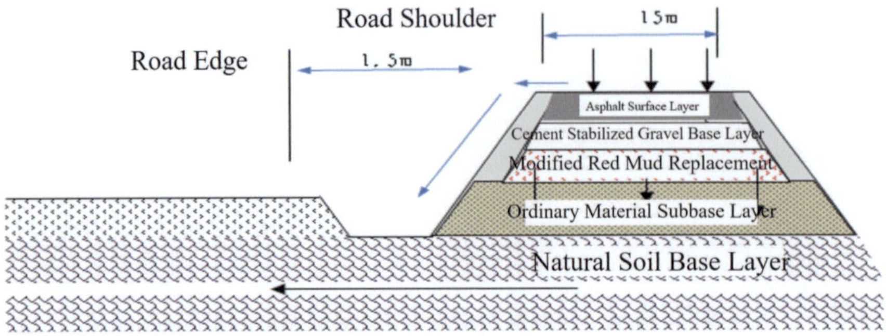

Fig. 6.32 Schematic diagram of modified red mud substitute road base structure and exposure scenario

infiltrate into the road surface are considered to flow into the side slope collection system or side ditch and infiltrate.

4. Impact on groundwater pH

The general road boundary for highway design is 100 mm, the ditch mouth is 1.2 m, and the shoulder is 0.75 ~ 3 m, so the shoulder to road boundary is calculated as about 1.5 m, see Fig. 6.32. There is generally a 1-m land use range outside the ditch mouth. Shandong highway design also often chooses permeable aggregate side ditches, and the infiltration coefficient is selected according to the empirical parameters provided by JTG/T-2012. After the rainwater infiltrates into the modified red mud layer, the $[H^+]$ is released and infiltrates into the groundwater layer, and then mixes with the rainwater that runs off to the valley and infiltrates.

a. Infiltration estimation

According to the "Highway Drainage Design Code" (JTG/T-2012), when surface water infiltrates into the road surface structure, the amount of water infiltrating into the asphalt pavement is calculated by the following formula:

$$Q_p = K_a B \tag{6.6}$$

where

- Q_p is the amount of surface water infiltrating per meter length of the driving lane [$m^3/(d\ m)$];
- K_a is the design infiltration rate of surface water per square meter of asphalt pavement [$m^3/(d\ m^2)$]; K_a is 0.15;
- B is the width of the one-way slope road surface. Assuming the road boundary to the shoulder, the 1-m land use range outside the highway ditch mouth, the infiltration rate reaching the groundwater due to rainfall is calculated by the following formula:

$$\text{the amount of surface water infiltrating } \mathrm{Q_i} = \mathrm{B_i} \times \mathrm{K_i}$$
$$\text{the amount of surface water infiltrating } \mathrm{Q_{other}} = \sum \mathrm{Q_i} \quad (6.7)$$

where

Q_i is the amount of surface water infiltrating per meter length of a certain type of road surface [m^3/(d m)];
K_i is the design infiltration rate of surface water per square meter of a certain type of road surface [m^3/(d m^2)];
B_i is the width of a certain type of road surface.

b. pH estimation

Based on the obtained rainfall infiltration amount, under the most unfavorable conditions, it is assumed that after the rainfall infiltrates into the modified red mud road base material layer, the alkaline solution that is leached directly enters the groundwater, or directly flows laterally into the highway's collection and drainage system. Then use the following formula to estimate the pH after the leachate from the modified red mud mixes with other rainwater reaching the groundwater or surface water. The estimation process does not calculate the evaporation amount.

$$\mathrm{pH_{estimation}} = -\log\left[\frac{\sum_{j=1}^{n}[H^+] * Q}{\sum_{j=1}^{n} Q_j}\right] \quad (6.8)$$

where: [H^+] is the hydrogen ion concentration ins each infiltration amount;
Q_j For each part of the infiltration volume.

Taking the highest pH of rainfall in Liaocheng City (i.e., the highest pH of rainfall in 2016 = 8.76) as an example, assuming the road boundary to the road shoulder is 1.5, the outer edge of the highway gutter is 1 m, the possible pH of groundwater after using modified red mud is calculated, as shown in Table 6.23.

The results show that the differences in pH estimated from different modified red mud blocks are not significant, basically increasing by 0.32 units on the original basis. The situation of the 14 modified red mud blocks is not listed one by one here. According to the rainfall pH range announced in the Liaocheng City Ecological Environment Quality Bulletin, it is estimated that if modified red mud is used as a substitute material to build a 1 km road, the change in groundwater pH can be seen in Table 6.24. The results show that if extreme rainfall pH is not considered, the use of modified red mud has little effect on groundwater pH.

5. Impact on surface water pH

Similarly, assuming that after the rainfall infiltrates into the modified red mud layer, the [H^+] released laterally discharges and merges with road surface runoff, slope

Table 6.23 Impact on groundwater pH after different modified red mud materials are used for subgrade materials

Modified red mud	Leaching stage	$[H^+]$ concentration mol/L	Estimated H^+ concentration mol/L	Estimated pH
XFHY201000-8	12.63	2.34E−13	8.9906E−10	9.05
XFHY201000-12	12.62	2.40E−13	8.9906E−10	9.05
XFHY201001-8	12.84	1.45E−13	8.9901E−10	9.05
XFHY201001-12	12.74	1.82E−13	8.9903E−10	9.05
XFHY201401-8	12.74	1.82E−13	8.9903E−10	9.05
XFHY201401-12	13	1.00E−13	8.9899E−10	9.05
XFHY201601-8	12.39	4.07E−13	8.9915E−10	9.05
XFHY201601-12	12.63	2.34E−13	8.9906E−10	9.05
XFHY201701-8	12.42	3.80E−13	8.9914E−10	9.05
XFHY201701-12	12.56	2.75E−13	8.9908E−10	9.05
XFHY201806-8	12.64	2.29E−13	8.9906E−10	9.05
XFHY201806-12	12.89	1.29 E−13	8.9901E−10	9.05
XFHY201905-8	11.79	1.62E−13	8.9978E−10	9.05
XFHY201905-12	12.48	3.31E−13	8.9911E−10	9.05

Note The highest pH of rainfall in Liaocheng City from 2016 to 2019 (pH = 8.73) was used for estimation

Table 6.24 Prediction of the impact on groundwater pH after using modified red mud

	Rainfall pH		Estimated pH of part of the rainfall entering groundwater after using modified red mud	
Year	pH range	pH average	pH range	pH average
2019	6.63–8.14	7.32	6.98–8.46	7.64
2018	7.18–8.50	7.67	7.50–8.82	7.99
2017	7.04–8.51	7.54	7.36–8.83	7.86
2016	6.26–8.73	7.84	6.58–9.05	8.16

drainage facilities, and gullies, etc. to collect rainfall, only road surface runoff drainage is considered here. The infiltration coefficient of the asphalt road surface is 0.15 $m^3/(d\ m^2)$, so the runoff coefficient is 0.85 $m^3/(d\ m^2)$, and 0.6 $m^3/(d\ m^2)$ is used for calculation here. The pH of surface water is estimated in the same way, and the results are shown in Table 6.25. Overall, the impact of the use of modified red mud on surface water is basically that the pH of the rainfall entering the surface system will increase by 0.1 units.

Table 6.25 Prediction of the impact on surface water pH after using modified red mud

	Rainfall pH		Estimated pH of part of the rainfall entering groundwater after using modified red mud	
Year	pH range	pH average	pH range	pH average
2019	6.63–8.14	7.32	6.73–8.24	7.42
2018	7.18–8.50	7.67	7.28–8.60	7.77
2017	7.04–8.51	7.54	7.14–8.61	7.64
2016	6.26–8.73	7.84	6.36–8.83	7.94

5. The impact of the modified red mud road use process on soil pH

(1) The impact of adding modified red mud leachate on soil pH

Due to the high content of alkaline substances in red mud, this part mainly discusses the possible impact of the use of modified red mud on soil pH. Take the modified red mud leachate with a pH exceeding 12 and add it to the farmland soil taken from Liaocheng City. After a week, measure the soil pH. The results are shown in Table 6.26.

The main types of soil in Liaocheng City are brown soil, tidal soil, saline soil, and wind-blown sand soil, among which tidal soil covers about 657,000 hectares,

Table 6.26 Possible impact of modified red mud leachate on soil pH

Leachate pH	Soil pH	Leachate pH	Soil pH	Leachate pH	Soil pH
12.35	8.39	12.63	8.26	12.40	8.23
12.14	8.29	12.62	8.42	12.38	8.48
12.31	8.42	12.84	8.5	12.65	8.38
12.02	8.41	12.74	8.45	12.06	8.42
12.35	8.45	13	8.42	12.81	8.47
12.14	8.41	12.74	8.39	12.29	8.12
12.24	8.37	12.39	8.36	12.49	8.19
12.29	8.38	12.63	8.44	12.30	8.42
12.23	8.36	12.42	8.41	12.42	8.29
12.12	8.34	12.56	8.42	12.33	8.47
12.15	8.38	12.89	8.29	12.00	8.01
Soil blank	7.87	–	7.87	–	7.87

accounting for 93.9% of the city's area, making it the largest type of soil in the city. Natural soil has the buffering capacity to resist and mitigate acid–base reaction changes. The soil used in the experiment is tidal soil, with a soil pH of 7.87. After adding different leachates to the soil, it was found that the soil pH increased by 0.14 ~ 0.63 units, with the highest being 8.47.

(2) The impact of the pH of the modified red mud leachate on the growth of soil flora and fauna:

During the long-term leaching experiment (NEN7375), a test was conducted on the impact of the modified red mud leachate on the growth of earthworms and Pak Choi.

a Earthworm growth experiment

The leachate of modified red mud with pH = 12 was added to the test soil, and the survival of the earthworms cultured in it was statistically analyzed. The survival rate statistics of the earthworms cultured for 7 and 14 days are shown in Table 6.27.

As can be seen from Table 6.27, the effects of the leachate of modified red mud on the survival rate of earthworms in the test soil are as follows: after 7 days of culture, except for one treatment where the survival rate of earthworms was 80%, the survival rate of earthworms in the other treatments was above 90%; after 14 days of culture, the survival rates of four treatments were 75.0%, 86.7%, 83.3% and 80.0%, respectively, and the survival rates of the remaining 10 treatments and blank earthworms were all above 90%.

According to the evaluation standards recommended by the International Organization for Standardization (ISO), the acute mortality rate (%) of earthworms is

Table 6.27 The impact of leachate pH on the survival rate of earthworms in the soil

Project	pH	7d	14d	7d	14d
Modified red mud leachate (%)	12.63	80.0	75	14	21
	12.62	100.0	86.7	0	15
	12.84	100.0	90	0	17
	12.74	93.3	93.3	12	12
	12.74	100.0	93.3	0	6
	13	96.7	96.7	6	6
	12.39	96.7	93.3	6	12
	12.63	90.0	90	10	10
	12.42	90.0	83.3	10	21
	12.56	93.3	93.3	12	12
	12.64	96.7	80	6	10
	12.89	96.7	93.3	6	17
	11.79	96.7	93.3	6	6
	12.48	96.7	90.0	6	17
Blank (%)	8.6	90	90	10	10

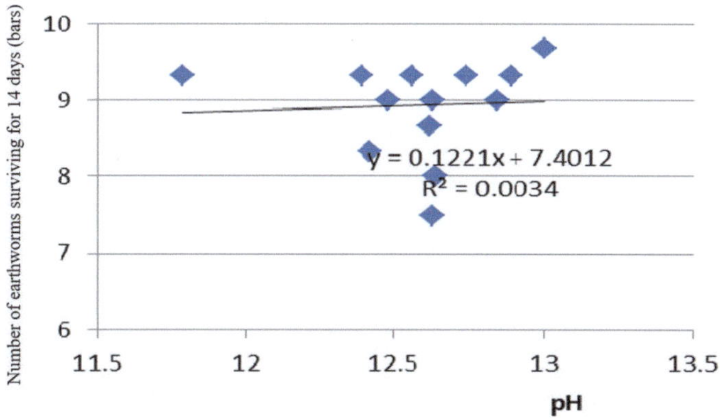

Fig. 6.33 Relationship between pH of leachate from modified red mud test block and earthworm survival rate

suggested to be used to divide the level of biological toxicity. Among them, if the mortality rate of earthworms is ≤ 10%, it is considered as micro-toxic or non-toxic; if 10% < the mortality rate of earthworms ≤ 30%, it is considered as low toxicity, and so on. The effect of the leachate of modified red mud on the earthworms in the test soil is mostly micro-toxic or non-toxic, and a small part is low toxicity. However, it can also be seen from the table that the two treatments with the highest pH (≥ 12.8) are micro-toxic or non-toxic. According to the regression trend analysis between pH and the number of earthworms surviving for 14 days (see Fig. 6.33). As can be seen from the figure, there is no correlation between the two, and the pH level does not have a significant correlation with the survival rate of earthworms.

b. Experiment on the germination and growth inhibition of Pak Choi:

The modified red mud leachate with a pH ≥ 12 was added to the test soil, and the germination and root growth of the Pak Choi planted in the soil were recorded. The results are shown in Table 6.28.

As can be seen from the table, when the modified red mud leachate (pH ≥ 12) is added to the test soil, the germination rate of the Pak Choi planted is all ≥ 90%, and the germination rate of the blank is 93.3 ~ 95%; the root length of the Pak Choi in the test soil ranges from 6.91 to 7.97 cm, some of which are inhibited, but not more than 10%. Therefore, it can be seen that the modified red mud leachate has little effect on the germination rate and root elongation of Pak Choi.

Table 6.28 The effect of modified red mud leachate on the growth of Pak Choi

Leachate pH	Germination rate (%)	STD	Root length	STD	Inhibition rate (%)
12.06	100	0.00	6.91	1.38	15.1
12.06	96.7	0.06	7.01	2.08	16.8
12.06	96.7	0.06	7.36	1.66	9.8
12.04	96.7	0.06	6.54	2.01	16.7
12.06	93.3	0.06	7.69	1.93	22.7
Blank 1	93.3	0.12	6.0	1.64	–
12.35	96.7	5.8	7.86	2.02	– 6.6
12.31	96.7	5.8	7.95	2.21	– 5.5
12.35	96.7	5.8	7.97	1.41	– 5.2
12.29	93.3	11.5	7.75	1.28	– 7.8
Blank 2	95.0	7.1	8.41	2.01	–

6.3.5 *Risk Analysis of Pollutant Dilution and Diffusion in Stabilized Red Mud*

Based on the analysis of exposure scenarios and environmental investigations, calculate the leaching concentration of hazardous substances after release into groundwater. The calculated leaching concentration is evaluated with the standard value of Class III groundwater as an indicator, representing the environmental risk of using red mud roads. The specific implementation process is: use of methods and environmental investigation → establishment of exposure scenarios → leaching concentration of hazardous substances in stabilized red mud → dilution attenuation model and parameters → calculation of leaching concentration → risk characterization.

After the modified red mud structure layer is leached by rainwater (solution), the hazardous substances in it are leached out with rainwater and migrate and diffuse into groundwater with the leaching solution, and then diffuse with the flow of groundwater, producing a dilution attenuation effect, represented by the dilution-attenuation factor (DAF). The dilution attenuation effect is divided into two processes, one is the dilution process of hazardous substances leaking from the landfill site, passing through the gas-filled zone into the groundwater mixing zone, represented by the dilution factor (DF); the second is the attenuation process of hazardous substances mixing with groundwater and migrating to the water extraction observation point, represented by the attenuation factor (AF). Since the analysis of the leaching concentration of hazardous substances in red mud road base entering groundwater is considered, only the process of DF is considered.

According to the hazardous substance dilution migration model established by the Water Resources Research Center of Texas State University, it can be applied to determine the dilution and attenuation process of hazardous substances migrating to groundwater under the condition of waste landfill disposal and polluted soil/site being drenched by rainwater. The model schematic is shown in Fig. 6.34.

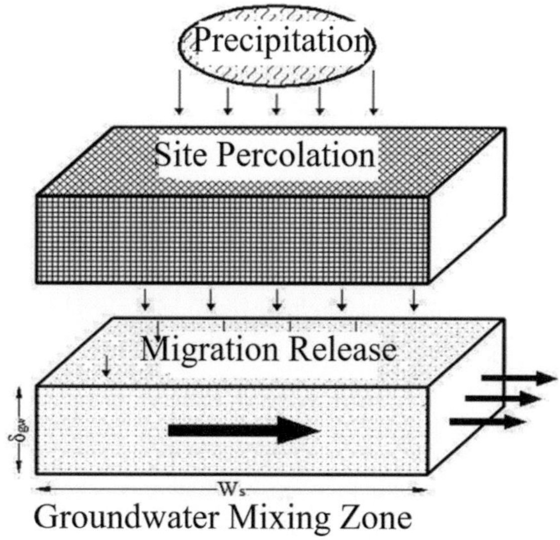

Fig. 6.34 Model of hazardous substance dilution process with precipitation infiltration

The dilution factor LDF after the hazardous substance enters the mixing zone is calculated by the following formula (6.9):

$$LDF = 1 + \frac{U_{gw}\delta_{gw}}{I_f W_s} \tag{6.9}$$

where U_{gw}—groundwater flow rate, cm/year; δ_{gw}—Thickness of the mixing zone, m; I_f—Soil layer permeation rate, cm/year; W_s—Lateral distance of the mixing zone, m.

The dilution factor DF is calculated by formula (6.10):

$$DF = \frac{C_{soil}}{C_{GW}} = \frac{LDF}{\left[\frac{\rho_b}{\theta_{ws}+K_d\rho_b+H\theta_{as}}\right]} \times \frac{L_1}{L_2} \tag{6.10}$$

where ρ_b—Soil density, t/m^3; θ_{ws}—Volumetric water content of the gas-filled zone, cm^3-water/cm^3-soil; θ_{as}—Air volume of the gas-filled zone, cm^3-air/g-soil; L_1—Thickness of the red mud layer, m; L_2—Depth from the bottom of the red mud layer to the groundwater level, m; K_d—Soil–water distribution coefficient, cm^3—water/g-soil; H—Henry's constant

The selection of calculation index parameters refers to the reference values in HJ25.3-2014 "Technical Guidelines for Risk Assessment of Polluted Sites" and is selected according to the actual working conditions of the road base. The calculated values of the leaching concentration C_i of hazardous substances when they reach groundwater and the standards for Class III groundwater pollutants are shown in Table 6.29.

Table 6.29 Leaching concentration values (C_i, mg/L) of hazardous substances (reaching groundwater)

Item	Hg	Pb	Cd	Mn	Cu	Zn	Be	Ba
C_i	1.2E−06	7.8E−04	1.3E−04	2.4E−04	2.5E−04	3.3E−04	2.9E−05	3.9E−06
Class III groundwater standard limit	≤ 0.001	≤ 0.05	≤ 0.01	≤ 0.1	≤ 1.0	≤ 1.0	≤ 0.0002	≤ 1.0

Item	Ni	As	Co	Se	Mo	Cr(VI)	Fluoride
C_i	6.6E–04	8.0E−05	2.4E−04	1.3E−04	3.7E−03	1.1E−03	5.1E−02
Class III groundwater standard limit	≤ 0.05	≤ 0.05	≤ 0.05	≤ 0.01	≤ 0.1	≤ 0.05	≤ 1.0

After the hazardous substances in the modified red mud are released by rainwater leaching (dissolution), the concentration reaching the groundwater after dilution by the soil layer is all lower than the standard limit of the groundwater environmental quality standard Class III water. The environmental risk of using stabilized red mud for road base filling is acceptable.

6.4 Chapter Summary

The environmental risk of using modified red mud as a substitute material for highway construction is evaluated using a two-level assessment method. The main conclusions are as follows:

(1) When all the pollutants in the leachate of the modified red mud are lower than the third level standard of the "Groundwater Quality Standard" (GB/T14848), the environmental risk of its long-term use as a substitute material for various road constructions is acceptable, then the road use of the modified red mud is not restricted.
(2) After assessing and determining that the environmental risk of the long-term road use of modified red mud under specific scenarios of different highways is acceptable, the applicable conditions that the leaching content of V, Cr^{6+}, Cr^{3+}, As^{3+}, Se^{4+}, Se^{6+}, Mo, F should meet.

Furthermore, in the actual road use process of modified solidified red mud, the pH of groundwater and surface water will slightly increase; when the leachate of modified red mud is applied to the soil, the soil pH will slightly increase. The pH of the leachate from the modified red mud has little inhibitory effect on the growth of earthworms and Pak Choi in the test soil. After the harmful substances in the modified red mud are released by rainwater leaching (dissolution), the concentration reaching the groundwater after dilution by the soil layer is lower than the standard limit, and its environmental risk is within an acceptable range.

Chapter 7
Long-Term Environmental Impact of Modified Solidified Red Mud Road Use

7.1 Overview

Based on the physicochemical activity of bulk industrial solid waste represented by red mud, the key technology research on the large-scale regeneration road use of bulk solid industrial waste is carried out, which can not only save a large amount of road filling soil and reduce engineering costs, but also provide a new attempt for the resource utilization of bulk industrial solid waste. However, the pollutants in red mud will be released into the environment during the long-term use of it as road filling material, posing a risk of pollution to groundwater, surface water, and seawater. Therefore, to protect the environment, the environmental impact of solid waste such as red mud in road engineering application and operation is a focus and difficulty of its large-scale utilization. This chapter mainly analyzes the environmental impact indicators of red mud road use, evaluates its pollution control effect, and studies the long-term environmental impact of modified solidified red mud road use.

The analysis of the controllability of road use pollution of modified solidified red mud road use materials is provided by the Chinese Academy of Environmental Sciences and the Shandong Analysis and Testing Center.

According to the road structure design, the subgrade part of the test section is divided into upper and lower parts. The upper subgrade is filled with modified Bayer process red mud, and the lower subgrade is filled with lime-improved soil. The thickness of the modified red mud upper subgrade is 40 cm, and it is filled and compacted in two layers during construction. The distance from the normal groundwater level to the bottom of the modified red mud subgrade is about 1.2 m. The modified red mud filling layer is located above the groundwater level by 0.5 m, and an isolation layer of not less than 0.5 m is set in the middle. The possibility of the modified red mud subgrade being soaked by groundwater is extremely small. Therefore, in the long-term use of the road, the main consideration is the impact and damage of precipitation on the subgrade. To prevent rainwater from seeping into the inner red mud structure layer of the subgrade through the side of the subgrade, forming a lateral seepage

Y. Cheng and X. Zhu, *Utilization and Environmental Impact Assessment of Red Mud Resources*, https://doi.org/10.1007/978-981-96-7411-4_7

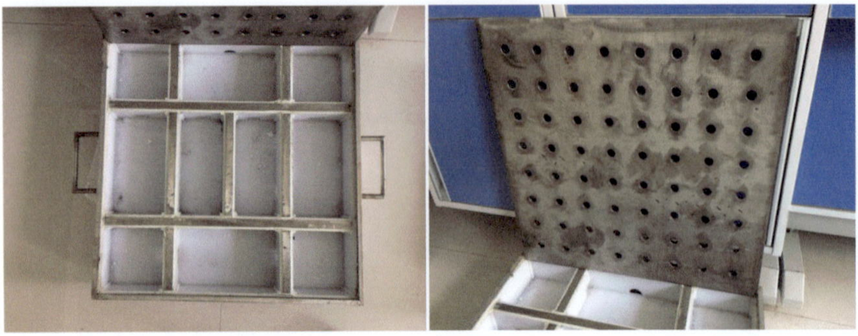

Fig. 7.1 Internal structure of the rainwater collection box

channel for pollutants, the upper subgrade of the modified red mud is isolated and sealed. The sealing treatment is located on both sides of the red mud, with a width of 80 cm. The material is lime soil with a mixing amount of 8%. After compaction, an isolation belt is formed on both sides of the red mud. During construction, the sealing soil and the modified red mud are spread and stabilized at the same time, then lime is added and mixed, and finally it is compacted synchronously with the internal modified red mud filler.

The pollutant collection device is specially customized by the environmental protection department, mainly composed of a collection box, a drain pipe, and a water storage device, used to collect the free water formed inside the subgrade under the influence of external precipitation. Inert materials are used inside the collection box to avoid interference with the concentration of harmful substances in the water sample, and its internal structure is shown in Fig. 7.1.

Two pollutant collection devices are buried in each of the modified red mud subgrade test section and the conventional material filling subgrade comparison section. The planar position of the collection box is located in a lane, with a horizontal distance of 2.75 m from the centerline of the road. Among them, the upper collection box in the test section is buried in the upper subgrade, the top of the box is flush with the top surface of the modified red mud subgrade, and the lower collection box is buried in the lower subgrade, the top of the box is flush with the bottom surface of the modified red mud subgrade; the upper collection box in the comparison section is buried in the upper subgrade, the top of the box is flush with the top surface of the lime soil upper subgrade, and the lower collection box is buried in the lower subgrade, the top of the box is flush with the bottom surface of the lime soil upper subgrade, as shown in Figs. 7.2 and 7.3.

There are four groundwater monitoring sampling wells set up for on-site environmental monitoring, which are located on both sides of the red mud test section and the conventional comparison section. The bottom of the monitoring well should reach below the groundwater level, and the groundwater samples in the well are regularly collected with a sampling pump, and the pH value and concentration of harmful substances in the water sample are tested according to environmental protection

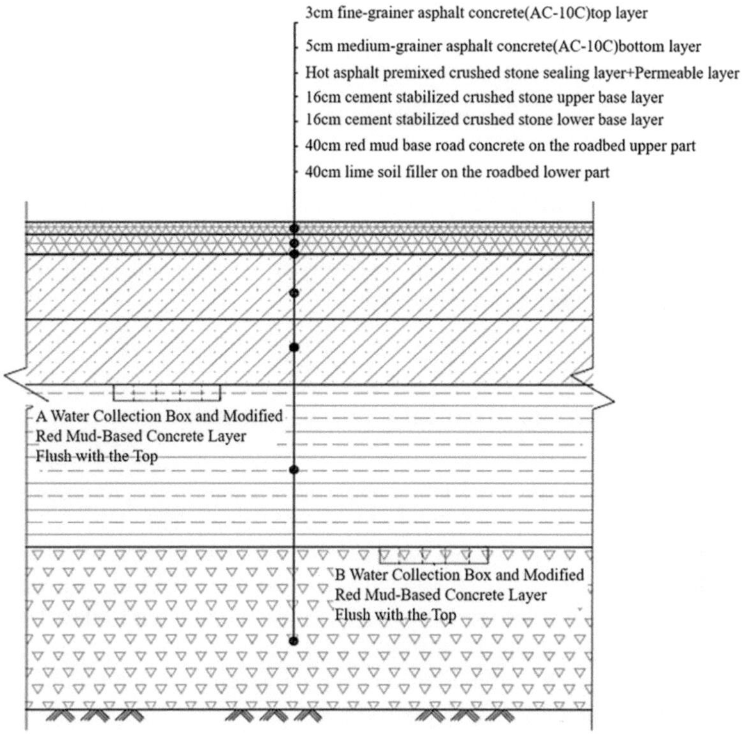

Fig. 7.2 Schematic diagram of the structure of the red mud subgrade test road

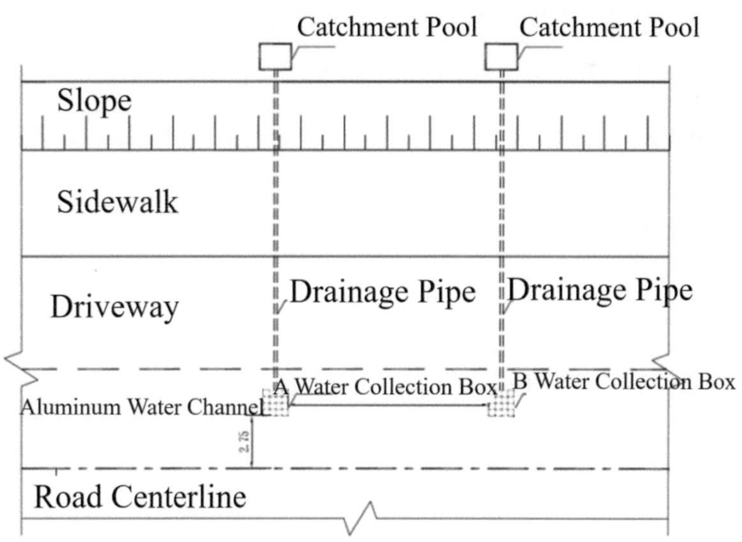

Fig. 7.3 Schematic diagram of the installation of the pollutant collection device

Table 7.1 Requirements for control indicators of radioactive nuclides in red mud road materials

Test item	Reference standard	Control indicator
Internal irradiation I_{Ra}	GB6566-2010	≤ 1.0
External irradiation I_r	GB6566-2010	≤ 1.3

standard methods. The monitoring wells were buried before the project started, and samples were taken for testing as a comparison blank sample. After the construction is completed, sampling tests are conducted every two weeks, and after two months, sampling tests are conducted once a month.

7.2 Road Use Environmental Control Standard

Before the modified solidified red mud is applied to roadbed filling, its environmental control indicators should comply with the specifications. The pollutant leaching concentration should be better than the Class III groundwater quality classification standard in the "Groundwater Quality Standard" (GB/T14848-9), and the pollutant content should be better than the Level I emission standard in the "Integrated Wastewater Discharge Standard" (GB8978-1996).

The radioactivity of the red mud road materials used in this study should comply with the provisions of Table 7.1 respectively.

In the leachate of solid waste prepared according to HJ/T 299-2007, the content of any hazardous component in the modified solidified red mud road materials should not exceed the concentration limit requirements listed in Table 7.2.

The sampling and sample preparation of red mud adopts the method specified in the "Technical Specification for Sampling and Sample Preparation of Industrial Solid Waste" (HJ/T 20-1998). The radioactive limit standard uses the detection method specified in the "Limit of Radioactive Nuclides in Building Materials" (GB 6566-2010). The leaching toxicity control standard is determined according to the method specified in the "Identification Standard for Leaching Toxicity of Hazardous Waste" (GB 5085.3-2007).

7.3 Environmental Monitoring and Evaluation of Modified and Solidified Red Mud Subgrade

7.3.1 Environmental Monitoring Plan

The structural schematic diagram of the two sections of the modified red mud subgrade test section is shown in Fig. 7.4. To monitor the leaching situation of pollutants in each layer of material inside the subgrade, soil samples are regularly taken

Table 7.2 Toxicity control indicators of leachate from modified red mud road materials

Serial number	Test item	Control indicator limit value (mg/L)	Test method
1	Total mercury	≤ 0.05	GB 5085.3-2007
2	Total chromium	≤ 1.5	GB 5085.3-2007
3	Hexavalent chromium	≤ 0.5	GB/T 15555.4-1995
4	Lead	≤ 1.0	GB 5085.3-2007
5	Cadmium	≤ 0.1	GB 5085.3-2007
6	Copper	≤ 1.0	GB 5085.3-2007
7	Zinc	≤ 5.0	GB 5085.3-2007
8	Beryllium	≤ 0.005	GB 5085.3-2007
9	Barium	≤ 1.0	GB 5085.3-2007
10	Nickel	≤ 1.0	GB 5085.3-2007
11	Arsenic	≤ 0.5	GB 5085.3-2007
12	Cobalt	≤ 0.05	GB 5085.3-2007
13	Selenium	≤ 0.01	GB 5085.3-2007
14	Manganese	≤ 2.0	GB 5085.3-2007
15	Inorganic fluorides	≤ 50.0	GB 5085.3-2007

from different layers (modified red mud layer, loess layer, and natural soil layer) after the completion of the test road, and the pH and concentration of hazardous substances of the subgrade material samples at different layers are tested according to the specified method to evaluate their impact on the surrounding environment.

A total of four groundwater monitoring sampling wells (Fig. 7.5) were set up for on-site environmental protection monitoring, which were respectively arranged on both sides of the red mud test section and the conventional comparison section. The monitoring wells of the modified red mud subgrade test section are numbered as SY-1# and SY-2#, and the monitoring wells of the conventional subgrade comparison section are numbered as DB-1# and DB-2#. The bottom of the monitoring well should reach below the groundwater level line, and the groundwater samples in the well are regularly collected with a sampling pump, and the pH value and concentration of hazardous substances of the water samples are tested according to the environmental protection standard method. The monitoring wells were buried before the project started, and samples were taken for testing as blank comparison samples. After the construction is completed, sampling tests are conducted every two weeks, and after two months, sampling tests are conducted once a month. After the construction of the red mud subgrade is completed, the groundwater samples in the four monitoring wells are collected at 14d, 28d, 42d, 56d, 90d, 120d, 150d, 180d, as shown in Fig. 7.6.

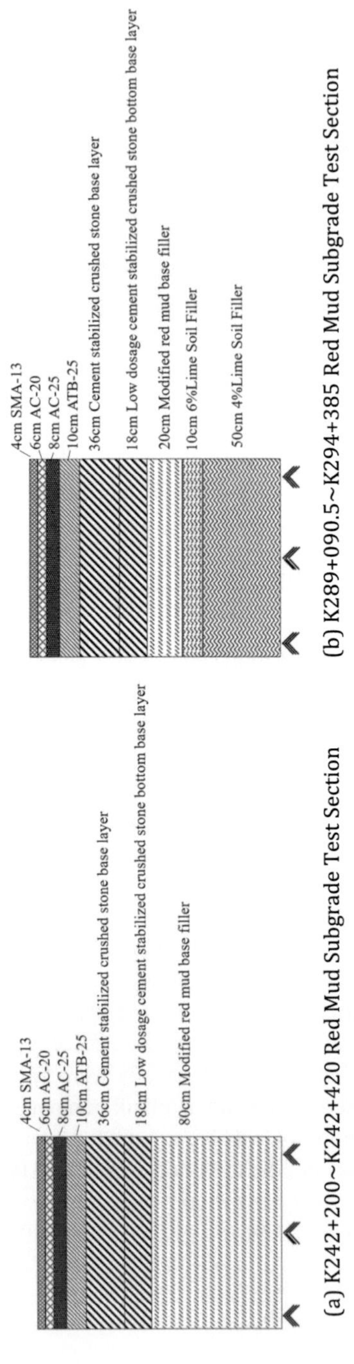

Fig. 7.4 Schematic diagram of the structure of the red mud subgrade test road

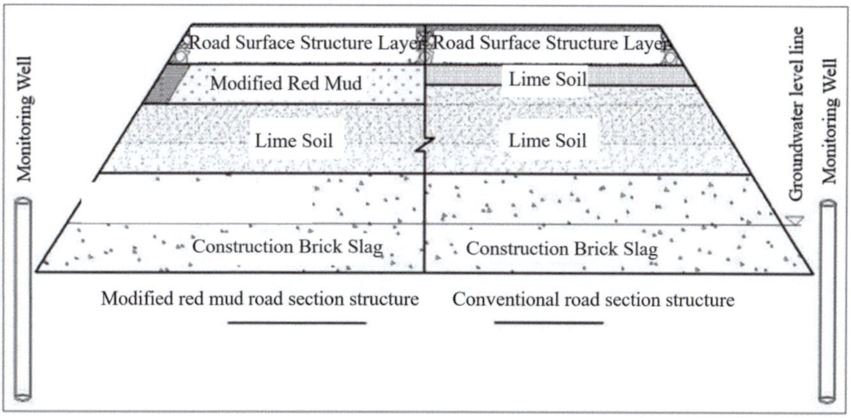

Fig. 7.5 Schematic diagram of the installation of groundwater monitoring wells

Fig. 7.6 Basis for the collection of water samples from the on-site monitoring wells

7.3.2 Environmental Monitoring Results and Evaluation

At the beginning of the leaching experiment, the first sample was taken every 3 h. From the second time, samples were taken every 24 h, and a total of 10 test samples were taken. The pH value of the leachate within 24 h was measured at the same time of each sampling. After 14d, the pH of the leachate of each group basically reached equilibrium, as shown in Table 7.3.

The trend of pH value changes in the standing solution is shown in Figs. 7.7 and 7.8. The pH range of the red mud standing solution is between 10.50 and 12.00, and the pH value of the on-site modified red mud is between 10.0 and 11.0. As the soaking time increases, the pH values of red mud at all ages show a gradually decreasing trend. Compared with other materials, the pH value of red mud is lower than that of cement and cement stabilized gravel.

As can be seen from the figure, the pH value of the cement stabilized gravel is significantly higher than that of other material samples, and the pH value of the modified red mud, especially the on-site modified red mud, is significantly lower than the predicted pH value of the red mud. This indicates that the modification treatment can effectively neutralize the alkaline components in the red mud, and suppress the leaching and diffusion of ions through stable solidification.

The results of pollutant detection in each layer of the modified red mud subgrade test Sect. 7.2 are shown in Figs. 7.9, 7.10, 7.11, 7.12, 7.13 and 7.14.

As shown in Fig. 7.10, the pH values of the modified red mud layer at K289 + 800, the 1.7 m deep raw soil layer, and the 0.5 m deep loam layer at K290 + 540 all show an increasing trend over time, and the pH values of all three locations are higher than the limit value of Class III groundwater.

As shown in the figure On July 11, the mercury content at the modified red mud layer at K289 + 800 has a decreasing trend over time. The mercury content in the 0.5 m deep loam layer at K290 + 540 and the 1.7 m deep loam layer at K289 + 800 were all below the detection value in three tests, and the mercury content at all three locations was below the Class III groundwater limit.

Table 7.3 Changes in pH value of different leachates over time

Test material	Time interval/h									
	3	24	48	96	120	144	170	193	218	316
Original red mud	11.90	11.85	11.70	11.35	11.05	10.95	10.90	10.80	10.75	10.40
Indoor modified red mud	11.95	11.45	11.85	11.40	11.10	11.05	10.65	10.70	10.70	10.05
On-site modified red mud	10.60	10.85	11.15	11.05	10.90	10.90	10.55	10.50	10.25	9.95
5% lime soil	10.45	11.10	11.10	11.25	10.75	10.75	10.30	10.15	10.35	9.80
Cement test block	12.55	13.00	12.85	12.95	12.95	12.90	12.85	12.90	12.75	12.65
Water stable gravel specimen	12.00	12.30	12.15	12.05	11.80	11.90	11.65	11.65	11.6	11.60

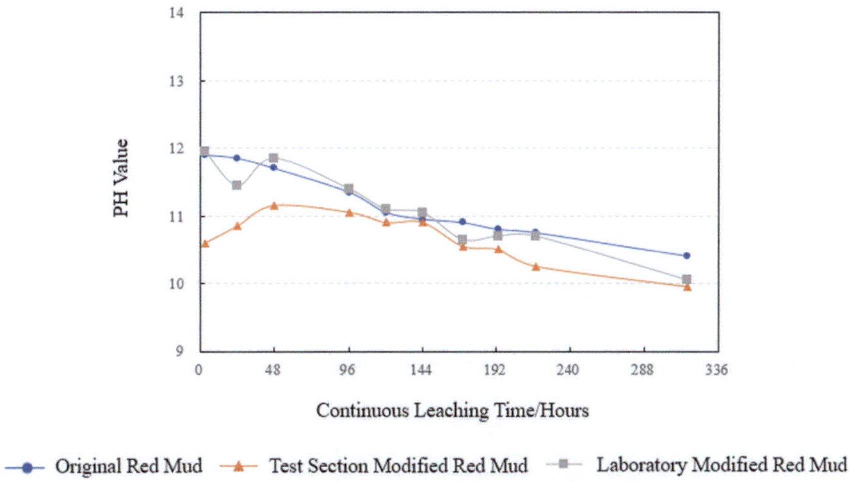

Fig. 7.7 Changes in pH values of red mud and modified red mud over time

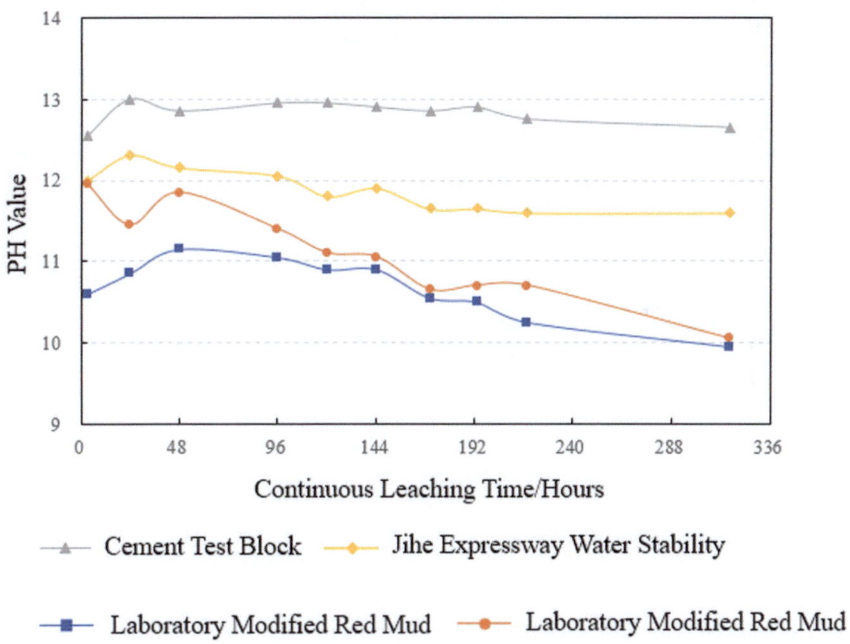

Fig. 7.8 Changes in pH values of different materials over time

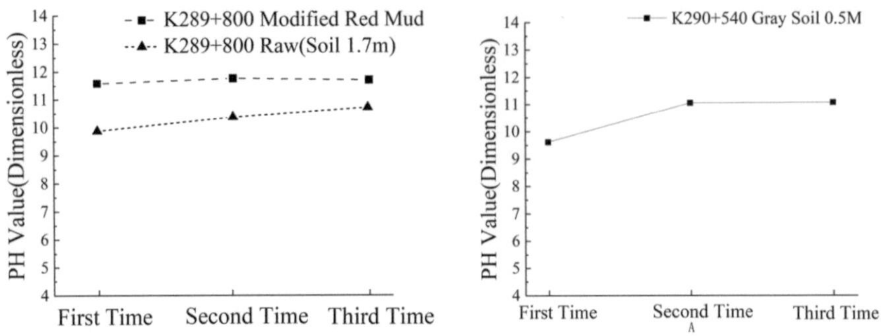

Fig. 7.9 pH detection results of each layer inside the subgrade

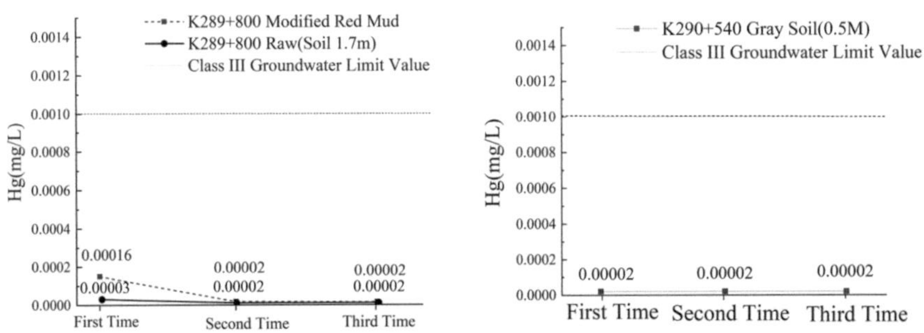

Fig. 7.10 Mercury concentration detection results of each layer inside the subgrade

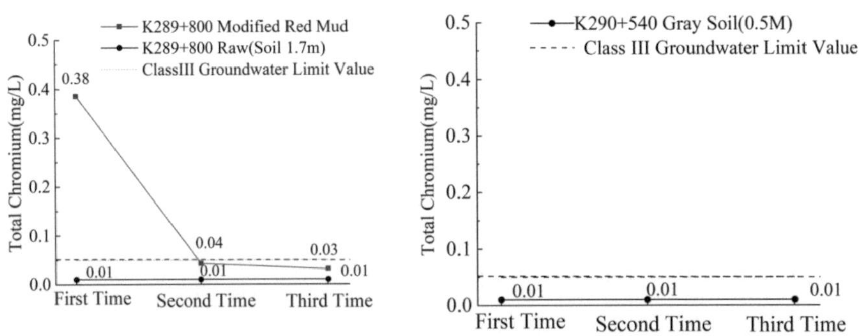

Fig. 7.11 Detection results of total chromium concentration in each layer inside the subgrade

As shown in Fig. 7.12, the total chromium content in the first test of the modified red mud layer at K289 + 800 was higher than the Class III groundwater limit. However, with the increase of time, the total chromium content gradually decreased and was lower than the Class III groundwater limit. The total chromium content in the 1.7 m deep loam layer at K289 + 800 and the 0.5 m deep loam layer at K290 +

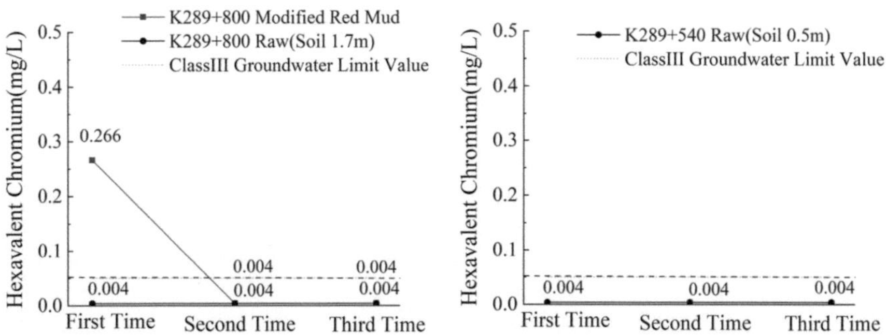

Fig. 7.12 Detection results of hexavalent chromium concentration in the red mud layer inside the subgrade

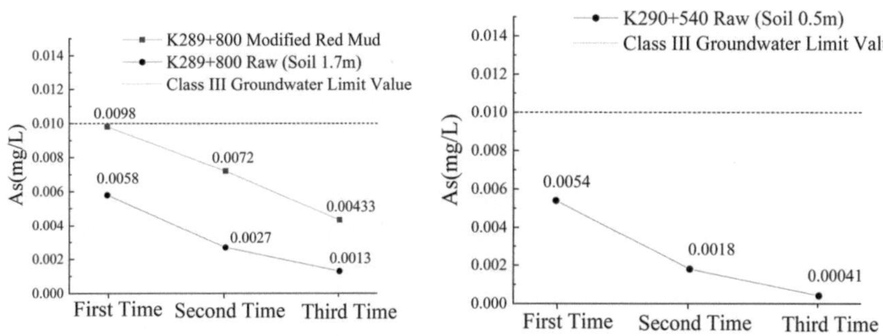

Fig. 7.13 Detection results of arsenic concentration in each layer inside the subgrade

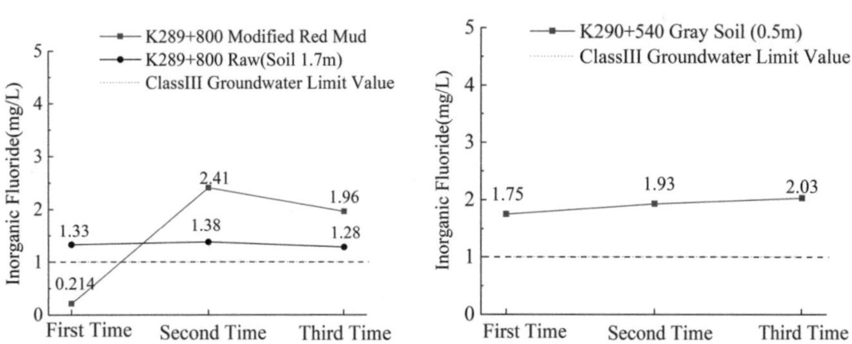

Fig. 7.14 Detection results of inorganic fluoride concentration in each layer inside the subgrade

540 were all below the detection value in three tests, and were lower than the Class III groundwater limit.

As shown in Fig. 7.13, the hexavalent chromium content in the modified red mud layer at K289 + 800 decreases with time, and the content starts to be lower than the Class III groundwater limit in the second test. The arsenic content in the 1.7 m deep loam layer at K289 + 800 and the 0.5 m deep loam layer at K290 + 540 were all below the detection value in three tests, and were lower than the Class III groundwater limit.

Figure 7.14 shows the detection results of inorganic fluoride concentration in each layer inside the subgrade. The arsenic content at the modified red mud layer at K289 + 800, the 1.7 m deep loam layer, and the 0.5 m deep loam layer at K290 + 540, all three locations have an increasing trend over time, and the arsenic content at all three locations is below the Class III groundwater limit.

The inorganic fluoride content at the modified red mud layer at K289 + 800, the 1.7 m deep loam layer, and the 0.5 m deep loam layer at K290 + 540, all three locations are higher than the Class III groundwater limit. The inorganic fluoride content at the modified red mud layer at K289 + 800 and the 1.7 m deep loam layer decreases with time, while the 0.5 m deep loam layer at K290 + 540 first decreases and then slightly increases.

In addition, beryllium was only tested in the first test, and the content in each soil layer under each working condition was below the detection value of 0.005 mg/L, but higher than the Class III groundwater limit of 0.002 mg/L; lead was only tested in the first test, and the content in each soil layer under each working condition was below the detection value of 0.05 mg/L, but higher than the Class III groundwater limit of 0.01 mg/L; some substances were found to have low concentrations through the detection data, and were lower than the Class III groundwater limit: copper, zinc, cadmium, barium, nickel, total silver, selenium.

The detection results of pollutants in each layer inside the subgrade of the modified red mud subgrade test Sect. 7.1 are shown in Figs. 7.15, 7.16, 7.17, 7.18, 7.19 and 7.20.

Under both conditions, the pH value first decreases and then increases over time, slightly higher than the Class III groundwater limit.

The arsenic content in each test under both conditions is below the Class III groundwater limit.

Under both conditions, the inorganic fluoride content increases over time. The loam layer at K242 + 080 was below the Class III groundwater limit in the first two tests, but exceeded the Class III groundwater limit in the third test, while the loam layer at K242 + 225 exceeded the Class III groundwater limit in all three tests. In addition, beryllium was only tested in the first test, and the content in each soil layer under each working condition was below the detection value of 0.005 mg/L, but higher than the Class III groundwater limit of 0.002 mg/L; lead was only tested in the first test, and the content in each soil layer under each working condition was below the detection value of 0.05 mg/L, but higher than the Class III groundwater limit of 0.01 mg/L; some substances were found to have low concentrations through the detection data, lower than the detection limit, and lower than the Class III groundwater limit: total chromium, hexavalent chromium, mercury, copper, zinc, cadmium, barium, nickel, total silver, selenium.

After testing, the pH value of the groundwater samples in the four monitoring wells has slightly decreased and then basically stabilized compared to before the project

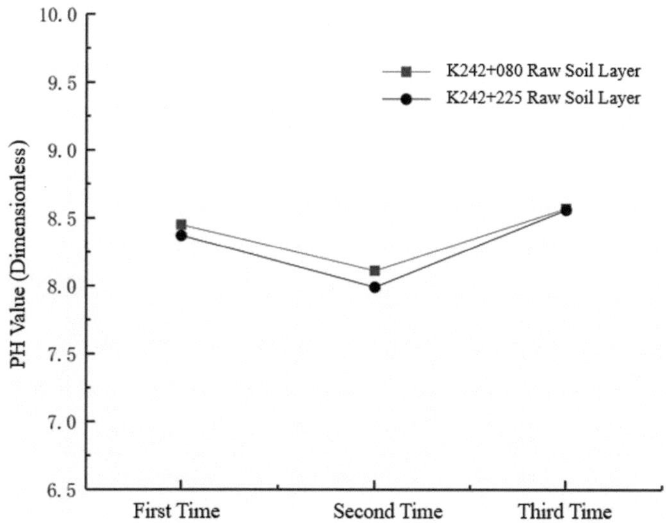

Fig. 7.15 pH monitoring results of the loam layer

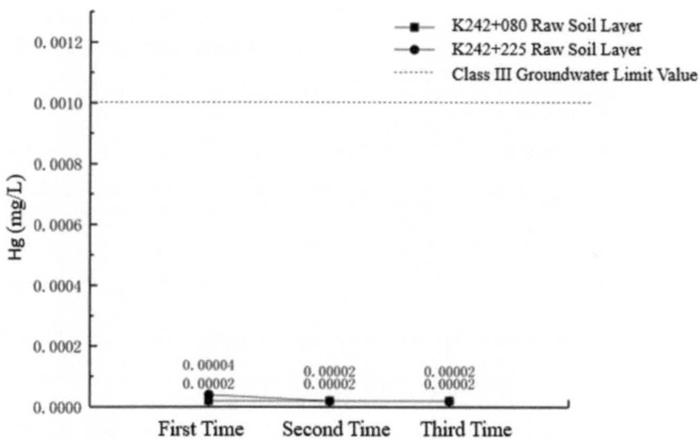

Fig. 7.16 Mercury concentration monitoring results of the loam layer

construction, with an overall distribution in the range of 7.0–8.0, and the modified red mud test section is slightly lower than the conventional subgrade comparison section, as shown in Fig. 7.21.

The changes in the concentration of hazardous substances can be roughly divided into four situations:

(1) Below the detection limit. The concentrations of cyanide, total mercury, total chromium, lead, cadmium, beryllium, selenium, silver, copper, antimony, and vanadium are all below the environmental standard detection limit;

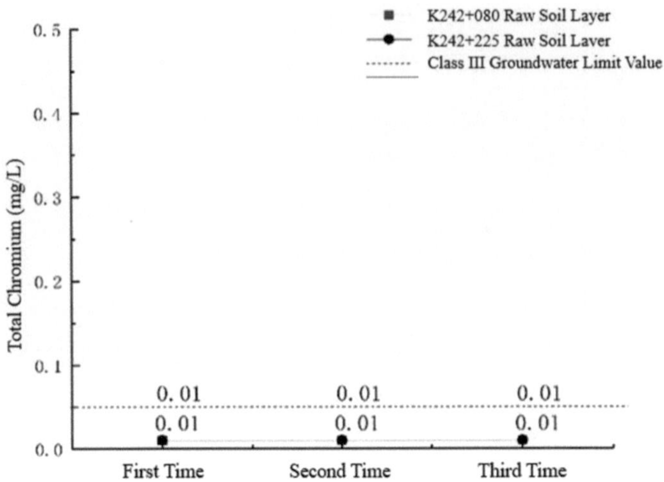

Fig. 7.17 Total chromium concentration monitoring results of the loam layer

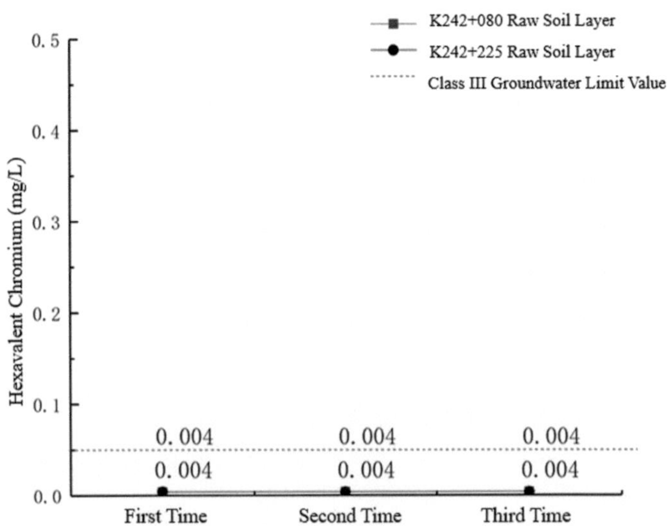

Fig. 7.18 Hexavalent chromium concentration monitoring results of the loam layer

(2) Increases then stabilizes. The concentrations of fluoride and barium are slightly higher than the blank samples before road construction, and basically stabilize after 14 ~ 28 days;
(3) Decreases and not detected. The concentrations of hexavalent chromium, arsenic, cobalt, and molybdenum gradually decrease after the road is completed, and are below the detection limit after 28 days;

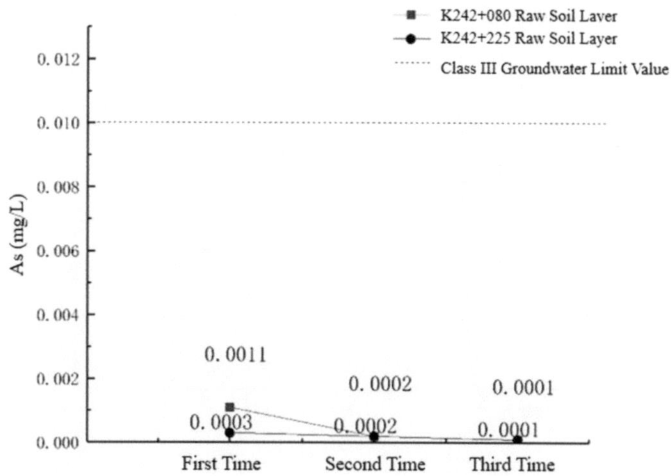

Fig. 7.19 Arsenic concentration monitoring results of the loam layer

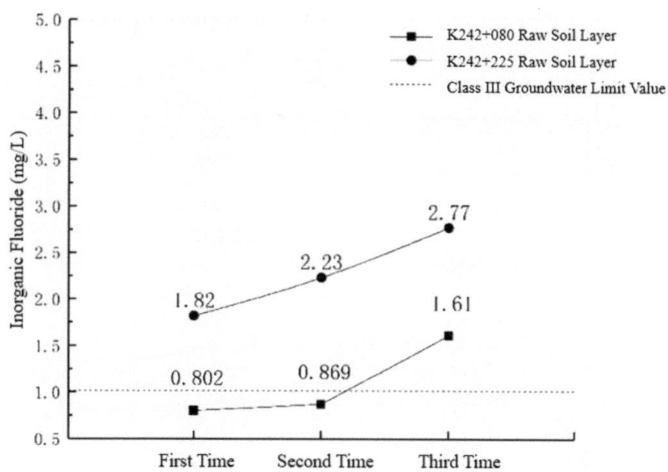

Fig. 7.20 Inorganic fluoride concentration monitoring results of the loam layer

(4) Small fluctuations. The concentration of hazardous substances fluctuates irregularly with time, but the overall fluctuation is not large, such as nickel and manganese. For situations (2) ~ (4), the concentration change curves of fluoride, hexavalent chromium, and nickel are selected as typical examples, as shown in Figs. 7.22, 7.23 and 7.24. It can be seen that the pH value and the trend of hazardous substance concentration changes in the groundwater of the red mud test road section and the conventional comparison road section are consistent, and they rise or fall and tend to stabilize at the same time, indicating that the

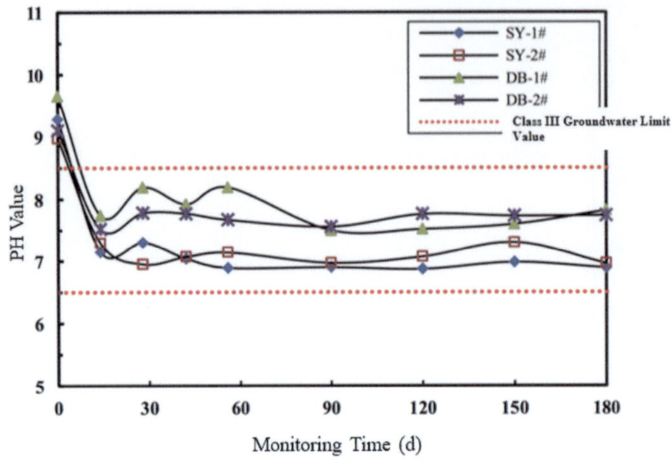

Fig. 7.21 Time-varying curve of pH value of groundwater samples

cause of this change may be affected by the construction process, not caused by the leaching of modified red mud.

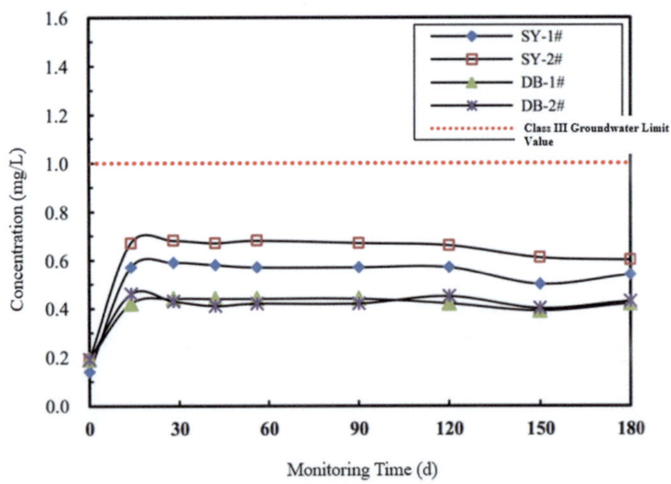

Fig. 7.22 The change curve of fluoride concentration in groundwater samples over time

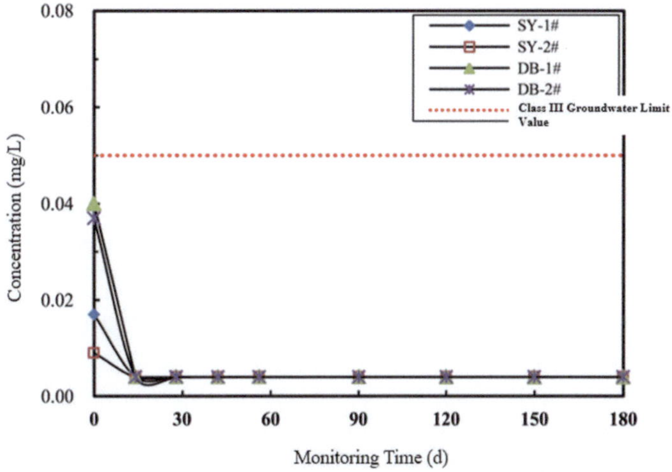

Fig. 7.23 The change curve of Cr^{6+} concentration in groundwater samples over time

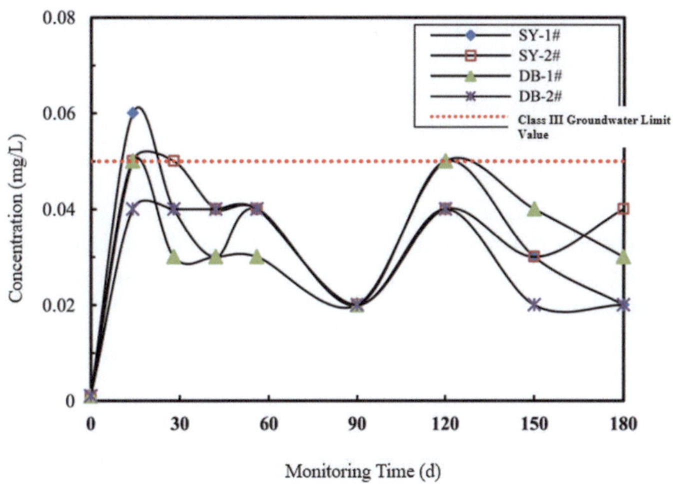

Fig. 7.24 The change curve of Ni concentration in groundwater samples over time

7.4 Environmental Reliability Measures for Modified Solidified Red Mud Roads

7.4.1 Reliable Environmental Performance

① The choice of modified materials should not increase the leaching amount of pollutants or characteristic pollutants

When modifying the original red mud, the selected modified materials are mainly composed of calcium-based materials + modified reductants. Because the co-disposal of waste in cement kilns can lead to higher metal components in cement products than conventional cement products. If the modified materials are not properly selected, they may introduce other pollutants, or the content of certain pollutants is high, but the purpose of solidifying characteristic pollutants is not achieved. Therefore, when selecting modified materials, the principle should be to not increase the content and leaching amount of characteristic pollutants F, V, Cr, As, Se, Mo in the original red mud sample.

② Optimize the modification process to ensure the stability of the modified red mud properties

Due to differences in the grade of alumina raw materials at different times and adjustments in process parameters, the pollutant composition in red mud may be unstable, and the components of the modified materials used may also differ, causing differences in the pollutant composition in the modified red mud. It is recommended to further optimize the modification process conditions to ensure the stability of the modified red mud properties, while ensuring that the leaching content of characteristic pollutants meets the applicable requirements and takes into account the leaching content of other pollutants.

③ Strengthen sampling and establish a pollutant file in modified red mud and modified materials

It is recommended to strengthen the sampling of each batch of modified red mud, and establish a file record of the content and leaching amount of the main pollutants in each component of red mud, modified red mud, and modified materials; according to the pollutant composition in red mud, it helps to quickly screen and determine suitable modified materials.

④ Establish a file for the section of the modified red mud road, and archive it in time for traceability

All sections of the modified red mud road should be well recorded and archived, recording the use date, quantity, range, pollutant content in the modified red mud, and the content of the main pollutants in the environmental media (groundwater, surface water and soil) during the implementation of the project.

7.4.2 Construction Process Management

The quality assurance of road construction with modified solidified red mud should strictly comply with the current relevant laws, regulations, and technical requirements of national, local, and industry standards.

① The planning, design, construction, and environmental protection management of roads using modified solidified red mud should comply with the relevant requirements of the latest revised version of the “Shandong Province Highway Administration Regulations”.

② The road surface design and construction of roads using modified red mud should meet the relevant provisions of the “Highway Asphalt Pavement Design Specification” (JTG D50-2017) and its latest version.

③ The engineering properties and testing of the soil used in road projects using modified red mud should comply with the relevant requirements of the “Highway Geotechnical Test Specification” (JTG 3430-2020) and its latest version.

④ In order to improve the durability of highways, reduce the damage of surface water and groundwater to the roadbed, road surface and structures, the expressways using modified red mud should design the highway drainage system according to the “Highway Drainage Design Specification” (JTG/T D33-2012) and its latest version. Highway drainage facilities should not be used for other non-highway drainage purposes.

⑤ Each stage of the design of the road construction project using modified red mud should pay attention to environmental protection design, and its environmental protection design should comply with the technical requirements of the “Highway Environmental Protection Design Code” (JTG B04-2010). The following are closely related to this assessment.

Article 3.2.2, paragraph 3 (5) stipulates “Take comprehensive measures to reduce the pollution of traffic noise, exhaust gas, wastewater, etc. to the environment”.

Article 3.2.4, paragraph 3 stipulates, “The newly established material yard should consider the impact of the location and number of the material yard on slope vegetation, river flow direction, and soil and water conservation”.

Article 3.2.4, paragraph 5 stipulates “The comprehensive drainage facilities of the subgrade and pavement should form a system and must not interfere with the local irrigation and drainage system”.

Article 4.2.4, paragraph 1, should demonstrate and determine the width of the highway subgrade cross-section, central divider, hard shoulder, and protective slope.

Article 4.2.7, when there is a lack of soil source or when the project requires, on the basis of technical and economic comparison, priority should be given to using industrial slag, blown (filled) sand or fly ash as fillers to build road embankments, reducing the occupation of land for soil extraction.

Article 5.3.2 Surface runoff from the road surface must not be directly discharged into drinking water bodies and aquaculture water bodies.

Article 5.3.3 Highways must not occupy the drinking water bodies of concentrated residential areas; when the edge of the subgrade is less than 100 m from the drinking water body and less than 20 m from the aquaculture water body, greening belts or other isolation and protection measures should be taken.

Article 5.3.4 When highways pass through lakes, reservoirs, wetlands and other surface runoff catchment areas, measures should be taken to prevent highways from blocking surface runoff.

Article 5.3.6 In the drinking water groundwater source protection area, no seepage structures that pollute the groundwater source should be set up.

Article 6.1.1 stipulates that highway construction projects should mainly prevent and control the environmental pollution of sound, gas, water, and solid waste. Among them, paragraph 3 is "Pollution of the water environment by domestic sewage, construction wastewater, and engineering waste in facilities along the highway".

Article 6.1.2 stipulates that the prevention and control of highway environmental pollution should mainly target environmentally sensitive points, among which the water environment sensitive points are drinking water source protection areas and water bodies with special requirements.

Article 6.2.4, paragraph 3 stipulates "When using construction waste or industrial waste for embankment, it should be wrapped in soil, not exposed, and greened in time".

Article 6.4 related requirements for water pollution prevention and control. Especially 6.4.2, when the highway passes through drinking water source areas and water bodies with higher requirements for water environmental quality, it should comply with the regulations. It is recommended that the road sections using modified red mud should avoid drinking water source areas and water bodies with higher requirements for water environmental quality. In addition, the sewage treatment of facilities along the line should comply with the provisions of Article 6.4.3, and the treatment of domestic sewage during the construction period should comply with the provisions of Article 6.4.4.

Article 8.2.1, paragraph 2 stipulates, "The drainage facilities of the subgrade and pavement should be systematically improved, form a system, should be intercepted and sent far away, and guided by the situation".

In addition, in addition to complying with the above-mentioned regulations and technical requirements, it should also comply with the relevant regulations and technical standards of the current national, local and industry.

7.4.3 Regular Environmental Monitoring

For the completed red mud subgrade, environmental impact monitoring should be carried out. The monitoring scope includes possible environmental impacts on groundwater and surface water. The main monitoring object is groundwater, and when necessary, it should also include surface water, soil, etc. The monitoring items should be determined based on the conclusions of the preliminary environmental investigation phase and the implementation phase work plan. Including site pollutants and site characteristic parameters determined by environmental investigation, the dangerous waste monitoring items that may be involved should refer to the relevant indicators in the "Identification Standards for Leaching Toxicity of Hazardous Waste" (GB5085.3-2007).

① Organization and implementation of monitoring work

Monitoring work generally includes information collection and sorting, monitoring plan preparation, monitoring point layout, sample collection and on-site analysis, sample laboratory analysis, data processing, monitoring report preparation, etc. Among them, point layout and sampling are the key points of monitoring work. Necessary technical and management measures should be taken in the process of sample collection, preparation, transportation, and analysis.

② Monitoring point layout

Determine the layout range according to the geographical location and site boundary determined by the on-site environmental investigation. For areas within the layout range with similar hydrogeological conditions, a random layout method can be used to set up monitoring points. Groundwater monitoring points should be in areas suspected of severe environmental impact, and points should also be considered downstream of groundwater runoff. Surface water monitoring points should be in areas suspected of severe environmental impact, and points should also be considered downstream or at the confluence of groundwater runoff. Soil monitoring points should be in areas suspected of severe environmental impact, and a zoning layout method should be used for layout.

③ Sample Collection

When sampling groundwater, the most representative samples should be obtained with the lowest sampling frequency according to the local hydrogeological conditions. When setting up monitoring wells, foreign water and fluids should be avoided, and seepage prevention measures should be taken at the ground wellhead. When sampling surface water, stirring of bottom sediments should be avoided. To reflect the hydraulic connection between surface water and groundwater, the sampling frequency and time of surface water should be as consistent as possible with groundwater sampling. Soil samples are generally collected by excavation, using simple tools such as shovels and spades, or by drilling. The sampling process should minimize soil disturbance and ensure that soil samples are not secondarily contaminated during the sampling process.

④ Sample Testing and Analysis

After sampling is completed, portable analytical instruments can be used for qualitative and semi-quantitative analysis of field samples. Quantitative analysis of groundwater, surface water, and soil samples in the laboratory should be strictly carried out in accordance with the requirements of relevant environmental protection standards.

7.5 Chapter Summary

This chapter carried out environmental monitoring and impact analysis of red mud road base based on the Ji-Qing Expressway and the experimental road in Binzhou Beihai Economic Development Zone. The following conclusions were drawn.

The changes in the concentration of hazardous substances can be roughly divided into four situations.

① Below the detection limit. The concentrations of cyanide, total mercury, total chromium, lead, cadmium, beryllium, selenium, silver, copper, antimony, and vanadium are all below the environmental protection standard detection limit;
② Increase and then stabilize. The concentrations of fluoride and barium are slightly higher than the blank samples before road construction, and basically stabilize after 14 ~ 28d;
③ Decrease and not detected. The concentrations of hexavalent chromium, arsenic, cobalt, and molybdenum gradually decrease after the road is completed, and are below the detection limit after 28d;
④ Slight fluctuations. The concentration of hazardous substances fluctuates irregularly with time, but the overall fluctuation range is not large, such as nickel and manganese.

Long-term environmental safety measures for the use of red mud roads have been proposed. These mainly include ensuring the stability of modified red mud properties, standardizing the construction and environmental protection design management of road sections using modified red mud, and paying attention to strengthening regular environmental monitoring.

Chapter 8
Post-evaluation of Environmental Impact of Modified and Solidified Red Mud Road Use

The key to environmental risk assessment is to determine the source release concentration and exposure concentration. This project uses the maximum leaching method (HJ/T-299) for environmental impact assessment to characterize the source release concentration of pollutants in red mud. Based on exposure scenario analysis and combined with environmental investigation, the exposure concentration of pollutants after release to the exposure point was simulated and calculated. Using the groundwater class III standard value as the evaluation standard, the environmental risk of the use of modified and solidified red mud roads was characterized. The main process includes: use mode and environmental investigation, establishment of exposure scenarios, concentration of characteristic pollutants in modified red mud, dilution attenuation model and parameters, calculation of exposure concentration and risk characterization. The brief process is shown in Fig. 8.1.

Based on the requirement of red mud for road use and considering economic rationality and construction convenience, a composite modification and solidification material based on efficient composite excitation and particle surface activation is used for harmless treatment of red mud. This technology uses phosphogypsum and cement as the main basic carrier materials, and also adds heavy metal ion reductants, complexing agents and other solidification modification materials to produce, and the red mud modifier and red mud are mixed in a ratio of 8:92. The modified red mud is used for road base construction. In view of the strong alkalinity of red mud, phosphogypsum is used as a component of the red mud modifier to adjust the alkaline substances in the red mud. At the same time, the Ca^{2+} in phosphogypsum can react with the F^- in red mud to form CaF_2, realizing the solidification of F^- in red mud. In addition, the solidifier adds reductants, complexing agents and other substances to reduce and solidify Cr^{6+} and other heavy metals in red mud.

Y. Cheng and X. Zhu, *Utilization and Environmental Impact Assessment of Red Mud Resources*, https://doi.org/10.1007/978-981-96-7411-4_8

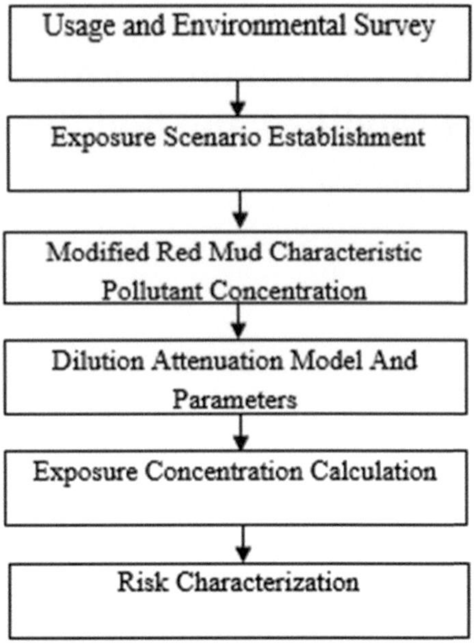

Fig. 8.1 Flowchart of environmental risk assessment

8.1 Preliminary Environmental Risk Assessment of Modified and Solidified Red Mud Road Use

In order to evaluate the leaching situation of pollution in red mud after modification, the leaching concentration of pollutants in raw red mud and modified red mud samples was detected (Table 8.1).

The indoor experiment results show that the modification has a significant inhibitory effect on the leaching of pollutants in red mud. The leaching concentrations of hexavalent chromium, selenium, inorganic fluorides, arsenic, and vanadium in red mud are reduced by 97.9%, 87%, 83.4%, 71.1%, and 69.1% respectively, indicating that the modification process has a good solidification and stabilization effect on red mud.

The foundation of this research is 1.0 m thick broken bricks, the upper layer of the broken bricks is 30 cm thick lime soil, this layer of lime soil is modified with quicklime, two fabrics and one film are laid on this layer of lime soil, the upper layer of the two fabrics and one film is 40 cm thick mature lime modified lime soil, the upper layer is 40 cm thick modified red mud fill layer, the outside of the modified red mud fill layer is lime soil wrapping structure. Above the modified red mud layer is the road surface structure layer, from bottom to top are two layers of 16 cm thick water-stable crushed stone base layer, hot asphalt premixed crushed stone sealing layer, two layers of asphalt concrete surface layer with thicknesses of 5 cm and 3 cm respectively. The thickness of the modified red mud fill layer from the normal groundwater level is about 1.2 m.

Table 8.1 Leaching of pollutants in red mud and modified red mud

Serial number	Test item	Standard basis	Test results	
			Red mud	Modified red mud (CPS-12)
1	Total mercury	< 0.00005	< 0.00005	–
2	Lead	< 0.02	< 0.02	–
3	Cadmium	< 0.004	< 0.004	–
4	Manganese	< 0.0005	< 0.0005	–
5	Total chromium	1.50	1.21	19.3%
6	Copper	0.003	< 0.001	–
7	Zinc	< 0.0001	< 0.0001	–
8	Beryllium	< 0.001	< 0.001	–
9	Barium	0.005	< 0.001	–
10	Nickel	0.0058	0.002	–
11	Arsenic	0.0097	0.0028	71.1%
12	Cobalt	0.0061	< 0.0005	–
13	Selenium	0.01	0.0013	87.0%
14	Molybdenum	0.16	0.12	37.5%
15	Silver	< 0.01	< 0.01	–
16	Antimony	0.03	< 0.02	33.3%
17	Vanadium	1.10	0.34	69.1%
18	Hexavalent Chromium	1.45	0.031	97.9%
19	Inorganic fluorides	16.1	2.68	83.4%

Note (1) method basis: HJ T-299; (2) there are 3 samples for each, and the leaching concentration of each pollutant in the table is the maximum value among the 3 samples; (3) < indicates below the detection limit

According to the research design requirements, the modified red mud fill layer is located above the groundwater level of 0.5 m, and there is no less than 0.5 m isolation layer in the middle. The thickness of the modified red mud fill layer from the normal groundwater level in this test section is about 1.2 m, and two fabrics and one film are laid in the middle to prevent the salt from the lower layer from infiltrating into the road base. The two fabrics and one film also have anti-seepage function. Therefore, there is no possibility of groundwater immersion in the modified red mud fill layer. The exposure scenario mainly considers the leaching (dissolution) action of precipitation through the road surface structure, causing harmful substances in the modified red mud layer to enter the groundwater with the leachate (water), thereby causing pollution to the groundwater. Therefore, the exposure scenario for this prediction assessment is the leaching (dissolution) of precipitation, and the environmental protection target is groundwater.

8.1.1 pH Value Environmental Impact Assessment

In environmental risk assessment, pH value is generally not included in the assessment category. However, considering the strong alkalinity of red mud, this project assessed and analyzed the possible environmental impacts of pH value when using red mud road.

According to the "Solid Waste Corrosiveness Determination Glass Electrode Method" (GBT 15555.12), the pH value of the crushed modified red mud sample was tested. The results show that the maximum value of the modified solidified red mud is 12.1. According to the previous pH value test results of cement concrete, the pH value is in the range of 11.7 ~ 12.2 (Table 8.2). It can be seen that the pH value of the modified solidified red mud is similar to that of cement concrete.

In actual modified solidified red mud roads, the red mud is compacted into a dense block structure. To study the release pattern of alkaline substances in the modified solidified red mud under the actual roadbed filling application conditions, the leaching method in the "Test of Diffusion Leaching Behavior of Inorganic Components in Block Waste or Building Materials" (NEN7375) was used to conduct a long-term experimental study on the pH value of the modified red mud sample block. The results show (Fig. 8.2) that the pH value of the leachate of the modified red mud sample block reached the highest value on the 1st and 9th days, about 8.7. After the 9th day, the pH value began to decrease, and after the 15th day, the pH value basically tended to balance, about 7.9.

Table 8.2 pH value test results of cement concrete sample leachate

Sample	1#	2#	3#	4#	5#	6#	7#	8#	9#	10#
pH	12.2	12.2	12.2	12.2	12.1	12.0	11.8	11.8	11.7	11.8

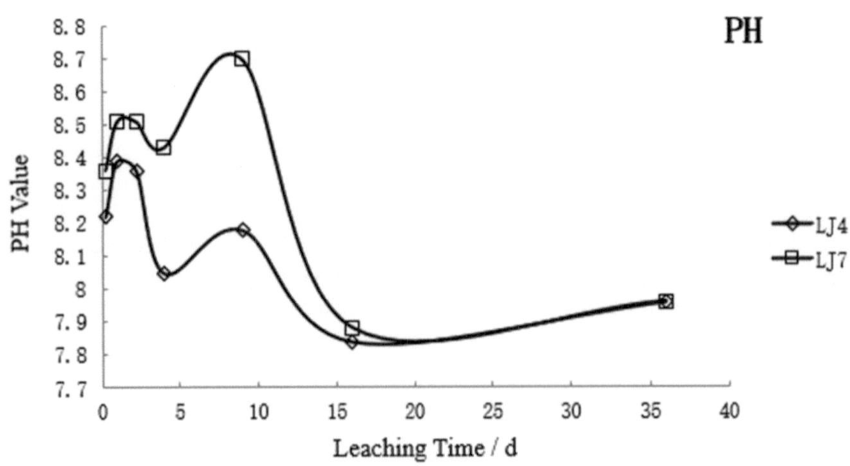

Fig. 8.2 pH test results of modified red mud sample block NEN7375 leaching method

In summary, the pH of the leachate of the modified red mud sample is basically the same as that of the cement concrete leachate, and when the NEN7375 leaching method, which is more in line with the actual application conditions, is used to test the pH value of the leachate of the sample block, the highest pH value is 8.7, which drops to 7.9 after 15 days and tends to balance. Therefore, the modified red mud sample will not have a significant adverse impact on the groundwater environment in actual applications.

8.1.2 Environmental Risk Assessment of Pollutants

After the modified solidified red mud roadbed is leached by rainwater (dissolved), the pollutants in it are dissolved by rainwater and migrate and diffuse into the groundwater with the leachate, and the pollutants entering the groundwater diffuse with the groundwater flow, that is, there is a dilution attenuation effect, which can be expressed by the dilution-attenuation factor (DAF), and the model schematic is shown in Fig. 8.3.

Hazardous components leak from the fill point, enter the groundwater mixing zone through the gas envelope (the groundwater in the fill area), and have a dilution effect on the hazardous components. The dilution factor is represented by DF (Dilution factor, DF). After mixing with groundwater, the harmful components migrate to the water well with the groundwater and also have an attenuation effect. Its attenuation dilution is represented by AF (Attenuation, AF). This research only considers the effect of DF.

According to the toxic substance dilution migration model established by the Water Resources Research Center of Texas State University (this model is used by Texas to determine the dilution attenuation multiples of toxic substances migrating to groundwater drinking water wells under the conditions of waste disposal and polluted soil/field rainwater leaching, and to formulate related pollution control standards (the model schematic is shown in Fig. 8.4), the calculation method of DF is as follows:

$$DF = \frac{C_{\text{soil}}}{C_{GW}} = \frac{LDF}{\left[\frac{\rho_b}{\theta_{ws} + K_d \rho_b + \dot{H}\theta_{\text{as}}}\right]} \times \frac{L_1}{L_2} \tag{8.1}$$

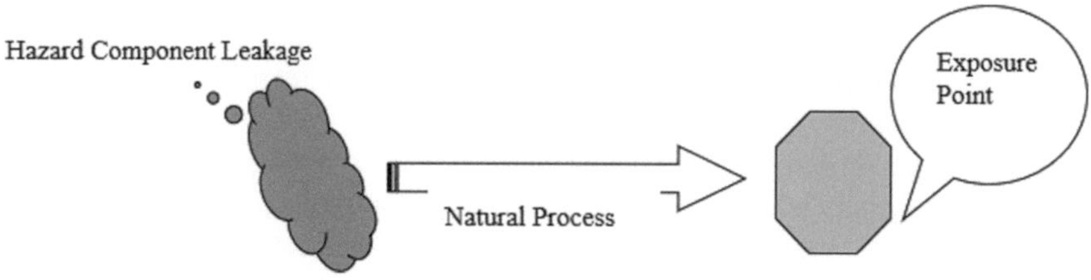

Fig. 8.3 Schematic of the dilution attenuation factor model

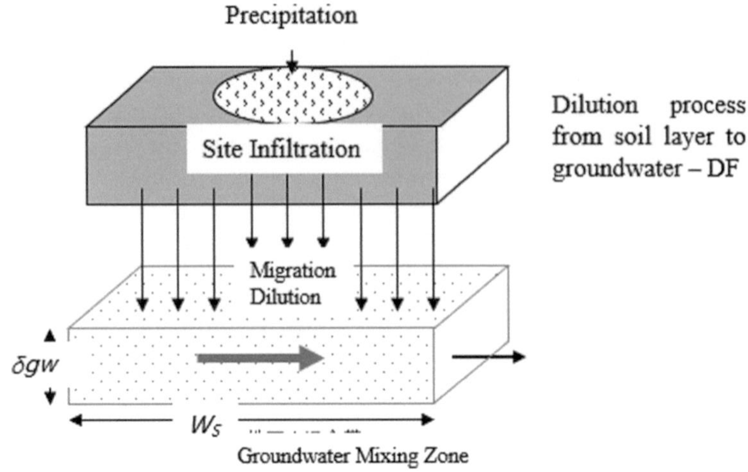

Fig. 8.4 Schematic of the soil layer leaching dilution process model

LDF is the dilution factor after the harmful components enter the mixing zone, and the calculation method is as follows:

$$LDF = 1 + \frac{U_{gw}\delta_{gw}}{I_f W_s} \tag{8.2}$$

In the formula, the value of the distribution coefficient Kd of the pollutant in the soil and water is referred to the recommended value in the Texas Risk Reduction Program (trrp). Some other main parameters directly refer to the reference values in the "Technical Guidelines for Risk Assessment of Polluted Sites (HJ25.3-2014)". And some other parameters are selected according to the actual situation of the Beihai Economic Development Zone in Binzhou, Shandong. For example: the depth of the seepage layer is the depth of the middle aquifer, which is 1.2 m; the soil permeability I_f is the net infiltration rate of the leachate into the groundwater during the rainfall process (Table 8.3). Regarding the infiltration problem of rainfall, there are many researches and model explanations in China, including Green-Ampt, Horton, Philip, Kostiakov models, etc. Due to the incomplete geotechnical hydrogeological data of the site, the EPA empirical model based on soil type rainfall infiltration was used. According to actual engineering data, the red mud roadbed is under lime soil and broken bricks, so the permeability is calculated according to sandy soil. The calculation results of DF and the concentration at the exposure point (Ci) are shown in Tables 8.4 and 8.5.

As can be seen from Table 8.5, after the pollutants in the modified red mud are released by rainwater leaching (dissolution), the concentration reaching the groundwater after dilution by the soil layer is all lower than the standard limit of Class III

Table 8.3 Main parameter meanings and reference values

Serial number	Code	Chinese name	Unit	Reference value
1	ρ_b	Soil density	Kg/m^3	1.5
2	θ_{ws}	Volumetric water content of the vadose zone (clay)	cm^3—water/cm^3—soil	0.12
3	θ_{as}	Volumetric air content of the vadose zone	cm^3—air/g—soil	0.26
4	L_1	Red mud thickness	m	0.4
5	L_2	Depth from the bottom of the red mud to the groundwater level	m	1.2
6	U_{gw}	Groundwater flow rate	cm/year	2000
7	I_f	Soil layer permeation rate	cm/year	3.89
8	δ_{gw}	Mixing zone thickness	m	3.4
9	W_s	Lateral distance of the mixing zone	m	24
10	K_d	Soil–water distribution coefficient	cm^3—water/g—soil	Log_{Kd}
11	H	Henry's constant	None	0

Table 8.4 Log_{Kd} and DF values of each pollutant

Serial number	Component	Log_{Kd}	DF
1	Hg	1.7	42.1
2	Pb	1.0	25.6
3	Cd	1.2	30.3
4	Mn	1.7	42.1
5	Cu	1.6	39.8
6	Zn	1.2	30.3
7	Be	1.4	35.0
8	Ba	1.0	25.6
9	Ni	1.2	30.3
10	As	1.4	35.0
11	Co	1.7	42.1
12	Se	0.34	9.9
13	Mo	1.3	32.7
14	Cr(VI)	1.1	27.9
15	Fluorine	2.2	54.0

Table 8.5 Exposure (to groundwater) point concentration (Ci, mg/L)

Item	Hg	Pb	Cd	Mn	Cu
Release concentration (C_0)	< 0.00005	< 0.02	< 0.004	< 0.01	< 0.01
DF	42.1	25.6	30.3	42.1	39.8
C	1.19E−06	7.81E−04	1.32E−04	2.38E−04	2.51E−04
Groundwater level III	≤ 0.001	≤ 0.05	≤ 0.01	≤ 0.1	≤ 1.0
Item	Zn	Be	Ba	Ni	As
Release concentration (C_0)	< 0.01	< 0.001	< 0.01	< 0.02	0.0028
DF	30.3	35.0	25.6	30.3	35.0
C_i	3.30E−04	2.86E−05	3.91E−04	6.60E−04	8.00E−05
Groundwater level III	≤ 1.0	≤ 0.0002	≤ 1.0	≤ 0.05	≤ 0.05
Item	Co	Se	Mo	Cr(VI)	F
Release concentration (C_0)	< 0.01	0.0013	0.12	0.031	2.68
DF	42.1	9.9	32.7	27.9	54.0
C_i	2.38E−04	1.31E−04	3.67E−03	1.11E−03	4.96E−02
Groundwater level III	≤ 0.05	≤ 0.01	≤ 0.1	≤ 0.05	≤ 1.0

water of groundwater environmental quality, the environmental risk of using modified red mud for the construction of the road base in Binzhou Beihai Economic Development Zone is acceptable.

Based on the laboratory test results of characteristic pollutants in the modified solidified red mud, the risk of the modified red mud to groundwater under the scenario of road base construction in Binzhou Beihai Economic Development Zone is assessed, and the following conclusions are drawn.

The composite red mud modification solidifier and implementation process have a good solidification and stabilization effect on the pollutants in the red mud. The use of modified solidified red mud in road base engineering will not have a significant adverse impact on the groundwater environment in actual applications. The environmental risk of groundwater pollution by heavy metals and other pollutants in the modified solidified red mud is acceptable.

In conclusion, the pollutants in the modified solidified red mud are affected by the source of the red mud and its modification process. When the source of the red mud or the modification process changes significantly, the potential environmental risks should be reassessed. The parameters for environmental risk assessment have strong regional characteristics, and there is uncertainty in the selection of parameters for the environmental risk assessment model and the assessment results. Based on the laboratory test results, after the completion of the modified red mud road utilization project, a daily monitoring plan for groundwater and a pollution emergency plan should be developed to prevent the road project from polluting the groundwater.

8.2 Post-evaluation of the Environmental Impact of Modified Solidified Red Mud Road Base

The post-evaluation of the environmental impact of the modified solidified red mud road is an important part of the utilization of red mud resources. The environmental impact assessment is carried out on the road sections that have been built using the modified solidified red mud road base in the infrastructure construction project in the Binzhou Beihai Economic Development Zone Vein Industrial Park. The groundwater, soil, and noise around the road of this project were mainly monitored.

In this study, modified solidified red mud was used as the road base material. When encountering rainy and snowy weather, the substances in the red mud may enter the regional groundwater, surface water, and soil through surface runoff with the rainwater, causing pollutants in the environment to exceed the standard. According to the collected red mud components and leachate for testing, their composition is shown in the Tables 8.6 and 8.7.

According to the test results of the red mud components and red mud leachate in the above table, in order to understand the pollution situation of red mud to the surrounding environment, 41 factors such as K^+, Na^+, Ca^{2+}, Mg^{2+}, iron, fluoride in the groundwater around the red mud road project; 9 factors such as pH, cadmium, mercury, arsenic, lead, chromium in the soil, and the noise environment around were monitored at the same time.

8.2.1 *Groundwater Environmental Quality Evaluation*

According to the actual road project situation, the groundwater status monitoring points shown in Fig. 8.5 were selected for the post-evaluation of the environmental impact of the modified solidified red mud road. The monitoring items are shown in Table 8.8.

According to the monitoring specifications of different projects, the specific analysis methods and detection limit values used in this post-evaluation of the environmental impact of the modified solidified red mud road are shown in Table 8.8.

Table 8.6 Experimental results of modified solidified red mud components

Test item	Al_2O_3	SiO_2	Fe_2O_3	CaO	MgO	K_2O	Na_2O	TiO_2	P	S
Red mud component	16.51	13.55	46.44	0.98	0.09	0.06	6.38	6.60	0.10	0.22

Table 8.7 Experimental results of modified solidified red mud leachate

Test item	pH	Fluoride (excluding calcium fluoride) (mg/L)
Red mud leachate	11.30	4.33

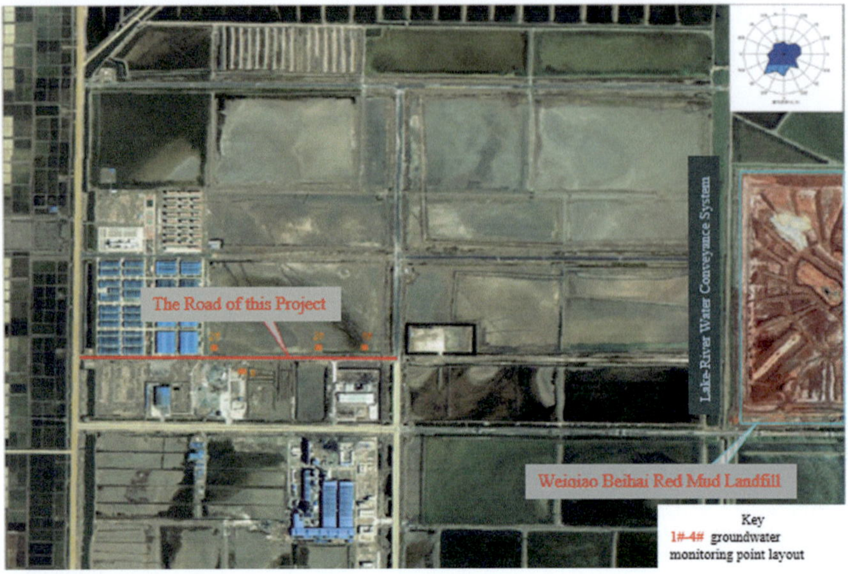

Fig. 8.5 Schematic diagram of groundwater monitoring points Scale 1:16,000

The monitoring results of the groundwater at each monitoring point were obtained by monitoring the on-site water quality environment, as shown in Table 8.9.

In order to carry out the post-evaluation of the environmental impact of modified solidified red mud for road use, the "Technical Regulations for the Application of Red Mud (Bayer Method) in Highway Engineering" (DB37/T 3559-2019) and the "Groundwater Quality Standard" (GB/T14848-2017) V class standard were used. The specific standard values of each evaluation factor are shown in Table 8.10.

The single factor index method is a common method for evaluating groundwater environmental quality, and the specific calculation formula is as follows:

$$P_i = \frac{C_i}{C_{i0}} \tag{8.3}$$

where

P_i Single factor index of pollutant I;
C_i Measured concentration value of pollutant$I$(mg/m^3);
C_{i0} Standard value of pollutant$I$(mg/m^3).

For pH value, its pollution index is calculated as follows:

$$S_j = \frac{7.0 - pH_j}{7.0 - pH_{sd}}, \left(pH_j \le 7.0\right) \tag{8.4}$$

Table 8.8 Overview of groundwater quality monitoring and analysis methods

Project name	Standard code	Standard name	Detection limit
pH	GB/T 5750.4-2006	Standard test method for sensory characteristics and physical indicators of drinking water 5.1 Glass electrode method	–
Total hardness		Standard test method for sensory characteristics and physical indicators of drinking water 7.1 EDTA titration method	1.0 mg/L
Total dissolved solids		Standard test method for sensory characteristics and physical indicators of drinking water 8.1 Weighing method	10 mg/L
Sulfate	HJ 84-2016	Water quality determination of inorganic anions (F^-, Cl^-, NO_2^-, Br^-, NO_3^-, PO_4^{3-}, SO_3^{2-}, SO_4^{2-}) ion chromatography method	0.018 mg/L
Chloride			0.007 mg/L
Nitrate nitrogen			0.004 mg/L
Fluoride	GB 7484-1987	Water quality determination of fluoride ion selective electrode method	0.05 mg/L
Iron	HJ 776-2015	Water Quality Determination of 32 Elements Inductively Coupled Plasma Emission Spectrometry	0.01 mg/L
Manganese			0.004 mg/L
Potassium			0.05 mg/L
Sodium			0.12 mg/L
Calcium			0.02 mg/L
Magnesium			0.003 mg/L
Aluminum			0.009 mg/L
Copper			0.006 mg/L
Zinc			0.004 mg/L
Volatile phenol	HJ 503-2009	Water quality determination of volatile phenol 4-Aminoantipyrine spectrophotometry Method 1 Extraction spectrophotometry	0.0003 mg/L
CO_3^{2-} HCO_3^-	DZ/T 0064. 49-1993	Underground water quality test method Titration method for determination of carbonate, bicarbonate and hydroxide	5 mg/L
Hexavalent chromium	GB/T 5750.6-2006	Standard test method for drinking water metal index 10.1 Diphenylcarbazide spectrophotometry	0.004 mg/L

(continued)

Table 8.8 (continued)

Project name	Standard code	Standard name	Detection limit
Oxygen consumption	GB/T 5750.7-2006	Standard test method for drinking water Organic matter comprehensive index 1.1 Acidic potassium permanganate titration method	0.05 mg/L
Ammonia nitrogen	GB/T 5750.5-2006	Standard test method for drinking water Inorganic non-metal index 9.1 Nessler's reagent spectrophotometry	0.02 mg/L
Nitrite nitrogen		Standard test method for drinking water Inorganic non-metal index 10.1 Diazotization spectrophotometry	0.001 mg/L
Cyanide		Standard test method for drinking water Inorganic non-metal index 4.1 Iso-nicotinic acid-pyridone spectrophotometry	0.002 mg/L
Sulfide		Standard test method for drinking water Inorganic non-metal index 6.1 N,N-diethyl-p-phenylenediamine spectrophotometry	0.005 mg/L
Total coliforms	GB/T 5750.12-2006	Standard test method for drinking water Microbial index 2.1 Multiple tube fermentation method	2 MPN/100 mL
Total bacterial count		Standard test method for drinking water Microbial index 1.1 Plate count method	1 CFU/mL
Mercury	HJ 694-2014	Water quality Determination of mercury, arsenic, selenium, bismuth and antimony Atomic fluorescence method	0.00004 mg/L
Arsenic			0.0003 mg/L
Selenium			0.0004 mg/L
Anionic surfactants	GB/T 5750.4-2006	Standard test method for drinking water Sensory characteristics and physical indicators 10.1 Methylene blue spectrophotometry	0.100 mg/L
Lead	HJ 700-2014	Measurement of 65 elements in water quality by inductively coupled plasma mass spectrometry	0.0009 mg/L
Cadmium			0.005 mg/L
Nickel			0.0006 mg/L
Cobalt			0.0003 mg/L
Thallium			0.0002 mg/L
Trichloromethane	HJ 810-2016	Water quality determination of volatile organic compounds—headspace/gas chromatography-mass spectrometry	1.1 μg/L
Carbon tetrachloride			0.8 μg/L
Benzene			0.8 μg/L
Toluene			1.0 μg/L

Table 8.9 Overview of groundwater monitoring results summary of groundwater monitoring results

Point number	Sampling time	pH	Total hardness	Total dissolved solids	Oxygen demand	Ammonia nitrogen	Nitrite nitrogen	Volatile phenol	Cyanide	Sulfide	Hexavalent chromium	Mercury	Arsenic	Selenium
1#	10.31	7.31	262	1.60×10^3	1.34	0.04	0.188	ND	ND	ND	ND	ND	0.0020	ND
2#	10.31	7.46	5.53×10^3	2.88×10^4	4.16	6.46	0.068	ND	ND	ND	ND	ND	0.0014	ND
3#	10.31	7.49	1.20×10^3	8.51×10^3	1.83	0.30	0.175	ND	ND	ND	ND	ND	0.0026	ND
4#	10.31	7.35	7.52×10^3	3.96×10^4	5.71	9.98	0.548	ND	ND	ND	ND	ND	0.0007	ND

Site number	Sampling time	Potassium	Sodium	Calcium	Magnesium	Fluoride	Chloride	Sulfate	Nitrate nitrogen	CO_3^{2-}	HCO^-	Lead	Cadmium
1#	10.31	39.7	470	73.9	17.3	1.34	772	103	3.59	ND	111	ND	ND
2#	10.31	157	7.88×10^3	571	976	0.85	1.48×10^4	2.53×10^3	0.537	ND	211	0.0012	ND
3#	10.31	110	2.47×10^3	165	182	0.90	4.46×10^3	441	0.348	ND	137	ND	ND
4#	10.31	360	1.15×10^4	936	1.20×10^3	0.36	2.06×10^4	4.19×10^3	9.03	ND	64	ND	MD

Position number	Sampling time	Iron	Manganese	Copper	Zinc	Aluminum	Nickel	Total coliforms	Total colony count	Anionic surfactants
1#	10.31	0.01	0.025	ND	ND	0.027	0.0007	ND	66,000	ND
2#	10.31	0.13	0.459	ND	0.006	ND	0.0024	240	19,000	ND
3#	10.31	0.05	0.342	ND	ND	0.037	0.0033	94	28,000	ND

(continued)

Table 8.9 (continued)

Position number	Sampling time	Iron	Manganese	Copper	Zinc	Aluminum	Nickel	Total coliforms	Total colony count	Anionic surfactants
4#	10.31	0.03	0.203	ND	ND	0.037	0.0030	ND	450	ND

Point number	Sampling time	Cobalt	Thallium	Chloroform	Carbon tetrachloride	Toluene	Benzene	Well depth (m)	Water depth (m)	Water temperature (°C)	Longitude (°)	Latitude (°)
1#	10.31	ND	ND	ND	ND	ND	ND	7.94	1.20	15.3	117.944582	38.054131
2#	10.31	0.0003	ND	ND	ND	ND	ND	7.80	1.65	15.7	117.941191	38.054425
3#	10.31	0.0012	ND	ND	ND	ND	ND	7.72	1.63	15.6	117.949139	38.054427
4#	10.31	0.0008	ND	ND	ND	ND	ND	3.45	2.10	15.4	117.950828	38.054425

Note "ND" indicates not detected (below detection limit)

Table 8.10 Groundwater evaluation standards

Evaluation factor	pH	Total hardness	Dissolved total solids	Ammonia nitrogen	Volatile phenol	Oxygen demand	Cyanide	Sulfate
Evaluation standard	< 2, > 12.5	> 650	> 2000	> 1.50	> 0.01	> 10	> 10.0	> 350
Evaluation factor	Chloride	Fluoride	Nitrate	Nitrite	total coliforms	Hexavalent chromium	Copper	Arsenic
Evaluation standard	> 350	> 2.0	> 30	> 4.8	> 100	> 0.10 (≤ 0.05)	> 1.50 (≤ 1.0)	> 0.05 (≤ 0.01)
Evaluation factor	Mercury	Lead	Cadmium	Iron	Manganese	Nickel	Zinc	Cobalt
Evaluation standard	> 0.002 (≤ 0.001)	> 0.10 (≤ 0.01)	> 0.01 (≤ 0.005)	> 2.0	> 1.50	> 0.10	> 5.00 (≤ 1.0)	> 0.10
Evaluation factor	Sulfide	Anionic surfactant	Benzene	Toluene	Selenium	Total colony count	Thallium	Sodium
Evaluation standard	> 0.10	> 0.3	> 0.12	> 1.4	> 0.1	> 1000	> 0.001	> 400
Evaluation factor	Chloroform	Carbon tetrachloride	Aluminum	–	–	–	–	–
Evaluation standard	> 0.3	> 0.05	> 0.5	–	–	–	–	–

Note The values in parentheses are from the DB37/T3559-2019 standard

$$S_j = \frac{pH_j - 7.0}{pH_{su} - 7.0}, \left(pH_j > 7.0\right) \tag{8.5}$$

where

S_j Standard index of pH;
pH_j pH value at point j;
pH_{sd} Lower limit of pH value specified in the groundwater quality standard;
pH_{su} Upper limit of pH value specified in the groundwater quality standard.

Based on the above evaluation methods, the results of the current status evaluation of groundwater quality for the infrastructure construction project of Binzhou Beihai Vein Industry Development Co., Ltd. Vein Industry Park have been calculated. As shown in Table 8.11, the total hardness, sulfate, dissolved total solids, chloride, sodium, total colony count, ammonia nitrogen, total coliforms monitoring factors in the regional groundwater exceed the minimum values specified in the "Groundwater Quality Standard" (GB/T14848-93) Class V standard, and the rest of the monitoring factors are all less than this standard minimum value. The groundwater quality belongs to Class V in the "Groundwater Quality Standard" (GB/T14848-93). In addition, all the items involved in the monitoring factors of this project in DB37/ T 3559-2019 Table 8.11 have not exceeded the standard, and meet the requirements for groundwater in this standard.

The post-evaluation of this project uses the historical monitoring data of the regional groundwater before the project was built (2015) for comparison. The historical monitoring layout is shown in Table 8.12, the historical monitoring data is shown in Table 8.13, and the evaluation results are shown in Table 8.14.

As can be seen from the above table, in 2015, the regional groundwater environment's dissolved total solids, permanganate index, total hardness, ammonia nitrogen, sulfate, chloride, etc., all exceeded the minimum values stipulated in the "Groundwater Quality Standard" (GB/T14848-93) Class V standard, and the rest of the indicators were all less than the minimum values stipulated in the Class V standard.

Comparing the two sets of data, the use of red mud modified materials as the road base did not have a negative impact on the regional groundwater. The factors such as dissolved total solids, permanganate index, total hardness, ammonia nitrogen, sulfate, chloride, etc., exceeding the minimum value of Class V standard in the "Groundwater Quality Standard" (GB/T14848-93) are mainly related to the regional hydrogeological chemical conditions. The groundwater in the area is Quaternary pore groundwater, the main source of recharge is seawater, the main discharge route is surface evaporation, the type of groundwater is chlorinated sodium type, the deep layer is chloride, sulfate-sodium type water, and the shallow brackish water has a higher degree of mineralization due to evaporation concentration.

Table 8.11 Overview of groundwater environmental quality current status evaluation results

Item	1#	2#	3#	4#
Total hardness	0.40	8.51	1.85	11.57
Dissolved total solids	0.30	14.40	4.26	19.80
Sulfate	0.29	7.23	1.26	11.97
Chloride	2.21	42.29	12.74	58.86
Nitrate nitrogen	0.12	0.02	0.01	0.30
Fluoride	0.67	0.43	0.45	0.18
Iron	0.01	0.07	0.03	0.02
Manganese	0.02	0.31	0.23	0.14
Sodium	1.18	19.70	6.18	28.75
Aluminum	0.05	–	0.07	0.07
Oxygen consumption	0.13	0.42	0.18	0.57
Ammonia nitrogen	0.03	4.31	0.20	6.65
Nitrite nitrogen	0.04	0.01	0.04	0.11
Total coliforms	–	2.40	0.94	–
Total colony count	66.00	19.00	28.00	0.45
Arsenic	0.04 (0.20)	0.03 (0.14)	0.05 (0.26)	0.01 (0.07)
Lead	–	0.012 (0.12)	–	–
Nickel	0.01 (0.04)	0.02 (0.12)	0.03 (0.17)	0.03 (0.15)
Cobalt	–	0.003	0.012	0.008
Zinc	–	0.001 (0.006)	–	–

Note Non-detection and non-standard factors are not evaluated: the results in parentheses are calculated according to the standard DB37/T3559-2019 (Table 8.8)

Table 8.12 Monitoring data collection situation (2015)

Source of quotation	Monitoring time	Monitoring location	Monitoring factors
Shandong Province analysis testing center	2015.10	1# Cha Jian village 2# East side of latitude twenty-five road 3# Red mud dumping ground 4# Dongfeng port 5# Fu Jia platform	pH, dissolved total solids, Permanganate index, total hardness, NH_3-N, sulfate, chloride, cyanide, fluoride, total coliforms, volatile phenol, nitrate, nitrite, sulfide, petroleum, hexavalent chromium, arsenic, mercury, cadmium, lead, copper, nickel, zinc

8.2.2 Soil Environmental Quality Evaluation

According to the actual engineering situation, the soil status monitoring points shown in Fig. 8.6 were selected for the post-evaluation of the environmental impact of the

Table 8.13 Historical groundwater monitoring data in 2015 (unit: mg/L, except pH)

Monitoring items	1# Cha Jian village	2# East side of latitude twenty-five road	3# Red mud dumping ground	4# Dongfeng port	5# Fu Jia platform
PH value (dimensionless)	8.01	7.74	8.21	8.14	8.75
Dissolved total solids	5.34 × 104	7.50 × 104	1.05 × 105	1.30 × 105	4.31 × 103
Permanganate index	13.1	10.9	28.4	22.8	3.24
Total hardness	3321	5147	7635	9913	1224
Ammonia nitrogen (NH_3-N)	4.55	3.37	0.05	0.02	0.02
Sulfate	2858	5114	6852	8628	240
Chloride	25,193	39,885	52,796	57,404	1936
Fluoride	0.41	0.94	0.91	0.07	0.12
Nitrate nitrogen	8.35	19	24.87	27.77	0.84
Cyanide	Not detected	Not detected	Not detected	Not detected	Not detected
Total coliforms	7	2	Not detected	2	Not detected
Nitrite nitrogen	Not detected	Not detected	Not detected	Not detected	Not detected
Sulfides	Not detected	Not detected	Not detected	Not detected	Not detected
Petroleum	Not detected	Not detected	Not detected	Not detected	Not detected
Hexavalent chromium	Not detected	Not detected	Not detected	Not detected	Not detected
Arsenic	Not detected	0.0007	0.002	0.0028	Not detected
Mercury	Not detected	Not detected	Not detected	Not detected	Not detected
Cadmium	Not detected	Not detected	Not detected	Not detected	Not detected
Lead	Not detected	Not detected	Not detected	Not detected	Not detected
Copper	0.02	0.016	0.018	0.008	0.008
Nickel	0.011	0.002	0.003	0.004	0.025
Zinc	0.088	0.036	0.014	0.014	0.07
Well depth m	30	30	–	–	40

(continued)

Table 8.13 (continued)

Monitoring items	1# Cha Jian village	2# East side of latitude twenty-five road	3# Red mud dumping ground	4# Dongfeng port	5# Fu Jia platform
Buried depth m	18	1.6	–	–	1.5
Water temperature °C	14.9	14.6	15.5	14.7	15.3

Table 8.14 Single factor evaluation results of each groundwater monitoring point in 2015

Monitoring Items	1# Cha Jian village	2# East side of latitude twenty-five road	3# Red mud dumping ground	4# Dongfeng port	5# Fu Jia platform
Total dissolved solids	26.7	52.5	65	21.6	19.2
Permanganate index	1.31	2.84	2.28	0.32	1.17
Total hardness	6.04	13.89	18.0	2.2	3.66
Ammonia nitrogen (NH_3-N)	8.9	0.1	0.04	0.04	5.3
Sulfate	8.2	19.58	24.65	0.69	12.94
Chloride	71.98	150.85	164.01	5.53	58.81
Nitrate nitrogen	0.01	0.03	0.00	0.00	0.03
Cyanide	–	–	–	–	0.09
Total coliforms	0.07	–	0.02	–	0.11
Arsenic	–	0.04	0.056	–	0.012
Copper	0.01	0.01	0.01	0.01	0.03
Nickel	0.11	0.03	0.04	0.25	0.12
Zinc	0.02	0.00	0.00	0.01	0.01
Fluoride	0.21	0.47	0.46	0.04	0.06

modified solidified red mud road, and the soil status monitoring points and monitoring project conditions are shown in Table 8.15, with a monitoring frequency of 1 sample per day.

Refer to the current specifications, the monitoring methods of different monitoring items in the soil environment are shown in Table 8.16.

According to different project monitoring specifications, the specific analysis methods and detection limit values used in this post-evaluation of the environmental impact of the modified solidified red mud road are shown in Table 8.17.

The soil around this project is guest soil from agricultural land sampling points, and construction has not yet started around the project, so chromium refers to the

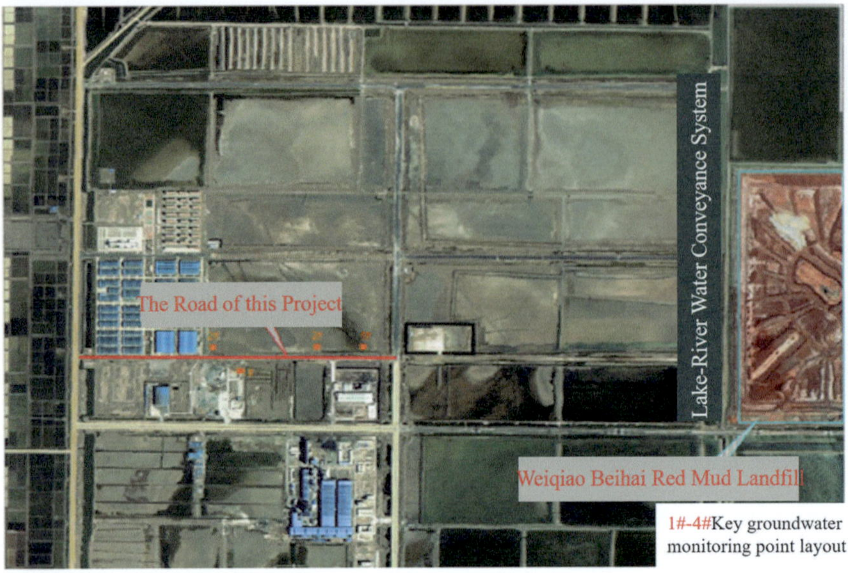

Fig. 8.6 Schematic diagram of soil monitoring points (scale 1:16,000)

Table 8.15 Soil monitoring overview

Number	Monitoring location	Sampling depth	Monitoring items
1#	South side of the road (within 200 m range)	0 ~ 20 cm	pH, cadmium, mercury, arsenic, lead, chromium, copper, nickel, zinc
2#	North side of the road (within 200 m range)		

"Soil Environmental Quality Agricultural Land Soil Pollution Risk Control Standard (Trial)" (GB15618-2018) pH > 7.5 Other land requirements standards, the remaining projects implement the "Soil Environmental Quality Construction Land Soil Pollution Risk Control Standard (Trial)" (GB36600-2018) second type of land requirements, the specific standard values are shown in Table 8.18.

In order to carry out the post-evaluation of the environmental impact of the modified solidified red mud road, the single factor index method is used to evaluate the current status of soil environmental quality.

$$S_i = \frac{C_i}{C_{si}} \tag{8.6}$$

where

S_i Single factor index of pollutant;

Table 8.16 Overview of soil monitoring analysis methods

Project name	Standard code	Standard name	Detection limit
pH	NY/T 1377-2007	Determination of soil pH	–
Mercury	HJ 680-2013	Determination of mercury, arsenic, selenium, bismuth, and antimony in soil and sediment Microwave digestion/atomic fluorescence method	0.002 mg/kg
Arsenic			0.01 mg/kg
Lead	GB/T 17141-1997	Determination of lead and cadmium in soil quality—graphite furnace atomic absorption spectrophotometry	0.1 mg/kg
Cadmium			0.01 mg/kg
Chromium	HJ 491-2009	Determination of total chromium in soil by flame atomic absorption spectrophotometry	4 mg/kg
Copper	HJ 803-2016	Determination of 12 metal elements in soil and sediment Aqua regia extraction-inductively coupled plasma mass spectrometry (HJ 803-2016)	–
Zinc			–
Nickel			–

Table 8.17 Soil environmental status monitoring results (pH dimensionless, others mg/kg)

Site number	Sampling depth (m)	Monitoring time	pH	Mercury	Arsenic	Lead	Cadmium	Chromium
1#	0 ~ 0.2	10.31	8.29	0.0122	10.5	20.6	0.16	64
2#	0 ~ 0.2	10.31	8.13	0.0119	8.87	21.0	0.13	7

Site number	Sampling depth (m)	Monitoring time	Copper	Nickel	Zinc
1#	0 ~ 0.2	11.11	18.7	62	28
2#	0 ~ 0.2	11.11	17.3	64	28

Table 8.18 Soil environmental quality standards (unit: mg/kg)

Item	Cadmium	Mercury	Arsenic	Lead	Copper	Nickel	Chromium
Standard value	65	38	60	800	18,000	006	250

C_i Concentration of pollutant, mg/kg;
C_{si} Evaluation standard value of pollutant, mg/kg.

Table 8.19 presents the results of the evaluation of the current status of soil quality in the Binzhou Beihai Economic Development Zone Vein Industrial Park. According to the evaluation results, the chromium monitoring data of points 1# and 2# can meet the standard requirements for other lands with pH > 7.5 in the "Soil Environmental Quality Agricultural Land Soil Pollution Risk Control Standard (Trial)" (GB15618-2018); the monitoring data of other factors can all meet the requirements of the second type of land standard in Table 1 of the "Soil Environmental Quality Construction Land Soil Pollution Risk Control Standard (Trial)" (GB36600-2018). At the same time,

Table 8.19 Overview of soil evaluation results

Serial number	Item	1#	2#
1	Mercury	0.0003	0.0003
2	Arsenic	0.175	0.149
3	Lead	0.026	0.026
4	Cadmium	0.0025	0.002
5	Chromium	0.256	0.228
6	Copper	0.001	0.0096
7	Nickel	0.069	0.071

Note No evaluation is made without standard factors

all soil monitoring data of this project meet the standard requirements in DB37/T 3559-2019 (implemented according to GB 36600). Since the soil around the Vein Industrial Park in Binzhou Beihai Economic Development Zone is newly transported in during the construction of the Vein Industrial Park, it is not comparable with the soil background value before the project construction, so this post-evaluation does not compare the soil monitoring data.

8.3 Summary of This Chapter

In order to understand the continuous impact of the red mud modified material subgrade of the completed road project on the surrounding environment, this study conducted a post-evaluation of the environmental impact on the completed road sections in the "Binzhou Beihai Vein Industry Development Co., Ltd. Vein Industry Park Infrastructure Construction Project". According to the dilution migration empirical model of pollutants, the actual working environment conditions of the Bayer method red mud subgrade were simulated, and the estimated leaching concentration of two kinds of red mud subgrade hazardous substances when reaching groundwater was lower than the limit value of Class III groundwater. It can be seen that the modified red mud subgrade will not have a negative impact on the quality of groundwater.

According to the groundwater monitoring data of this project, the total hardness, sulfate, total dissolved solids, chloride, sodium, total bacterial count, ammonia, and total coliforms in the regional groundwater exceed the minimum value of Class V standard in the "Groundwater Quality Standard" (GB/T14848-93), and the rest of the monitoring factors are all less than the minimum value of the substandard. The groundwater quality belongs to Class V in the "Groundwater Quality Standard" (GB/T14848-93). All the items involved in the monitoring factors of this project have not exceeded the standard, and meet the requirements of this standard for groundwater quality. At the same time, compared with the regional groundwater monitoring data in 2015, the total dissolved solids, permanganate index, total hardness, ammonia nitrogen, sulfate, chloride, etc. all exceed the minimum value stipulated in the Class V standard of the "Groundwater Quality Standard" (GB/T14848-93). Therefore, the use of red mud modified materials as the subgrade in this project has not had a negative impact on the regional groundwater. The monitoring factors exceed the minimum value stipulated in Class V of the "Groundwater Quality Standard" (GB/T14848-93), which is mainly related to the regional hydrogeological chemical conditions. The groundwater in the area is Quaternary pore groundwater, the recharge source is mainly seawater, the discharge route is mainly surface evaporation, the groundwater type is chloride-sodium type, the deep layer is chloride, sulfate-sodium type water, and the shallow brackish water is due to evaporation concentration. The mineralization degree is high. According to the soil monitoring data of this project, chromium can meet the standard requirements for other uses in the "Soil Environmental Quality Agricultural Soil Pollution Risk Control Standard (Trial)" (GB15618-2018) where pH > 7.5; the monitoring data of other factors can all meet the "Soil Environmental Quality Construction Land Soil Pollution Risk Control Standard (Trial)" (GB36600-2018) Table 1 screening value second class land standard requirements. At the same time, all the soil factor monitoring data of this project meet the standard requirements in DB37/T 3559-2019.

References

1. Batemonghe. Research on the development of highways in China[J]. Northern Economy,2013,(3): 60–61.
2. The Ministry of Transport releases the national highway network plan: the total scale will be about 461,000 kilometers by 2035[N]. People's Network, 2022–07–25.
3. A.R. Hind, S.K. Bhargava, S.C. Grocott, The surface chemistry of Bayer process solids: a review, Colloids Surf, A: Physicochem. Eng. Asp. 146 (1999) 359–374.
4. Sun Zhaoyun, Cheng Yu. Current situation of red mud stockpiling and comprehensive utilization in Shandong Province[J]. Shandong Transportation Technology,2017,(02):99–100+107.
5. K. J. Bayer, German Patent No: 2150677, 1972.
6. Yang Shaowen, Cao Yaohua, Li Qing. Current situation and progress of comprehensive utilization of red mud from alumina production[J]. Mineral Protection and Utilization,1999,(06):46–49.
7. A. Kumar, S. Kumar, Development of paving blocks from synergistic use of red mud and fly ash using geopolymerization, Constr. Build. Mater. 38 (2013) 865–871.
8. Z.Sun, Y. Cheng, Present situation of red mud stockpiling and comprehensive utilization in Shandong province, Shandong Transp. Technol. 02(2017) 99–100+107.
9. S. Kumar, R. Kumar, A. Bandopadhyay, Innovative Methodologies for the Utilization of Wastes from Metallurgical and Allied Industries, Resour. Conserv. Recycl. 48(2006) 301–314.
10. Li Henglong. Development and practice of dry stacking technology for red mud in Western Australia[J]. Nonferrous Metals (Smelting Section).1993, (4):42~45.
11. Jiao Zhanzhong. Preliminary exploration of the stacking age of red mud dump in Shandong Aluminum Plant[J]. Light Metals,1989,(06):12–16.
12. Wang Litang. Effective ways to utilize red mud[J]. World Nonferrous Metals,1998,(08):43–45+39.
13. J.M. Amezaga, T.S. Rötting, P.L. Younger, et al., A Rich Vein? Mining and the Pursuit of Sustainability, Environ. Sci. Technol. 451(2011) 21–26.
14. Á. D. Anton, O. Klebercz, Á. Magyar, et al., Geochemical recovery of the Torna–Marcal river system after the Ajka red mud spill, Hungary, Environ. Sci.: Processes Impacts, 16(2014) 2677.
15. N. Uzinger, M. Rékási, Á.D. Anton, et al., Monitoring the clean-up operation of agricultural fields flooded with red mud in Hungary, Environ Sci. Pollut. Res. 23 (2016) 23573.
16. Zhang Qinghe, Zhuang Jianming, Wang Longqian. Current status and application of comprehensive utilization of red mud in plastic production[J]. Comprehensive Utilization of Mineral Resources, 1994,(01):37–40.
17. Liu Zigao, Yang Changshi, Cheng Zonghao, et al. Treatment and utilization of Bayer red mud[J]. Journal of Nonferrous Metals,1997,(01):43–47.
18. Zhang Zhengyong. Research and development of red mud resource utilization technology[J]. Guangzhou Chemical Industry,2013,41(06):29–31.

Y. Cheng and X. Zhu, *Utilization and Environmental Impact Assessment of Red Mud Resources*, https://doi.org/10.1007/978-981-96-7411-4

19. Z. Pan, Y. Zhang, Z. Xu, Strength development and microstructure of hardened cement paste blended with red mud, J. Wuhan Univ. Technol. 241(2009) 161–165.
20. M.D.A. Thomas, R.D Hooton, A. Scott, et al., The effect of supplementary cementitious materials on chloride binding in hardened cement paste, Cem. Concr. Res. 421(2012) 1–7.
21. R.C.O. Romano, H.M. Bernardo, M.H. Maciel, et al., Hydration of Portland cement with red mud as mineral addition, J. Therm. Anal. Calorim. 11(2017) 1–14.
22. Liang Naixing, Zhang Dengliang. Study on the mechanical properties of cement red mud concrete[J]. Journal of Xi'an Highway Transportation University,1995,(03):1–4.
23. W.C. Tang, Z. Wang, Y. Liu, et al., Influence of red mud on fresh and hardened properties of self-compacting concrete, Constr. Build. Mater. 178 (2018) 288–300.
24. Yang Jiakuan, Hou Jian, Qi Bo, et al. Pilot production and industrialization of alumina red mud non-fired bricks[J]. Environmental Engineering, 2006, 24(4):52–55.
25. Wang Xiaofeng, Lin Haiyan, Yang Yongfu, et al. Preparation of sintered bricks using Bayer red mud[J]. New Building Materials, 2007, 34(7):24–26.
26. Guo Hui, Guan Xuemao, Ma Xiaoe, et al. Experimental study on red mud environmental wall bricks[J]. China Ceramic Industry, 2007, 14(6):14–17.
27. S. Ordóñez, Catalytic Applications of Red Mud, an Aluminum Industry Waste: A Review, Appl. Catal., B 84(2008) 732–733.
28. Xu Zhifang, Su Ailing, Zhang Xinfeng, et al. Comprehensive recovery and utilization status of alumina red mud[J]. Shandong Metallurgy, 2010, 32(3):8–12.
29. Q. Liu, R. Xin, C. Li, et al. Application of red mud as a basic catalyst for biodiesel production, Acta Sci. Circumstantiae 25(2013) 823–829.
30. Guan Jianhong. Experimental study on recovering iron from red mud by pulsating high gradient magnetic separator[J]. Nonferrous Metals Science and Engineering, 2000, 14(4):15–18.
31. B. Mishra, A. Staley, D. Kirkpatrick, Recovery of value-added products from red mud, Miner. Metall. Process. 19(2002) 87–94.
32. D. Zhu, T. Chen, P. Jian, et al., Recovery of Iron From High-Iron Red Mud by Reduction Roasting With Adding Sodium Salt, J. Iron. Steel. Res. 19(2012) 1–5.
33. Liu Wanchao, Yang Jiakuan, Xiao Bo. Extraction of iron from Bayer red mud and preparation of building materials from residues[J]. Journal of Nonferrous Metals, 2008, 18(1):187–192.
34. Gao Jianyang, Chen Yuhai, Zheng Xia. Analysis of factors affecting the preparation of sponge iron from Bayer red mud[J]. Shandong Metallurgy, 2010, 32(4):37–40.
35. Fan Xianfeng, Xie Min, Xiao Ying, et al. Research on flotation process of Bayer alumina red mud[J]. Mining and Metallurgy, 1992(2):18–22.
36. Zhou Qiusheng, Fan Kuangsheng, Li Xiaobin, et al. Recovery of alumina from high iron red mud by sintering method[J]. Journal of Central South University (Natural Science Edition), 2008, 39(1):92–97.
37. Lu Guilin, Chi Songjiang, Bi Shiwen. Leaching of alumina and iron oxide from red mud[J]. Journal of Materials and Metallurgy, 2010, 9(1):000031–67.
38. Wang Keqin, Wang Hao, Li Shenghu. Research on recovering aluminum from red mud by hydrochloric acid leaching[J]. Nonferrous Metals (Smelting Section), 2012(7):16–18.
39. Chen Dingmao. Recovery of titanium dioxide from red mud[J]. Industry and Environment, 1994(3):42–45.
40. S. Agatzini-Leonardou, P. Oustadakis, P.E. Tsakiridis, et al., Titanium leaching from red mud by diluted sulfuric acid at atmospheric pressure, J. Hazard. Mater. 157(2008) 579–586.
41. Zhang Jiangjuan. Preliminary study on the recovery of titanium dioxide from red mud[J]. China Resources Comprehensive Utilization, 2003(1):28–30.
42. Jiang Pingguo, Wang Hongzhen. Study on the leaching of titanium from red mud[J]. China Nonferrous Metallurgy, 2008(2).
43. Chang Jun, Shao Yanhai, Li Shuo, et al. Current status and progress of comprehensive recovery of valuable metal elements from red mud[J]. Mining and Metallurgy, 2017, 26(3):59–63.
44. Wang Purong, Dai Huixin, Gao Likun, et al. Current status of scandium recovery and extraction[J]. Rare Metals,2012,36(03):501–506.

45. Abhilash, S. Sinha, M.K. Sinha, et al., Extraction of lanthanum and cerium from Indian red mud, Int. J. Miner. Process. 127(2014) 70–73.
46. Fan Yanjin, Liang Huanlong, Huang Taiyuan, et al. Experimental study on leaching red mud with titanium white waste acid[J]. Hydrometallurgy,2014,33(05):375–377.
47. Zhan Haihong, Liang Huanlong, Fan Yanjin, et al. Neutralization and precipitation of scandium in the second stage of sulfuric acid method titanium white waste acid[J]. Nonferrous Metals (Smelting Section),2014,(08):45–47+59.
48. Han Yi. Study on the preparation and application of red mud adsorbent[D]. Beijing University of Chemical Technology,2004.
49. Wu Jianfeng, Luo Wenhui, Xu Xiaohong, et al. Preparation and structure and properties of red mud ceramic insulation brick[J]. Journal of Wuhan University of Technology, 2008, 30(5):15–18.
50. Wei Jingjing. Study on the preparation of red mud porous ceramic material and its adsorption of heavy metal Cr(Vi)[D].Guangxi University,2014.
51. Wei Hongshan, Ma Xiaoe, Guan Xuemao, et al. Study on lightweight insulation ceramic material with high content of Bayer red mud[J]. Light Metals, 2017(4):16–19.
52. V.M. Sglavo, S. Maurina, A. Conci, et al., Bauxite 'red mud' in the ceramic industry. Part 2. Production of clay-based ceramics, J. Eur. Ceram. Soc. 20(2000) 245–252.
53. Wang Ronghua, Tang Longxiang, Liu Chunhua. Study on red mud filled modified polypropylene[J]. Modern Plastics Processing and Applications, 2016, 28(6):12–15.
54. Liu Lizhu, Gao Yushi, Wang Yong, et al. Feasibility study on the preparation of plastic-based composite sheet using fly ash and other inorganic solid wastes[J]. Green Technology, 2017(16).
55. Zhang Peixin, Yan Jiaqiang. Study on red mud microcrystalline glass[J]. Nonferrous Metals Engineering, 2000, 52(4):77–79.
56. Yang Huizhi, Sun Hongwei, Chen Changping, et al. Study on the preparation of red mud microcrystalline glass by sintering method[J]. Light Metals, 2007(4):22–24.
57. Zhang Shengbin. Research on gold slag-red mud microcrystalline glass[D]. Shandong Jianzhu University, 2016.
58. J. Alvarez, R. Rosal, H. Sastre, et al., Characterization and deactivation studies of an activated sulfided red mud used as hydrogenation catalyst, Appl. Catal., A 167(1998) 215–223.
59. J. Álvarez, S. Ordóñez, R. Rosal, et al., A new method for enhancing the performance of red mud as a hydrogenation catalyst, Appl. Catal., A 180(1999) 399–409.
60. S. Ordóñez, H. Sastre, F.V. DíEz, Characterisation and deactivation studies of sulfided red mud used as catalyst for the hydrodechlorination of tetrachloroethylene, Appl. Catal., B 29(2001) 263–273.
61. S. Ordóñez, H. Sastre, F.V. Díez, Catalytic hydrodechlorination of tetrachloroethylene over red mud, J. Hazard. Mater. 81(2001) 103.
62. G.W. Land, Trials of Additives for Sulfur-Dioxide Removal in Industrial Plants, J. Eng. Gas Turbines Power 92(1970) 30.
63. Chen Yi, Li Junqi, Huang Fang, et al. Study on the performance of Bayer red mud absorbing SO 2 waste gas[J]. Journal of Guizhou University of Industry: Natural Science Edition, 2007, 36(4):30–32.
64. G. Jones, G. Joshi, M. Clark, et al., Carbon Capture and the Aluminium Industry: Preliminary Studies, Environ. Chem. 3(2006) 297–303.
65. V. S Yadav, M. Prasad, J. Khan, et al., Sequestration of carbon dioxide (CO2) using red mud, J. Hazard. Mater. 176(2010) 1044.
66. S. Smiljanić, I. Smičiklas, A. Perić-Grujić, et al., Rinsed and thermally treated red mud sorbents for aqueous Ni2+ ions, Chem. Eng. J. 162(2010) 75–83.
67. H.S. Altundoğan, S. Altundoğan, F. Tümen, et al., Arsenic removal from aqueous solutions by adsorption on red mud, Waste Manage. 20(2000) 761–767.
68. R.C Sahu, R. Patel, B.C. Ray, Adsorption of Zn(II) on activated red mud: Neutralized by CO2, Desalination 266(2011) 93–97.
69. Y. Ma, C. Lin, Y. Jiang, et al. Competitive removal of water-borne copper, zinc and cadmium by a CaCO3-dominated red mud, J. Hazard. Mater. 172(2009) 1288.

70. Hao Xiaowei, Huang Yizong, Cui Yanshan, et al. The effect of red mud on the chemical forms and bioavailability of Pb and Zn in contaminated soil[J]. Journal of Environmental Engineering, 2010, 4(6):1431–1435.
71. Li Darong, Yang Weiwei, Li Qiujun, et al. Research on the improvement of lead and cadmium contaminated soil by silkworm sand and red mud[J]. Soil Bulletin,2015,46(04):977–984.
72. C.W. Gray, S.J. Dunham, P.G. Dennis, et al., Field evaluation of in situ remediation of a heavy metal contaminated soil using lime and red-mud, Environ. Pollut. 142(2006) 530–539.
73. Fan Meirong, Luo Lin, Liao Yulin, et al. Application of red mud in soil heavy metal pollution control and agricultural production[J]. Soil Bulletin, 2010(6):1531–1536.
74. Cai Delong, Zhao Shili. Experimental study on red mud silicon fertilizer in corn[J]. Regional Research and Development, 1997(3):83–84.
75. K.E. Snars, R.J. Gilkes, M.T.F. Wong, The liming effect of bauxite processing residue (red mud) on sandy soils, Soil Res. 42 (2004) 321–328.
76. Qi Jianzhao, Yang Jiakuan, Wang Mei, et al. Experimental study on red mud as road base material[J]. Highway Traffic Science and Technology, 2005, 22(6):30–33.
77. Qi Jianzhao. Experimental study on red mud road material[D]. Huazhong University of Science and Technology, 2005.
78. Chen Fan. Comparison of the performance and strength mechanism of red mud road base materials of different years[D]. Huazhong University of Science and Technology, 2007.
79. Bao Huiming, Mou Kali. Red mud modified asphalt viscosity and micro characteristics[J]. Journal of Henan University of Science and Technology: Natural Science Edition, 2017, 38(5):47–50.
80. H. Zhang, H. Li, Y. Zhang, et al., Performance enhancement of porous asphalt pavement using red mud as alternative filler, Constr. Build. Mater. 160 (2018) 707–713.
81. Yang Nanru. The physicochemical basis of alkali-aggregate reaction materials (I)[J].Journal of Silicate,1996,(02):209–215.
82. Yang Zhaojuan, Xiang Lan. Review on the comprehensive utilization of phosphogypsum[J]. Inorganic Salt Industry,2007,(01):8–10.
83. JTG T-F20–2015, Technical details for highway pavement base construction[S].
84. Jiang Tongchuan. Orthogonal experimental design[M]. Jinan: Shandong Science and Technology Press, 1985.
85. Yao Zhanyong, Ai Yizhong, Shang Qingsen, et al. Study on the mix ratio of Yellow River silt (silt) loess[J]. Rock and Soil Mechanics,2008,(07):1943–1948.
86. Pang Hao et al. Econometrics[M]. Beijing: Science Press, 2006.
87. Feng Zhongju, Xie Yongli. Theoretical calculation of the optimum moisture content and maximum dry density of the standard compaction test[J]. Journal of Chang'an University: Natural Science Edition, 2002, 22(2):10–13.
88. JTG E51–2009, Test procedures for inorganic binder stabilized materials in highway engineering[S].

MIX
Papier aus verantwortungsvollen Quellen
Paper from responsible sources
FSC® C105338
FSC
www.fsc.org

If you have any concerns about our products,
you can contact us on
ProductSafety@springernature.com

In case Publisher is established outside the EU,
the EU authorized representative is:
Springer Nature Customer Service Center GmbH
Europaplatz 3, 69115 Heidelberg, Germany

Printed by Libri Plureos GmbH
in Hamburg, Germany